Birkhäuser

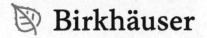

Studies in Universal Logic

This series is devoted to the universal approach to logic and the development of a general theory of logics. It covers topics such as global set-ups for fundamental theorems of logic and frameworks for the study of logics, in particular logical matrices, Kripke structures, combination of logics, categorical logic, abstract proof theory, consequence operators, and algebraic logic. It includes also books with historical and philosophical discussions about the nature and scope of logic. Three types of books will appear in the series: graduate textbooks, research monographs, and volumes with contributed papers.

More information about this series at http://www.springer.com/series/7391

Saloua Chatti

Arabic Logic from al-Fārābī to Averroes

A Study of the Early Arabic Categorical, Modal, and Hypothetical Syllogistics

Birkhäuser

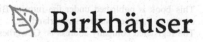

Saloua Chatti
Faculty of Human and Social Sciences
University of Tunis
Tunis, Tunisia

ISSN 2297-0282 ISSN 2297-0290 (electronic)
Studies in Universal Logic
ISBN 978-3-030-27465-8 ISBN 978-3-030-27466-5 (eBook)
https://doi.org/10.1007/978-3-030-27466-5

This book is published under the imprint Birkhäuser, www.birkhauser-science.com by the registered company Springer Nature Switzerland AG
The registered company address is: Gewerbestrasse 11, 6330 Cham, Switzerland

For my son Yassine
For all my family

Acknowledgements

This work would not have seen the light if I had not been invited by Prof. Jean-Yves Beziau to present a tutorial on Arabic Logic in his fourth Congress and School on Universal Logic (Unilog 2013, Rio de Janeiro). I thank him for his invitation to that huge event and for his interest in Arabic logic.

This book is the continuation of the works that I have done in the field of Arabic logic since I became interested in it. I feel that the analysis and the precise study of Arabic logic is highly indispensable to all scholars interested in the history of logic in general, since it is still not well known, in the western world as well as in the Arabic one, despite the seminal contributions of Prof. Nicholas Rescher whose work helped discover and analyze the writings of many logicians of that tradition. He was followed by many contemporary researchers who were interested in its various aspects. I myself focused a lot on categorical and modal logics and presented some conferences on these topics on many occasions. But in this tutorial, I wished to give a more complete image of a tradition which lasted several centuries. This is why I presented in the tutorial three different parts of these systems, which are the categorical logic, the modal logic, and the hypothetical logic. My analysis of these different theories focuses mainly on the original Arabic writings, in order to be as faithful as possible to the texts. I translated most of the quotations that I have given in this examination but I also used some good and well-known English translations such as those of Professors Tony Street, Nabil Shehaby, and Wilfrid Hodges, among others.

I am especially grateful to Prof. Wilfrid Hodges for his precious and invaluable help and advice and for all the various knowledgeable and detailed commentaries on this book, which made me improve it in several ways. Special thanks also to Prof. Jean-Yves Beziau, for all his help and valuable advice and suggestions. I am also very grateful to Prof. Stephen Read and to all the participants to the different events I attended, namely, Pr. Tony Street, Pr. Paul Thom, Pr. Riccardo Strobino, Pr. John N. Martin, Pr. Graham Priest, Pr. Hans Smessaert, Pr. Dany Jaspers,

Dr. Fabien Schang, Dr. Lorenz Demey, Dr. Alessio Moretti, and Dr. Amirouche Moktefi, for their various, valuable, and helpful remarks, commentaries, and questions. I thank also my Tunisian colleague Prof. Mokdad Arfa, who provided me with some useful documentation.

Contents

Chapter 1
General Introduction

Arabic logic started with the translations of the Aristotelian and Greek texts. These translations were first made in Syria from Greek to Syriac, then Arabic during the Umayyad Empire for some treatises, but the most important amount of translations was made during the Abbasid Empire [Baghdad, 750–1258 AD], starting from the reign of Abu Jaafar al-Manṣūr [754–775 AD], then Hārūn al Rashīd [786–809 AD] and al-Ma'mūn [813–833 AD] who founded an academy devoted to translation called "Beit al-Ḥikma" (literally: "the House of Wisdom") [830 AD] (see [127], 140). These interests concerned first some sciences like medicine and astronomy, but given the close relation between these sciences and logic, they extended to the philosophical and logical writings, which were especially encouraged by al-Ma'amūn ([127], 140). The translations were followed by commentaries of the whole Aristotelian corpus together with Porphyry's *Isagoge* and the writings of Plato, Plotin, and the Aristotelian commentators such as Alexander of Aphrodisias, Galen, Theophrastus, John Philoponus, and others. These commentaries are described in detail in Nicholas Rescher's *The development of Arabic Logic*, a book that has been translated in Arabic by Mohamed Mahrān, who added in his introduction of it some more information on the authors and their writings and the general conditions of the birth of Arabic logic. They were made by a number of

S. Chatti, *Arabic Logic from al-Fārābī to Averroes*, Studies in Universal Logic,
https://doi.org/10.1007/978-3-030-27466-5_1

scholars starting from Ibn al-Muqaffa‘ and Ya‘aqūb ibn Isḥāq al-Kindī [805–873, AD] among others. More details on translations will be given below.

I will first briefly present these logicians who are listed and grouped by period in Nicholas Rescher's book, and then I will focus on the early[1] authors in this tradition, namely, al-Fārābī [873–950, AD], Avicenna (Ibn Sīnā) [980–1037, AD], and Averroes (Ibn Rushd) [1126–1198, AD], since one has to study them before turning to their followers,[2] who are numerous in particular in the eastern areas. These authors' contributions are not equally important as we will show below, since Averroes, who is voluntarily very faithful to Aristotle, does not seem to add much to his logic, while Avicenna builds some logical systems which depart from Aristotle in multiple and various respects. As to al-Fārābī, he is mainly influenced by the Greek commentators like Alexander of Aphrodisias, and his contribution is the result of his development of the major achievements of these commentators and their Syriac translators (see [150], 529).

Thus, the focus on the classical early authors does not mean that I consider the later developments as being of no interest. On the contrary, these later developments, as exemplified by Avicenna's followers and as shown by some scholars such as Tony Street and Khaled El-Rouayheb, are very important and worth studying but since they are very much influenced by their predecessors, we need to study the early authors first. Anyway, I will sometimes refer to some of the achievements of later logicians when possible.

The problems that I raise are the following: what are the Arabic logicians' contributions? How do they define the main concepts and the syllogism itself? What methods do they use in their respective syllogistics? Are they different from Aristotle's ones? What are the characteristics of their logical systems? To what extent do these systems differ from Aristotle's and his followers' syllogistics, whether categorical or modal?

To answer these questions, I will study the systems of the authors mentioned and will analyze their categorical logic as well as their modal and hypothetical logics. My attempt to provide such an analysis does not ignore, however, the other studies that have been made by many authors, since the seminal work of Nicholas Rescher. What I want to do is to focus exclusively on the technical details of these systems

---

[1]Note that historically, the very first Arabic logicians are al-Kindī and presumably Ibn al-Muqaffa‘, who wrote a treatise entitled "*al-Mantiq*". Cristina D'Ancona says that we don't know if it is Muhammed Ibn al-Muqaffa' or rather his father Abdullah who translated "Porphyry's *Isagoge*, the *Categories*, *De Interpretatione* and *Prior Analytics*" ([62], Sect. 2), while N. Rescher [126] reports that it is the son Muhammed ibn Abdullah Ibn al-Muqaffa (d.c.800 A.D.) who is probably a translator or an author of a commentary in logic, and W. Hodges says that it must be Abdullah Ibn al-Muqaffa (the father) and the famous translator of *Kalilah wa Dimna* who wrote the commentary in logic entitled "al-Mantiq" (personal communication). As to al-Kindī, N. Rescher [126, Chap. 2, 15–27] evokes a treatise of al-Kindī on Aristotle's Organon entitled "On the objectives (or: subject-materials) of Aristotle in each of his treatises" (*Kitāb fī aghrāḍ Aristūtālīs fī kull wāhid min kutubihi*).

[2]More information about the authors of the post-classical periods, such as the post-Avicennans can be found in the studies of Tony Street and Khaled El-Rouayheb, for instance, among others.

by relying on the *original* Arabic writings and examining them in detail, using for that purpose the modern tools and symbolisms (especially the symbolisms of First-Order Logic and of Modal Logic).

However, I will not examine the analogical and inductive arguments which were parts of these author's systems too. For one thing, it would extend the topic of the present study, which focuses on deductive logic exemplified by the categorical and modal logics together with the hypothetical one. Second, although the inductive and analogical arguments are indeed parts of the systems studied, they do not have the same importance than deductive logic in these systems. Third, these arguments are very different from the deductive ones and they deserve another study on their own, in order to determine their characteristics and applications, which are very large in the Arabic tradition, since many analogical arguments, for instance, are very much used in theological and legal studies. The fourth reason is that the analogical and inductive arguments could be seen as methods that could be used in everyday life, in theology or in the empirical sciences, but they are not really parts of logic in its narrow sense, i.e., in the sense of the study of the *valid arguments*. For the analogical and inductive arguments are not valid as the deductive arguments are, given that they do not systematically and necessarily lead to true conclusions. This would be a reason to exclude them from logic *stricto* sensu.

Now, one could also want to determine if deductive logic in the Arabic tradition has had or not an influence on medieval logic, apart from the common Greek heritage. This influence has been variously appreciated, in particular, by the Western scholars. On the one hand, Carl Prantl thinks that Arabic logic had a great influence on the Western medieval logic because the "Western logic called *logica modernorum*, the so-called theories of the properties of terms, that is, supposition theory etc. from the twelfth century entered into the Latin world from translations of Byzantine and Arabic logical works (see Prantl 1867)" ([107], 180). On the other hand, Henrik Lagerlund says that L. M. de Rijk "showed in the 1960s [that this opinion] was completely wrong" ([107], 180). Both authors rely on purely historical details regarding the translations and the transmission of the Arabic writings. I note, however, that it is hard to determine exactly the influence of Arabic logic as long as the Arabic writings have not been properly studied; and it seems that at the time that L. M. De Rijk wrote his books (that is, in 1962, 1967, and 1970), the studies of Arabic logic in the Western world were quite rare. It is also hard to deny this influence as L. M. De Rijk seems to do given the many links and common features between Arabic philosophy in general and medieval philosophy. Furthermore, the studies of the translations themselves have been improved by some recent researches such as that of Charles Burnett, for instance, who reports more Arabic works translated than other scholars. Thus, he says that the whole of Averroes was known and also that al-Fārābī's *De interpretatione* and *Categories* were known in the Latin world along with other texts, as witnessed by the following quotation: "Moreover, references in Albert the Great and brief surviving fragments show that, aside from Averroes' commentaries, al-Fārābī's summaries of at least the *Categories* and the *De interpretatione*, and his commentaries of the *Prior* and the *Posterior Analytics* were known in Latin in the Middle Ages. Moreover, a summary of the *Posterior*

Analytics had been included in the Arabic Encyclopedia known as the *Brethren of Purity*, and was translated into Latin with an attribution to al-Kindī" (see [48], 601). He refers to both Grignaschi (1972) for the translation of al-Fārābī and to A. Nagy (1897) for the *Brethren of Purity* and al-Kindī (see [48], 601, notes 12 and 13). This contrasts with what other people such as Dag N. Hasse claim, namely, that "only a few works of Arabic logic were translated into Latin ..." among which "the *Isagoge* part of Avicenna's *Shifā*" and mainly "al-Ghazālī's *Intentions of the Philosophers*" (*Maqāṣid al-falāsifa*), along with "al-Fārābī's *Enumeration of the sciences*" (see [80], Sect. 3) and Averroes's commentaries. Al-Fārābī's other treatises are not cited by this author and only a small part of Avicenna's *Shifā* is said to have been translated. This difference can explain why some people deny the influence of Arabic logic or give it a very little importance; but the work of Charles Burnett and the historians that he cites is very important in that it makes it possible to justify the links between some Arabic logicians and some Medieval ones.

In addition, some medieval logicians such as Gerard of Cremona were able to read Arabic, and even to translate the Arabic texts, as shown by Charles Burnett in his article where he says the following: "And yet we find Gerard of Cremona translating the latter work *from Arabic* in the same century, and the thirteenth century all Averroes's Middle Commentaries on the *Organon* were translated" (see [48], 598, emphasis added). Many treatises of the Arabic authors such as Avicenna and al-Fārābī are also said to have been translated ([48], 598).[3] So the knowledge of Arabic must also be taken into account when one considers the influence of Arabic logic on the Latin West.

Anyway, I wish to show the specificities of Arabic logic by *studying the texts themselves*, without focusing that much on the eventual influences from one side or the other. For that purpose, I will analyze the Arabic theories, concepts, inferences, and arguments. This is why I will focus, as I have stressed earlier, on the original writings of the Arabic authors together with some comments made by eastern and western scholars. I will try to follow as faithfully as possible the original texts but also to use the symbolism of modern logic to render the different propositions and arguments as clearly as possible. I don't focus that much on historiography, in the sense of searching for the different influences and sources of the Arabic authors, whether Greek or other, even if these sources are indicated when necessary; rather my study is first and foremost theoretical in the sense that I will analyze the theories themselves in order to clarify them and to compare them with Aristotle's theories and eventually his commentators' ones, when they are evoked. My study does not focus immediately on the eventual influences that Arabic logic may have had on medieval logicians, but it can make it possible for some *other* researchers interested in the historical links and the transmissions from one tradition to another, to

[3]Charles Burnett evokes the treatise of al-Fārābī entitled "*On the Syllogism*" [presumably *al-Qiyās*], and also some parts of *al-Shifā*, by Ibn Sīnā, plus *The aims of the Philosophers* of al-Ghazālī.

compare between these traditions because it clarifies the doctrines and the methods used by the Arabic authors.

I will thus use the modern symbolism of propositional, predicate, and modal logics.[4] This practice is justified by the fact that the modern symbolism helps understand and state precisely the different propositions and syllogisms of whatever kind. Some concepts used by Avicenna, for instance, need to be interpreted in order to validate the syllogisms used. Although this has some inconvenience, because Arabic logic, as Aristotelian and traditional logic, is less formal than modern logic and relies on the ordinary language in its way of expressing the propositions and the syllogisms or arguments, it can be useful to clarify the ideas involved. Despite the fact that a clear-cut correspondence between traditional logic and modern logic is probably not realistic, this kind of examination can nevertheless have many advantages. On the one hand, it helps clarify the concepts used by the traditional logicians; on the other hand, it is a very reliable instrument to test the validity of the arguments held by these logicians.

I wish to give some precisions on my method. Unlike the studies on Arabic logic which try to relate it to the general philosophy endorsed by the various authors, including their epistemology, their metaphysics, and their theology, I will focus much more on the *technical aspects and features of the logics* themselves. Furthermore, among the technical features, I will focus on some aspects that have not been very much studied by the other scholars interested in the subject. As an example of these aspects, I can mention the theory of oppositions, whether categorical or modal, which is very rich, particularly in Avicenna's frame, and which has some distinct characteristics that differentiate it from the purely Aristotelian theory. Another example is the hypothetical syllogistic as well as the modal logic, which present some interest and deserve some more studies, given its complexity and its originality in particular in Avicenna's frame. Modal syllogistic has indeed been studied by a number of scholars, who brought out its distinct features and its links with Aristotle's modal syllogistic.[5] But it still deserves some more examination given the complexity of the propositions involved and the multiple interpretations that these propositions give rise to.

On the other hand, it seems that many concepts, methods, and even inferences that one can find in the writings of the medieval logicians are close to those that we

---

[4]In the Western area, many people use the modern symbolisms to interpret the traditional texts as we will see in the whole book. In the Arabic area, see [75] for an application of modern methods to Arabic logic.

[5]See, for instance, Tony Street in his "An outline of Avicenna's syllogistic" *Archiv für Geschichte der Philosophie*, 84 (2), ([136], 129–160), Paul Thom in his paper "Logic and metaphysics in Avicenna's modal syllogistic" ([143], 283–295). Wilfrid Hodges too studied modal logic, in many of his papers and books, for instance, Wilfrid Hodges "Ibn Sīna on modes", *Ibāra* ii.4, 2010 ([81]), "Ibn Sina's alethic Modal Logic" ([93]), see also [83] and *Mathematical Background to the Logic of Avicenna*, 2014 ([92]). See also Henrik Lagerlund and Allan Bäck, who published papers on the modal syllogistic of Avicenna, e.g., Allan Bäck "Avicenna's conception of the modalities", *Vivarium* XXX, 2, 217–255, 1992 ([39]), and Henrik Lagerlund "Avicenna and Tūsi on modal logic", in *History and Philosophy of Logic*, 30: 3, 227–239, 2009 ([105]) (see also [41] for islamic logic in general).

find in Arabic logic. This closeness might be due to the fact that both of them are influenced by the Greek commentators of Aristotle. But is this influence the whole story? Isn't there any direct link between the Arabic tradition and the medieval one? Maybe no straightforward answer can be given to that question, but the study of the theories themselves, their common features as well as their differences could help determine the missed links between both traditions. This is why my aim in this research is to study the theories themselves in order to determine their specificities and their results. In this respect, I wish to clarify the doctrines defended in the systems chosen and to examine them in the light of modern logic. This kind of research may be helpful in the sense that it can make the comparison with other traditions easier, although it is not purely historical, in the sense that I don't try to determine exactly who read whom,[6] and whether or not the translations from Arabic to Latin have been made or not, or were available or not at these times. It focuses on the original texts and the theories studied by using a symbolism that helps determine precisely the nature of the propositions, concepts, and inferences endorsed by the authors and the differences between these theories and the Aristotelian ones, and eventually some of his commentators, in particular, those who are explicitly cited.

Finally, I have to mention a specific difficulty characteristic to the study of ancient texts in general and Arabic texts in particular, namely, the degree of faithfulness and adequacy of what is published under the names of these authors. These problems are well known. For one thing, some publications seem to be falsely attributed to authors like al-Fārābī, for instance, and one has to be careful by comparing several editions of the same text and retain what is common to them. Then the texts may contain some corrupt passages and contain words or sentences badly written, which the publisher must interpret in order to make them coherent. This also requires not only a comparison between the available editions in order to warrant as much faithfulness to the author as possible but also some knowledge of the logical doctrines endorsed by these authors; for only a good knowledge of the logical rules, definitions and concepts can help correct some incongruities that are sometimes present in the edited texts.

Also in some cases, the texts are lost, which makes the study of that particular author incomplete anyway, although one can sometimes rely on the quotations found in the texts of other authors who knew him. But the information remains fragmentary and insufficient if the original text of that specific author is not available.

[6]The transmissions between Arabic logic and other traditions, in particular, Greek and Western Medieval, have been studied by Zimmermann (1972) ([150]), who focuses mainly on al-Fārābī and his Greek and Syriac predecessors. See also A. Hasnawi and W. Hodges (2016) ([79]) and J. Brumberg-Chaumont ([46]) who studies the transmissions of Ancient logic to both Arabic logic and Western Medieval logic.

References

15. Al-Fārābī, Abū Naṣr. 1988. *Kitāb al Qiyās*. In *al-Manṭiqiyāt li-al-Fārābi*, vol. 1, texts published by Mohamed Teki Danesh Pazuh, Edition Qom, 115–151.
16. Al-Fārābī, Abū Naṣr. 1988. *al-Qiyās al-Ṣaghīr*. In *al-Manṭiqiyāt li-al-Fārābi*, vol. 1, texts published by Mohamed Teki Danesh Pazuh, Edition Qom, 152–194.
17. Al-Fārābī, Abū Naṣr. 1988. *Kitāb al-Burhān*. In *al-Manṭiqiyāt li-al-Fārābī*, vol. 1, texts published by Mohamed Teki Danesh Pazuh, Edition Qom, 267–349.
18. Al-Fārābī, Abū Naṣr. 1988. *Mā yanbaghī an yuqaddama qabla ta'allum al-falsafa*. In *al-Mantiqiyyāt li-al-Fārābī*, texts published by Mohamed Teki Danesh Pazuh, Edition Qom, 1–10.
19. Al Fārābī, Abū Naṣr. 1990. *Kitāb al Ḥurūf*, Dar al Machriq, Beirut.
20. Antonelli, Aldo. 2008. Non-monotonic logic. In *Stanford encyclopedia of philosophy*, ed. E. N. Zalta. http://plato.stanford.edu/entries/logic-nonmonotonic/.
21. Aristote. 1971. *Premiers Analytiques*, Translated by J. Tricot, Librairie philosophie J. Vrin, Paris.
22. Aristotle. 1991. *Categories*. In *The complete works of aristotle*, vol. 1, ed. Jonathan Barnes. The Revised Oxford Edition.
23. Aristotle. 1991. *De Interpretatione*. In *The complete works of aristotle*, vol. 1, ed. Jonathan Barnes. The Revised Oxford Edition.
24. Aristotle. 1991. *Prior analytics*. In *The complete works of aristotle*, vol. 1, ed. Jonathan Barnes. The Revised Oxford Edition.
25. Arnault, Antoine and Pierre Nicole. 1970. *La logique ou l'art de penser*. Editions Flammarion.
26. Averroes. 1982. *Talkhīṣ Manṭiq Arisṭu* (*Paraphrase de la logique d'Aristote*), vol. 1: *Kitāb Al-Maqūlāt* (pp. 3–77), *Kitāb al-'Ibāra* (pp. 81–141), *Kitāb al-Qiyās* (pp. 143–366), edited by Gérard Jehamy, Manshūrāt al-Jāmi a al-lubnānīya, al-Maktaba al-sharqiyya, Beirut.
27. Averroes. 1982. *Kitāb al-Jadal*. In *Talkhiṣ Manṭiq Arisṭu*, vol. 2, ed. Gérard Jehamy, Manshūrāt al-Jāmi a al-lubnānīya, al-Maktaba al-sharqiyya Beirut, 499–661.
39. Bäck, Allan. 1992. Avicenna's conception of the modalities. *Vivarium* XXX (2): 217–255.
41. Black, Dedorah. 1998. Logic in Islamic Philosophy‖, Routledge. Available on line in http://www.muslimphilosophy.com/ip/rep/H017.htm#H017SECT2.
46. Brumberg Chaumont and Julie. 2016. The legacy of ancient logic in the middle ages. In *The Cambridge Companion to Medieval Logic*, ed. Catarina Dutilh Novaes and Stephen Read, 19–44. Cambridge University Press.
48. Burnett, Charles. 2004. The translations of Arabic works on logic into Latin in the middle ages and the renaissance. In *Handbook of the history of logic*, vol. 1, ed. Dov Gabbay and John Woods, 597–606. Elsevier BV.
62. Couturat, Louis. 1901. *La logique de Leibniz*, Georg Olms Verlagsbuchhandlung Hildesheim, New Edition (1969).
75. Fakhoury, Adel. 1981. *Mantiq al 'Arab min wijhati naḍar al mantiq al ḥadīth* (in Arabic) (*Arabic logic from the point of view of modern logic*), 2nd Edn., Beirut.
79. Hasnawi, Ahmed and Wilfrid Hodges. 2016. Arabic logic up to Avicenna. In *The cambridge companion to medieval logic*, ed. Catarina Dutilh Novaes and Stephen Read, 45–66. Cambridge: Cambridge university Press.
80. Hasse, Dag Nickolaus. 2014. Influence of Arabic and Islamic thought on the Latin West. In *Stanford encyclopedia of philosophy*, ed. Edward N. Zalta. http://plato.stanford.edu/entries/arabic-islamic-influence/#Log.
81. Hodges, Wilfrid. 2010. Ibn Sīnā on modes, '*Ibārah* ii.4'. http://wilfridhodges.co.uk/arabic07.pdf.
92. Hodges, Wilfrid. forthcoming. *Mathematical background to the logic of Avicenna*. http://wilfridhodges.co.uk/arabic44.pdf.

93. Hodges, Wilfrid. 2010. *Ibn Sīnā's Alethic Modal Logic*, to appear. http://wilfridhodges.co.uk/arabic47.pdf.
83. Hodges, Wilfrid. 2012. 'Ibn Sīna's Modal Logic', plus 'Permanent and Necessary in Ibn Sīna', presented in the workshop *Modal Logic in the Middle Ages*, University of St-Andrews. http://wilfridhodges.co.uk/arabic20a.pdf.
105. Lagerlund, Henrik. 2009. Avicenna and Tūsi on modal logic. *History and Philosophy of Logic* 30 (3): 227–239.
107. Lagerlund, Henrik. 2012. Arabic logic and its influence. *al-Mukhatabat*, no. 1, pp. 175–183.
127. Rescher, Nicholas. 1964. *The development of arabic logic*. University of Pittsburgh Press [Arabic translation by Mohamed Mahrān, Dar el Ma'ārif, Cairo (1985)].
126. Rescher, Nicholas. 1966. *Galen and the Syllogism*. Pittsburgh: University of Pittsburgh Press.
136. Street, Tony. 2002. An outline of Avicenna's Syllogistic. *Archiv für Geschichte der Philosophie* 84 (2): 129–160.
143. Thom, Paul. 2008. al-Fārābī on indefinite and privative names. *Arabic Sciences and Philosophy* 18 (2): 193–209.
150. Zimmermann, F. W. 1972. Some observations on Al-Fārābī and logical tradition. In *Islamic philosophy and the classical tradition, essays presented by his friends and pupils to Richard Walzer on his seventieth birthday*, ed. S. M. Stern, Albert Hourani and Vivian Brown, Cassirer, Oxford, 517–546.

Chapter 2
The Rise of Arabic Logic: Authors, Translations, Topics

As reported by Cristina d'Ancona in her article "Greek Sources in Arabic and Islamic Philosophy" (2013) ([64]), the translations of the Greek corpus started "before the rise of Islam" ([64], Sect. 1) in Syria, by translations from Greek to Syriac, in the fourth and the fifth centuries (AD) and were made by some "theological schools of Edessa and Nisibi" ([64], Sect. 1). But it is during the Umayyad Empire first and particularly during the Abbasid Empire that the translations from Greek into Syriac and afterward from Syriac into Arabic really developed and grew. Under the Umayyad Empire (661–750, AD), some translations have been made in particular by the Christian Syriacs, for instance, Porphyry's *Isagoge* ([64], Sect. 1). But the greatest amount of translations has been made under the Abbasid Empire, starting from the reign of al-Manṣūr (r. 754–775, AD) and including the reign of Hārūn al Rashīd (r. 786–809, AD) and the reign of al-Ma'mūn (r. 813–833, AD).[1] The own secretary of al-Manṣūr, namely, Abdullah Ibn al-Muqaffaʻ (d. 756), is said to have translated "Porphyry's *Isagoge*, the *Categories*, *De Interpretatione* and *Prior Analytics*" ([64], Sect. 2), although some historians attribute these translations rather to his son Muhammed ibn Abdullah Ibn al-Muqaffaʻ. Apart from these very first translations, Cristina d'Ancona claims that there are three major periods and trends with regard to the translations.

The first one concerns the translations made by the so-called "circle of al-Kindī" ([64], Sect. 3), under the reign of al-Ma'mūn whose "leader" ([64], Sect. 3) was Abu Yaʻqūb ibn Isḥāq al-Kindī (805–873, AD), the very first Arabic philosopher. It is in that period that the so-called *Beyt al-Ḥikma* (literally the House of Wisdom) has been founded (that was on 830 AD, according to N. Rescher ([127], 140)). This group of authors translated Aristotle's *Metaphysics* plus the *Prior Analytics* (translated by Ibn al-Bitrīq) ([64], note 32) *Sophistici Elenchi*, and other philosophical treatises ([64], Sect. 3). Some of Alexander of Aphrodisias' commentaries (for instance, "the first book of Alexander's commentary on the De gen. corr…." ([64], Sect. 3)), together with some writings of Plotinus, Proclus and John

[1] Ibn Khaldūn evokes only two of these, namely, Abu Jaʻfar Al-Manṣūr ([100], 74) and al-Ma'mūn ([100], 75), the latter being the most interested in sciences in general and the most encouraging to the whole translation process ([100], 75).

S. Chatti, *Arabic Logic from al-Fārābī to Averroes*, Studies in Universal Logic,
https://doi.org/10.1007/978-3-030-27466-5_2

Philoponus (for instance, the "Physics") among other things ([64], Sect. 3) were also translated at that period.

The second trend concerns the translations made by Ḥunayn ibn Isḥāq (809–877, AD), his son Isḥāq ibn Ḥunayn (845–910, AD), and their collaborators. They translated almost all the treatises, both from Greek into Syriac and from Syriac into Arabic. We will talk here mainly about the logical treatises. Among these treatises, we can cite *De Interpretatione*, translated into Syriac by Ḥunayn, and from Syriac to Arabic by his son Isḥāq, the *Prior Analytics*, from Greek into Syriac "partly by Ḥunayn and partly by Isḥāq, and into Arabic by a certain Tayadurus" ([64], Sect. 4), who has been identified "with Tadhari ibn Basil Akhi Istafan" ([64], note 57) by some historians. His translation is the one that is used by Abderrahman Badawi in his recent edition of Aristotle's *Organon* (1980) [40]. In this period too, the *Posterior Analytics* and the *Topîcs* have been translated into Syriac by Isḥāq and for part of the former by his father. The translation into Arabic of the *Topics* has been made by "Abu 'Uthman al-Dimashqī (books I–VII) and by Ibrahim ibn 'Abdallah (book VIII)" ([64], Sect. 4). Some of Alexander's treatises such as *On the principle of the All* and *On Intellect* have been translated by Isḥāq into Arabic ([64], Sect. 4), together with Porphyry's *Isagoge*, and a commentary on the *Categories* which have been translated by Abu 'Uthman al-Dimashqī for the former and by Ḥunayn for the latter ([64], Sect. 4). Many of Galen's works have also been translated by Ḥunayn and one of his pupils.

The third trend concerns the translations made by "Abu Bishr Mattā ibn Yūnus, Yaḥiā ibn 'Ādī and the Baghdad Aristotelians" ([64], Sect. 5). The first author translated, for instance, "the Syriac version of the *Posterior Analytics* made by Isḥāq ibn Ḥunayn and the Syriac translation of the *Poetics*" ([64], Sect. 5) plus other Aristotelian treatises. Yaḥiā ibn 'Ādī translated "the Syriac version of the *Topics* made by Isḥāq ibn Ḥunayn" ([64], Sect. 5).

As mentioned by A. Hasnawi and W. Hodges ([79], Sect. 2.1), these works of the Aristotelian Greek commentators, such as "Themistius (d.c. 388) and … the commentaries of the members of the late Neoplatonic school (fifth-sixth centuries)" which have been translated have had an influence on the Arabic commentaries, in particular, those of our three authors. For among these commentaries, the two authors evoke "glosses on the *Categories* which reflect Simplicius' commentary on this treatise" ([79], 47); they say that "traces [of Alexander's commentary on *Prior Analytics*] are visible in the works of al-Fārābī, Avicenna and Averroes" ([79], 47–48) and add that "the same can be said about Themistius' paraphrase of *Prior Analytics*" ([79], 48). So by studying the texts, one should find the close links between these Greek commentaries and the Arabic ones, but this influence does not seem to be the whole story, in particular, in Avicenna's writings, which contain many original and rich developments, which are different from the theories presented by his predecessors.

Nicholas Rescher says in *The Development of Arabic Logic* [127] that these translations gave rise to various kinds of commentaries of the Aristotelian treatises as well as Porphyry's *Isagoge*. These commentaries use the writings of the Greek commentators such as Alexander of Aphrodisias and Galen, among others, who presumably transmitted the Stoics' heritage to the Arabic authors. However, it seems that no commentary of the Stoics' treatises is really made by the Arabic logicians in the way the commentaries of the Aristotelian treatises are made.[2] They only include the study of the hypothetical syllogisms, characteristic of the Stoïc tradition, inside their respective correspondents of the *Prior Analytics* or even the *Categories*, as in al-Fārābī's case. Their own studies of these kinds of syllogisms are different. Avicenna gives them much importance and constructs a whole system involving these syllogisms and other complex ones mixing between categorical and hypothetical propositions, while Averroes, for instance, treats them as really secondary, as shown by the simple fact that he devotes only a few pages of his *Kitāb al-Qiyās* to the analysis of these arguments. Al-Fārābī provides in his *al-Maqūlāt* and a short part of his *al-Qiyās* an analysis of the hypothetical syllogisms, but he treats them in a relatively elaborate way.

The commentaries were of different kinds, oscillating from short commentaries looking more like summaries to very long commentaries, which include quotations of the Aristotelian text, mentions of the ancient criticisms, and personal answers to these criticisms, in order, most of the time, to defend the Aristotelian positions against its critics. They also are attempts to make the Aristotelian claims coherent and to explain them as clearly as possible by means of examples and detailed analyses of the different ideas defended by Aristotle. However, these commentaries lead, in the final analysis, to distinct opinions, which are sometimes significantly different from that of Aristotle, despite their alleged faithfulness to the texts. This is clear in al-Fārābī's commentaries which introduce many relatively elaborate distinctions, under the influence of Alexander of Aphrodisias and presumably the Syriac translators of the Aristotelian corpus (see [150], 520ff for details). Avicenna is the most original logician in this respect in so far as he does not only comment on Aristotle but presents what he considers as a new system, more elaborate and complex than Aristotle's one, although influenced by it, given that the general background from which he starts is Aristotelian.

However, before analyzing in detail these commentaries and treatises, I will first briefly present the main logicians mentioned by Nicholas Rescher in the above text, and then I will focus on the three authors already mentioned, namely, al-Fārābī (873–950, AD), Avicenna (980–1037, AD), and Averroes (1126–1198, AD). This choice is motivated, not only by the historical predominance of these three authors, which is undisputable, but also and above all, by the fact that the *technical features* of their systems are not really well known until now, both in the Arabic world and in the western area, although there are some very good analyses of these logics as such, that have been made by some western and eastern authors. My aim is to provide an analysis based on the *original Arabic* texts in order to clarify the theories and the concepts and to offer to the readers a faithful and non-deviated

[2] See, for instance, Tony Street "Arabic Logic" where he notes that "...although it is clear that Stoïc logic filtered through to scholars working in Islamic law and theology, there is no tradition of translating Stoic works and commenting on them comparable to that devoted to Peripatetics works,..." ([137], 526–527).

interpretation of these logical works. For that purpose, I will use the modern logical symbolism because it is a very efficient tool to check the validity of the rules and arguments.

The authors are classified chronologically by Rescher ([127], 149, 150, 154, 177, 186, 189). The classification includes almost all the Arabic philosophers and scientists, even those who were much more interested by mathematics or other sciences than by logic as such. For instance, al-Khawarizmi, who is included in Rescher's classification, was above all a mathematician and an astronomer, much more than a logician. He was the creator of Algebra, as witnessed by the title of his seminal contribution in Algebra, which is *Kitāb al mukhtaṣar fi ḥisāb al-jabr wa-l-muqābala* (The Compendious Book on Calculation by Completion and Balancing) written under the reign of al-Ma'mūn (813–833). So among the authors listed by Rescher, I will mention only those who wrote in the field of logic.

In the first list of the period (800–900, AD) provided by Rescher, we can retain the names of Ibn al-Muqaffa' (750–815, AD) who wrote a short commentary of Aristotle's writings, al-Kindī, who was a philosopher and a commentator, Ḥunayn Ibn Isḥāq (809–877, AD), Isḥāq Ibn Ḥunayn (845–910, AD) who both are very important translators ([127], 149–150), while the second period (900–1000, AD) comprises, among others, Abū bishr Mattā (870–940, AD), who was mainly a translator, al-Fārābī (873–950, AD), who is a philosopher and a commentator, and Ikhwēn Assafā (The Brethren of Purity, 970–1030, AD) ([127], 154–155).

In the third period (1000–1100, AD), we can find figures such as Ibn Sīnā (Avicenna) (980–1037, AD) and al-Ghazālī (1059–1111, AD) ([127], 177). The former is almost unanimously considered as the greatest logician in the Arabic-speaking world, whereas the latter who was influenced in many respects by Avicenna was widely known in the Medieval West. In the fourth period (1100–1200, AD), the authors cited are mainly Andalusian as, for instance, Ibn Bāja (1090–1140, AD) who was an astronomer, philosopher, and physicist, and was known under the name of Avempace in the Latin West, Ibn Rushd (Averroes) (1126–1198, AD), and Ibn Maimūn (known under the name of Maïmonide in the West) (1135–1204, AD) ([127], 186). The fifth list (1200–1300, AD) comprises, mainly, though not only Afḍal al-Dīn al-Khūnajī (1194–1249, AD), Athīr al-Dīn al-Abhārī (1200–1265, AD), who wrote a very influential and popular textbook on logic, Nasir-eddin at-Tusi (1201–1274, AD), who commented on Avicenna's *al-Ishārāt wa al-Tanbīhāt*, and Najmeddine al-Qazwīnī al-Kātibī (1220–1276 or 1292, AD) ([127]), who was a follower of Avicenna. We can also cite Fakhreddin al-Rāzī, who was a very important post-Avicennan logician in the eastern area.

Al-Khūnajī was a logician and was known to have generalized, applied to logic and made popular the distinction between the conceptions and assents which is still considered until now as describing the two main topics of logic. This distinction can be found in many Avicennan treatises, and it is at the heart of the subsequent studies in Arabic logic. Nasir-eddin at-Tusi is a follower of Avicenna and has commented on his texts, making his own theory intimately related to that of Avicenna. Najmeddin al-Kātibī al-Qazwīni was a logician and he wrote a book called

Al-Risāla al-shamsīya (The Solar Epistle), where he develops the temporal analysis initiated by Avicenna.

Most logicians of the second period (900–1000), belonged to the so-called "School of Baghdad" which was concerned by the commentaries and the teaching of the Aristotelian Corpus. The members of the school of Baghdad were mainly Christian. But al-Fārābī who was also member of that school was Muslim. The Muslim scholars became interested in logic later and taught it in Arabic. They made commentaries of the whole Aristotelian Corpus including *Posterior Analytics, Topics*, and so on. They were also interested in modal logic and the hypothetical syllogisms whose study was in general included in the *Prior Analytics*. They also studied analogical and inductive arguments; some of them (*Ikhwan aṣṣafā*) used this kind of arguments in theological discussions ([127], 164).

The main texts translated are the treatises of the Organon. The so-called "Arabic Organon" contains

1. Porphyry's Isagoge,
2. Categories,
3. De Interpretatione,
4. Prior Analytics,
5. Posterior Analytics,
6. Topics,
7. Sophistical Refutations,
8. Rhetorics, and
9. Poetics ([127], 133).

However, the North African historian Ibn Khaldūn [1332–1406 AD] complains, in his *al-Muqaddima* (*Prolegomena*), about the fact that some of these treatises were almost abandoned in later periods for nobody would study them and logic became a pure formal discipline having nothing to do with science in general or metaphysics. He also says that in these times, the commentaries of Aristotle's treatises have been abandoned in favor of textbooks and summaries which do not contain any new theory. He himself did not produce any new logical system, but he was interested in the evolution of the research in that field, for he wrote a chapter devoted to logic in his *al-Muqaddima* ([100], Sect. VI, Chap. 22) where he evaluated the latest developments of logic and of its teaching in the Arabic area. Ibn Khaldūn mentions eight treatises to which a ninth one (Porphyry's Isagoge) has been added by "the Greek philosophers (*ḥukamā al-yūnāniyyīn*)" ([100], 94). Four of these treatises (*Categories*, *De Interpretatione*, *Prior Analytics* and *Posterior Analytics*) ([100], 93) are devoted to the study of the "form of the syllogism" ([100], 92), while the other four (namely, *Topics*, *Sophistical Refutations*, *Rhetorics*, and *Poetics*) were devoted to the study of "their matter" ([100], 92), Porphyry's *Isagoge* being devoted to the study of the "five predicables" ([100], 94). These books have been studied with great attention until the period of Averroes, but later on, Ibn Khaldūn says that some authors changed the field by introducing some significant modifications such as the suppression of the *Categories*, considered as "nonessential" ([100], 94). In addition, they included inside

De Interpretatione the analysis of conversions, which was part of the *Topics* in Ancient logic ([100], 94) and removed from logic the five following treatises: *Posterior Analytics* (*al-Burhān*), *Topics* (*al-Jadal*), *Rhetorics* (*al-Khaṭāba*), *Poetics* (*al-Shi'r*), and *Sophistical Refutations* (*al-Mughālaṭa*) ([100], 94–95). According to him, these treatises are fundamental in logic; this is why he deplores the fact that they were no more studied, after Averroes. This had, according to him, the following consequence: logic was no more viewed as a tool for science, but as an independent art. Ibn Khaldûn evokes two authors in this context: Fakhr al-Dīn ibn al-Khaṭīb[3] and Afḍal al-Dīn al-Khūnajī, who both wrote very influential books, which were taught in the eastern areas "until this [i.e. his own] time" ([100], 95). Ibn Khaldūn's complaints may be understood in the light of his own view about logic, its significance, and its utility. For according to him, logic is first and foremost a tool for other sciences, and this is why one has to start studying it before any other science. It is as a tool for other sciences that logic is the most useful, not as an independent discipline.

Ibn Khaldūn's report acknowledges, however, a *significant move* in the way logicians viewed their own field. For in the new conception endorsed by al-Khunāji and Fakhreddin al-Rāzī, logic appears as an independent and formal discipline, having its own subject matter and worth studying for itself, independently of its alleged relations with other sciences and its usefulness for them. This new conception of logic as an independent and formal discipline makes it closer to the modern conception, as K. El-Rouayheb rightly notes when he says: "Ibn Khaldūn himself lamented this development, but the resulting narrower view of the scope of *manṭiq* made it much closer to the contemporary understanding of 'logic' than the earlier Peripatetic conception of it as a discipline that covers all the books of the *Organon*" ([73], 67). This new conception departs significantly from the traditional one, which mixes between logic proper and its theoretical or practical applications and could ultimately be due to the influence of Avicenna on these authors and their followers.

Anyway, in the first period which is mainly our concern here, the topics studied were rather various and rich. The three authors were interested in the categorical syllogistic together with modal logic and the hypothetical syllogisms which were in general included into the *Prior Analytics*, and sometimes even in the *Categories*.

Before entering into the details of their doctrines, let us first consider the aims of the three logicians considered, since as we will see below, their contributions are significantly different, so that it could be useful and enlightening to first see what they wanted to do with their logical systems in order to explain the differences that we can find in these systems.

Let us start with al-Fārābī. According to al-Fārābī, logic is first and foremost a tool, which is useful to reach the truth. But his aim in studying logic has much to do with teaching, since he seems to believe that Aristotle's logical corpus and the commentaries that have been made on it by the Greek authors, whether the

[3]This author is Fakhreddin al-Rāzī.

peripatetics or the neoplatonists arrived at maturity and can be taught without significant changes. Thus, in his book *al-Ḥurūf*, he says:

"… after dialectical methods have been justified to the greatest extent possible and have almost become scientific. Things proceed in this manner until philosophy attains the condition it was in at the time of Plato. They continue to be engaged in these matters until things become settled where they were during Aristotle's time. Theoretical science is completed, the mathematical methods are all distinguished, theoretical philosophy and universal practical philosophy are perfected, and they cease to contain any object of examination. It becomes an art that is *only learned and taught*, and it is taught both to a select audience and commonly to all. Select instruction proceeds by demonstrative methods only, whereas common instruction, which is public, proceeds by dialectical." ([19], 151.14–152.3, emphasis added).

This focus on teaching and also dialectical methods can be seen through his own treatment of logic and his use of it, as we will see below, since he does not really aim at constructing a new system, different from Aristotle's one. He seems to consider logic more as a tool to reach the truth in other sciences or fields than as a science by itself, which could be studied for its own interest. For instance, in the book *Ihṣā al-'Ulūm*, he says what follows:

"The rules of logic which are the tools by which the intellected things where the reason could make errors or might omit to perceive are examined (*yumtaḥanu bihā*)" ([3], 68).

So logic is seen as an instrument that can help the intellect to correct errors and reach the truth, when the things examined are not obvious and not sufficiently clear. Its main purpose is thus to verify the truth of our knowledge and to help reaching the truth in various fields and domains. This verification could be made by one person for herself or for other people, for instance, in teaching or in discussions, for it is by following the rules of logic that one can check and verify the correction of the deductions that lead from true premises to a conclusion.

Al-Fārābī was educated within the Aristotelian tradition as Majid Fakhry says in what follows:

"In the field of logic, al-Fārābī's standing was unmatched. He was the first logician to break with the Syriac (Jacobite-Nestorian) tradition, which flourished at Antioch, Edessa and Qinnesrin, and refused for religious reasons to proceed beyond the first four parts of the Aristotelian logical corpus, i.e. the Categories, Peri hermeneias, the first part of *Analytica Priora* and the *Isagoge* or *Introduction to the Categories*, written by Porphyry of Tyre. His logical output *covered the whole Organon*, together with the *Rhetorica* and *Poetica*, as well as the Isagoge of Porphyry, in the form of paraphrases or large commentaries" ([74], 154, emphasis added).

He did thus know all the Aristotelian treatises and he commented in particular on *Peri Hermeneias* and on *Prior Analytics*. Majid Fakhry adds that:

Abū Naṣr al-Fārābī himself reports that he received instruction from Yuhanna Ibn Haylān up to the end of *Analytica Posteriora* (*Kitāb al-Burhān*). What came after the "existential moods" used to be called the unread part, until it was read then. The rule, thereafter, once the responsibility devolved upon Muslim teachers, was to read what one was able to read of the existential moods. Abū Naṣr states that he read up to the end of *Analytica Posteriora* [*Kitāb al-Burhān*]" (From Ibn Usaybi a, "*Uyūn al-anbā*", trans. in [74], 159).

He was thus a trained commentator whose aim was not that much to construct a new system of his own. Rather, he relied on the already available writings of Aristotle and his followers and focused on the clarification of Aristotle's doctrines, although by doing so, he did arrive at some original ideas that can be seen as improvements of the doctrine, since he criticized some points in Aristotle's doctrine, especially in modal logic, and added many precisions in categorical logic. In hypothetical logic, his contribution is also independent of Aristotle's text, since this whole field is not that much Aristotelian, but related to the Stoïc logic.

As to Avicenna, he was very prolific in logic, and did not hesitate to express his disagreement with some of Aristotle's doctrines and to modify them in a novel way. He did not hesitate to build new systems that cannot be found in the writings of Aristotle or in his commentators, whether Greek or Arabic. His own view with regard to logic is different from al-Fārābī's view, for according to him, logic is a science on its own, even if it is also a tool for other sciences. As we can see from the passages below, he freely claimed his departure from Aristotle in some fields and the changes that he felt necessary to introduce:

> "There is nothing of account to be found in the books of the ancients which we did not include in this book of ours; if it is not found in the place where it is Customary to record it, then it will be found in another place which I thought more appropriate for it. To this I added some of the things which I perceived through my own reflection and whose Validity I Determined through my own theoretical analysis, especially in Physics and Metaphysics—and even in Logic, if you will; for although it is Customary to prolong [the discussion on] the first principles of logic with material that does not belong to Logic but only to the philosophical discipline—I mean the First Philosophy—I avoided mentioning any of that [in Logic] and wasting thereby time, and deferred to it in its [proper] place…
>
> ….
>
> I also wrote [another book . . .] in which I presented philosophy as it is naturally [perceived] and as required by an unbiased view which neither takes into account in [this book] the views of colleagues in the discipline nor takes precautions here against creating schisms among them as is done elsewhere; this is my book on Eastern philosophy. But as for the present book, it is more elaborate and more accommodating to my Peripatetic colleagues. Whoever wants the truth [stated] without indirection, he should seek the former book; whoever wants the truth [stated] in a way which is somewhat conciliatory to colleagues, elaborates a lot, and alludes to things which, had they been perceived, there would have been no need for the other book, then he should read the present book" (*Madkhal* 10.1–17, trans. [78], 43–45).

As we will see below, Avicenna's systems are the most original and innovative ones in the Arabic early tradition. His aim in logic was not so much to be faithful to Aristotle or to comment on his writings; rather, it was to study logic on its own by considering it as a science, not only as a tool. This is why he created his own categorical, modal, and hypothetical systems, which, in many respects, depart clearly from Aristotle's syllogistic, even if the general background of Avicenna's investigations remains Aristotelian. Thus starting from Aristotle's syllogistic, and taking it as its main basis, he felt free to introduce some changes in his analysis of the categorical as well as the modal propositions and moods. These changes affected the analysis of propositions and their logical relations and consequently the moods stated, whether in categorical or in modal logic. His hypothetical logic has no counterpart in any of Aristotle's writings and even in his commentators's ones. As Dimitri Gutas says,

> "Avicenna's rationalist empiricism is the main reason why he strove in his philosophy on the one hand to perfect and fine-tune logical method and on the other to study the human soul and cognitive processes at an almost unprecedented level of sophistication and precision. In section after section and chapter after chapter in numerous works he analyses not only questions of formal logic but also the very conditions operative in the process of Guessing Correctly and hitting upon the middle term: how one can work for it and where to look for it, and what the apparatus and operations of the soul are that bring it about" ([78], 376).

Averroes can be seen as the author who departs the most from Avicenna, for his aim is primarily and mainly to remain as faithful as possible to Aristotle, and to defend his theories against all the criticisms, misunderstandings, false commentaries, etc. that could have been made by his followers. As one of the contemporary authors says, Averroes' aim is to return back to the authentic Aristotelian writings, and "to restore the authentic doctrine of Aristotle" ([69], 51), that is, to get rid of all what has been added to this theory, which is seen by Averroes as some kind of distortion of the original doctrine.

These different aims can explain the different features of the three theories that we will analyze in this book. If logic is mainly seen as a tool, or if it is seen as essentially a defense of Aristotle's doctrine and nothing else, this cannot lead to the same results as the view that logic is a science by itself that is not necessarily dependent on Aristotle's findings or on his specific doctrine, even if it remains close to it. So if Avicenna was innovative unlike his predecessor and his successor, it is surely because of his vision of logic and his (relative) independence toward Aristotle. And if Averroes did not introduce a significant change in logic, it is precisely because he did not want to do that, given his will to "restore" Aristotle's logic. Similarly, al-Fārābī's view led him to consider logic as an already well-known theory that one could teach and that does not need to be significantly improved.

Let us now analyze these theories, starting from categorical logic.

References

3. Al-Fārābī, Abū Naṣr. 1968. *Iḥṣā al-'Ulūm*, ed. Uthman Amin, Maktabat al-anjelu al-misriyya, Cairo.
19. Al-Fārābī, Abū Naṣr. 1990. *Kitāb al Ḥurūf*. Beirut: Dar al Machriq.
40. Badawi, Abderrahman. 1980. *Manṭiq Arisṭu*, vols. 1 and 2, Dar al Kalam, Beirut.
63. Czeżowski, Tadeusz. 1955. On certain peculiarities of singular propositions. *Mind* 64: 287–308.

64. D'Ancona, Cristina. 2013. Greek sources in Arabic and Islamic philosophy. In *Stanford encyclopedia of philosophy*, ed. Edward N. Zalta, http://plato.stanford.edu/entries/arabic-islamic-greek/.
69. Elamrani-Jamal, Abdelali. 1995. Ibn Rušd et les Premiers Analytiques d'Aristote: Aperçu sur un problème de syllogistique modale. *Arabic Sciences and Philosophy* 5: 51–74.
73. El-Rouayheb, Khaled. 2016. Arabic logic after Avicenna. In *The Cambridge Companion to Medieval Logic*, eds. Dutilh Novaes, Catarina, and Read, Stephen, 67–93. Cambridge University Press.
74. Fakhry, Majid. 2002. *Al-Fārābī, founder of Islamic neoplatonism, his life, works and influence*. Oxford: Oneworld.
78. Gutas, Dimitri. 2014. *Avicenna and the aristotelian tradition: introduction to reading Avicenna's philosophical works*, 2nd ed. Leiden: Brill.
79. Hasnawi, Ahmed, and Hodges, Wilfrid. 2016. Arabic logic up to Avicenna. In *The Cambridge companion to medieval logic*, ed. Catarina Dutilh Novaes and Stephen Read, 45–66. Cambridge: Cambridge University Press.
100. Ibn Khaldūn, Abdurrahmān. 2005. *Al-Muqaddima*, eds. Abdessalam Chaddadi, Beyt al Funūn wa al-'Ulūm wa al-'Ādāb, Casablanca, Morocco.
127. Rescher, Nicholas, *The development of Arabic Logic*, University of Pittsburgh Press, Arabic translation by Mohamed Mahrān, Dar el Ma'ārif, Cairo, (1964), Arabic translation (1985).
125. Rescher, Nicholas. 1992. *Studies in the history of Arabic logic*. University of Pittsburg Press, (1963); Arabic translation by Mohamed Mahrān, Cairo (1992).
137. Street, Tony. 2004. Arabic logic. In *Handbook of the history of logic*, vol. 1, eds. Gabbay, Dov, and Woods, John. Elsevier, BV.
150. Zimmermann, F. W. 1972. Some observations on Al-Fārābī and logical tradition. In *Islamic Philosophy and the Classical Tradition, Essays presented by his friends and pupils to Richard Walzer on his seventieth birthday*, eds. S. M. Stern, Albert Hourani and Vivian Brown, 517–546. Cassirer: Oxford.

Chapter 3
Categorical Logic

3.1 Conceptions and Assents

The Arabic logicians traditionally divide logic into two parts as witnessed by Tony Street in his article "Arabic and Islamic Philosophy of Language and Logic" ([139]). These two parts are the following:

1. The study of "conceptions" (*Taṣawwer*) (translation Street 2008; this word is translated as "Conceptualisations", in W. Hodges and T. A. Druart "Al-Fārābī's Philosophy of Logic and Language" ([90]).
2. The study of "Assents" (*Taṣdīq*) (translation Street [139]).

This division of logic into conceptions and assents can first be found in al-Fārābī. It is also present in Avicenna's text as we will show below. In al-Fārābī, this distinction is made in relation with teaching, for instance, in the book entitled *al-Alfāẓ al-musta'mala fi al-manṭiq*, where al-Fārābī says what follows:

> "For everything that can be learned by a discourse, the learner must necessarily go through three situations (*aḥwāl thalātha*): the first one is to conceive (*taṣawwur*) this thing and to understand the meaning of what he heard from the teacher, that is, the meaning that the teacher intended by the discourse. The second thing is to *assent* (*an yaqa'a lahu al-taṣdīq*) *to what he conceived* or understood from the discourse of the teacher. And the third one is to memorize what has been conceived and has been assented to." ([4], 87, emphasis added)

About this opinion on both conceptualization (*taṣawwur*) and assent (*taṣdīq*), Wilfrid Hodges comments in this way: "… in another passage on teaching, he adds that 'We can seek *taṣdīq* either of simple things or of compound'. For al-Fārābī a proposition is always compound, so he is telling us that non-propositional concepts can be assented to." ([90], Sect. 8), and he adds:

> "The view that both propositions and non-propositional concepts can be true runs fairly deep in al-Fārābī's thinking, although he recognises that not everybody agrees with it ([*Commentary on De Interpretatione*] 52.13f). For example this view allows him to think of definition of non-propositional concepts and demonstration of propositions as overlapping

S. Chatti, *Arabic Logic from al-Fārābī to Averroes*, Studies in Universal Logic,
https://doi.org/10.1007/978-3-030-27466-5_3

procedures; there can be definitions that are identical with demonstrations except in the order of their parts ([*Demonstration*] 47.11). Avicenna a hundred years later found it essential to distinguish between non-propositional and propositional concepts. Provocatively he used al-Fārābī's own terminology of *taṣawwur* and *taṣdīq* to fix the distinction; for Avicenna any concept can be conceptualised, but only propositions can go on to be verified" ([90], Sect. 8)

This distinction means that logic studies first what expresses the concepts, that is, the names and the verbs and their characteristics, second what expresses the propositions and the arguments. But as the passages above show, in al-Farabi's view, both procedures are not sharply distinguished, since he says that the pupil "assents to what he conceived," so that it seems that concepts can also be something that one can assent to, while in Avicenna's view, one can only assent to propositions, not to concepts. So the distinction as it is defined by al-Fārābī contains some confusion related to the fact that concepts should not be objects of assent, but only of conception, which W. Hodges noted in the passages cited above, which Avicenna does not commit.

Since logic is concerned with reasoning, and reasoning is related primarily with propositions which contain names and verbs expressing various kinds of concepts and are themselves involved in syllogisms, which lead from premises to a conclusion, it seems natural to start the study of the arguments by analyzing the components of the propositions which constitute them. But this characterization of logic is not the only one in the Arabic area. It has been endorsed by some logicians who contest Avicenna's characterization according to which logic is the study of the so-called "Second Intentions".

In Avicenna's writings, this distinction is explained as follows:

"A thing is knowable in two ways: one of them is for the thing to be merely Conceived so that when the name is uttered, its meaning becomes present in the mind without there being truth or falsity, as when someone says "man" or "do this!"...The second is for the Conception to be accompanied with Assent, so that if someone says to you, for example, every whiteness is an accident" you do not only have a conception of the meaning of this statement, but also assent to it being so." (*al-Shifā, al Madkhal*, 17, cited by [139], Sect. 2.1)

The first is the study of single words (classified in several classes) by means of definitions. The second is the study of propositions, their truth values, and the way they can be proved. This includes the study of the different arguments, notably the syllogistic arguments, whether categorical, modal, or hypothetical. According to K. El-Rouayheb ([72], 70), the view that Conceptions and Assents are the subject matter of logic was endorsed by al-Khūnājī against the rival view that the subject matter of logic is Second Intentions" which was endorsed by Avicenna, particularly in his "*Eisagoge* and *Metaphysics* of *al-Shifā*" ([72], 70). Tony Street, who evokes K. El-Rouayheb ([72]) and A. I. Sabra ([131]) says what follows:

"Avicenna's doctrine on the subject matter of logic was not adopted by the majority of logicians who followed him (pace Sabra (1980) 757). Quite the contrary, Khūnajī argued in the second quarter of the thirteenth century that the subject matter of logic was Conceptions and Assents, a claim that was energetically resisted by the remaining Avicennan purists like Tūsī. A recent study has clarified what is at issue in this debate (El-Rouayheb (2012))" ([139], Sect. 2.1.3)

Avicenna himself evokes this view about logic both in his Metaphysics and in his *Eisagoge* (*al-Madkhal*) of *al-Shifā*, according to K. El-Rouayheb ([72], 75, note 15). He does not talk about it in his logical treatises such as *al-Qiyās* (the correspondent of *Prior analytics*) or *al-Ishārāt wa-al-tanbīhāt*. But before examining what he says in these logical treatises, let us see what is meant by "Second Intentions". According to K. El-Rouayheb, this concept means, in Avicenna's view "what accrues to first intentions due to the latter's existence in the mind" ([72], 74). Second intentions are some sort of second-level concepts, that is, concepts applying to other ones, such as the concept "universal" which applies to the concept "animal" which itself corresponds "to entities in the real world" ([139], Sect. 2.1.2). Thus, the second intentions studied by logic, in this view, are the concepts that are properties of other concepts, not those that are properties of real objects. In his *Metaphysics*, Avicenna evokes explicitly this view and explains it as follows: "The subject matter of logic, as you know, is given by the secondary intelligible meanings, based on the first intelligible meanings, with regard to how it is possible to pass by means of them from the known to the unknown, not in so far as they are intelligible and possess intellectual existence ([an existence] which does not depend on matter at all, or depends on an incorporated matter)" (Avicenna, *Metaphysics*, p. 7, cited in [139], Sect. 2.1.2). Note that, in this passage, he suggests that the study of secondary intentions is what makes the mind "pass from the known to the unknown," that is, arrive to an unknown conclusion from known premises, which is exactly what all logical arguments do. He thus introduces almost implicitly the notions of deduction and of reasoning in describing the subject matter of logic.

However, in *al-Ishārāt*, for instance, Avicenna gives a slightly different and very general definition:

> "What is meant by logic, for men, is that it is a regulative (*qanūnīya*) tool whose use prevents his mind from making errors (*'an yaḍalla fī fikrihi*)" ([36], 117).

where logic is seen as an Organon much more than as the study of "Second Intentions". His follower Tūsī comments on this definition by saying that logic is, according to Avicenna "a science by itself (*'ilmun bi nafsihi*) and a tool with regard to other sciences..." ([36], 117, note (1)).

In other treatises, Avicenna presents almost the same definition, for he says in *al-Najāt*, for instance, what follows: "... for [logic] is the tool that prevents the mind from errors in what men conceive and assent to and it is what leads to the true convictions by providing their reasons and by following its methods" ([32], 3) and he goes on talking precisely about the so-called Conceptions and Assents, in the chapter that just follows that quotation. For instance, he says that "Every knowledge and science is either conceptions or assents. The conception is the first science and is acquired by the term (*ḥadd*)... like our conception of the essence of men" ([32], 3). This quotation shows that the idea of Conceptions and Assents can be found in Avicenna's text itself and is not foreign to Avicenna, but it is applied here to all sciences, not only to logic, although the treatise itself is devoted to the study of logic. In *Mantiq al-mashriqiyyīn*, Avicenna entitles the first chapter "On the science of logic" and explains that title by saying that logic is "the first art in conceptions and assents (*al-fann al-'awel fi-al-taṣawer wa-al-taṣdīq*)" ([31], 9). This means once again that the division above is not foreign to Avicenna's thought.

However, although the notions of conception and assent are indeed present in Avicenna's texts, these notions are not specifically applied to logic in Avicenna's view. This is, according to T. Street, what distinguishes Avicenna's view about logic from Khūnajī's one, for "All knowledge, according to Avicenna, is either Conception or Assent... What the later logicians in the line of Fakhreddīn al-Rāzī did was make Conceptions and Assents the subject matter of logic. We know that Khūnajī was the first to do this thanks to a report in the *Qistās al-Afkār* of Shamseddīn as-Samarḳandī (d.c. 1310)." ([139], Sect. 2.1.3). So, what applied to all sciences in Avicenna's view applies in the later one only to logic. According to Tony Street, the shift from the first view to the later one is due to some weaknesses of Avicenna's view which he expresses as follows: "The claim that Avicenna's identification of secondary intelligibles as logic's subject matter is *inaccurate* and *too narrow* to achieve what he hopes it can" ([139], Sect. 2.1.3 my emphasis). This narrowness is also noticed by K. El-Rouayheb who reports an objection made by some people and "endorsed by Kātibī," according to which "second intentions are not exhaustive of the subject matter of logic" ([72], 74). The reason evoked is that some concepts studied by logic are first intentions rather than second intentions, for he says, reporting Katibi's opinion: "The logician 'investigates' (*yabḥathu 'an*) concepts such as 'differentia' and 'genus' and, crucially, these are intrinsic accidents of first intentions. It follows that the subject matter of logic includes first intentions as well as second intentions." ([72], 74).

This rival view became the most popular one and is still present in the Arabic traditional logic books, for instance, the famous Kātibī's book *Al-Risāla al-Shamsīyya* and "even much later Arabic handbooks on logic,... for example *Sullam al-'ulūm* by the Mughāl scholar Muḥibbullāh Bihārī (d. 1707)" ([72], 78). Almost all of the textbooks start by the study of concepts, then the propositions, and finally the syllogistic arguments. The view that logic is the study of Conceptions and Assents seems then to be largely accepted by Arabic logicians as claimed both by El-Rouayheb and by M. Mahrān in ([127], introduction, 40). However, K. El-Rouayheb says that, in Khūnajī's view, "the subject matter of logic is presented as being 'the objects of conceptions and assents' (*al- ma'lūmāt al-tasawwurīya wa-al-tasdīḳīya*)" ([72], 71) rather than merely conceptions and assents. The shift between "objects of conceptions and assents" and "conceptions and assents" *simpliciter* occurred, according to him, in the writings of "Khūnajī's students, Ibn Wāṣil and Sirāj al-Dīn al-Urmāwi (d. 1283)." ([72], 71, note 5). It was followed later on by several scholars.

By the study of conceptions, what is meant is the clarification of the meanings of simple words and complex ones. The aims are mainly (among others):

1. to distinguish between general and particular words.
2. to indicate how the word signifies (by "equivalence" (*muṭābaḳa*) or "inclusion" (*tadhammun*) or "implication" (*iltizām*).

3. to distinguish between singular and complex words.
4. to examine the characteristics of names, verbs, and particles (M. Mahrān, [127], introduction, 47).

These aims are accomplished in distinct ways by the various scholars, for the classifications of words and their functions are not exactly the same from author to author, as mentioned by M. Mahrān in his introduction to Rescher's book (see [127], 41–49). The Arabic analyses of the concepts extend the Aristotelian background to include some Stoic features and characteristics, according to M. Mahrān, who mentions that "the idea of 'signification' and 'significant' comes from the Stoics' logic" ([127], p. 43).

The analyses may also involve some grammatical distinctions specific to the Arabic language. For instance, the complex singular words are typically compound names, used very frequently by the Arabic people, such as "Abd al Malik" ([127], 44). As to the distinction between names, verbs, and particles, it can be found in Sībawayhī's[1] grammatical treatise entitled *al-Kitāb* (I: 12), but it is applied to logic by al-Fārābī, for instance (see [7], 22). Al-Fārābī provides a complete listing of logical words in his treatise entitled *Al-Alfaẓ al-musta'amala fī al-Manṭiq*. Although the particles are used both by the grammarians and the logicians,[2] al-Fārābī stresses the fact that his own use is logical, for he privileges the logical meanings of these words, i.e., the ways they are used by logicians in their treatises. He even refers to the Greek tradition in his classification of particles, for, he says, the Arabic grammarians do not really classify them, because they do not give to each category a specific name ([4], 42).

The classification presented by al-Fārābī includes some particles related directly to logic such as "*Al*" ("The"), "*Kull*" ("Every" and "all"), and "*Ba'd*" ("some") ([4], 44), that is, the quantifiers and some articles, plus the particles of negation such as "*Laysa*" and "*lā*" ("not" and "no") or those expressing the logical connectives such as "*Immā*" ("or"), "*In kāna*", "*idhā*" ("when" and "in case"), "*Lākin*" and "*'Illā 'anna*" ("but" and "however"), "*Fa 'idhan*" ("therefore"), and so on ([4], 8–9).[3] This classification could be seen as an anticipation of the Medieval studies and listings of the so-called syncategoremata, which were very popular in the Middle Ages. These listings were different from one logician to another but all of the distinction between syncategoremata and categoremata can be found in all treatises.[4]

Some other particles are related to the Aristotelian categories, since they help characterize the category of the predicate, whether the quantity, the quality, or the

[1]Sībawayhī is considered as the founder of Arabic grammar. His distinction has been challenged by some other grammarians, but it resisted all the criticisms and seems to be still admitted nowadays by the contemporary grammarians.

[2]For a general study of the relation between logic and grammar in al-Fārābī's frame, see [147] and see also ([5], 80). For the relation between grammar and logic in the Arabic tradition, see [148] and [149].

[3]See the whole classification and its analysis in [52].

[4]For a general study of this topic, see [102]. For the classification presented by Peter of Spain, see [133].

substance. Others determine the aim of doing or of saying something. They are thus related to the final cause, for the final causes even in Aristotle's *Posterior Analytics* are expressed in some syllogisms by the middle term. So we may include this kind of words inside traditional logic, although this inclusion might seem odd for modern and contemporary logicians.

Note, however, that the modal words, that is, "*mumkin*" (possible), "*muḥāl*" or "*mumtani*" (impossible), and "*wājib*" or "*darūrī*" (necessary) are not cited by al-Fārābī, which means that the words expressing the modalities are not considered as particles. This omission may be explained by the fact that these words are merely adverbs and may also be verbs. For this reason, they cannot be classified as particles, which are neither verbs nor names (adverbs could be seen as a kind of names).

Now, what about the assents? This part of logic is the study of propositions and arguments. It is the most important part of the logical texts, even in Arabic logic, since the treatises corresponding to *De Interpretatione* (called *al-'Ibāra* in Arabic) are devoted to the classification of the several kinds of propositions and their relations, while those corresponding to *Prior Analytics* (called *al-Qiyās*, in Arabic) are devoted to the study of the different syllogisms, that is, the deductive arguments, whether categorical or modal or hypothetical. The inductive and analogical arguments are also studied by the different Arabic logicians in their respective *al-Qiyās*.

The propositions are either true or false, which means that Arabic logic is bivalent. They may be elementary (*basīṭa)* or complex (*murakkaba*). The complex propositions are either predicative (*ḥamlīya*) or hypothetical (*sharṭīya*). There are many kinds of predicative propositions, which are either singular or indefinite or quantified. The quantified propositions are either universal or particular, the indefinites are not quantified, but they are generally treated as particulars in al-Fārābī's and Avicenna's frames, while in Averroes' one, they are ambiguous (sometimes particular, sometimes universal). The syllogistics presented in al-Fārābī's and Averroes' frames focus on the predicative propositions and admit three figures and the same valid moods as Aristotle in each figure. While both of them devote only a few pages to the hypothetical syllogisms, Avicenna departs from his predecessor and his successor by defending a whole theory of hypothetical syllogisms, using the two main kinds of hypothetical propositions and containing exactly the same moods and figures than the categorical syllogistic.

The hypothetical propositions may be either conditional (or connected = *muttaṣila*) (e.g., "If the sun rises, it is daytime") or disjunctive (or separated = *munfaṣila*) (e.g., "a number is either odd or even"). The study of these complex propositions includes elements of the Stoic logic, since the five Stoics' Indemonstrables and their variants are cited in al-Fārābī's frame, while Averroes cites most of them. As to Avicenna, he develops a whole hypothetical syllogistic which takes a considerable part of his *al-Qiyās* and is summarized in other treatises such as *al-Najāt* and *al-Ishārāt wa-t-Tanbīhāt*. We will present it in our Chap. 5.

The reasonings are classified into three kinds which are the following:

1. Syllogistic reasonings (*Qiyās*),
2. Inductive reasonings (*Istiqrā*), and
3. Analogical reasonings (*Tamthīl*).

The syllogistic reasonings include the categorical syllogisms as well as the modal ones and the hypothetical syllogisms in their two versions. We will present all those that are treated in al-Fārābī's, Avicenna's, and Averroes' frames in Chap. 5 . However, we will not study the analogical and inductive arguments, for they deserve another examination.

3.2 The Oppositions Between Categorical Propositions

As we said above, the predicative propositions may be either singular or indefinite or quantified. The singular propositions contain a particular subject and a general predicate. They may be affirmative or negative. Both the indefinite and the quantified propositions contain general terms as subjects and predicates, the difference between them being that the indefinite propositions do not contain any quantifier while the quantified ones contain one of these words.

Their relations are determined by means of their truth values, which in turn are fixed by considering the matter modalities of the propositions.

These matter modalities are the following:

1. The matter necessity, which means that the predicate is satisfied by all of the subject individuals, as in the following: "Men are animals."
2. The matter possibility, that is, the fact that the predicate is satisfied by some of the subject individuals, but not all, e.g., "Men are writers."
3. The matter impossibility, i.e., the predicate is not satisfied by any of the subject individuals, as in: "Men are stones".

So all material sentences containing concrete subjects and predicates pertain to one matter or another depending on whether the predicate is satisfied or not by the subject individuals, and in case it is satisfied, on whether it is satisfied by all of them or only by some of them. These matters are determined by a set of truth values corresponding to each kind of quantified propositions, whether affirmative or negative. The table below shows the truth values of the four quantified propositions in the three matters:

Necessary matter	Possible matter	Impossible matter
Every A is B: True	Every A is B: False	Every A is B: False
Some A is B: True	Some A is B: True	Some A is B: False
No A is B: False	No As is B: False	No A is B: True
Not every A is B: False	Not every A is B: True	Not every A is B: True

These matter modalities are used in all the frames considered here, starting from that of al-Fārābī. They help determine the relations between the quantified propositions in a precise way, for all the propositions considered have a determined value in all the cases considered. Avicenna, in particular, provides the table presented above and shows in *al-'Ibāra* that these truth values justify in a semantic way the whole set of logical oppositions between the quantified propositions.

We have to note, however, that al-Fārābī is not the first author to evoke and analyze these matter modalities, for they can be found in the writings of the Greek commentators, such as Alexander of Aphrodisias, Ammonius, and Stephanus and also in the writings of some earlier Arabic logicians, as many authors have shown. For instance, F. W. Zimmermann says that "The list of "necessary", "possible", and "impossible" is based, according to Ammonius (215.11 ff), on the argument that "a predicate must hold of a subject either always or never or just occasionally" ([150], note 8, 538). He adds that we find similar terminologies in authors like Ibn al-Muqaffa' and Ikhwān aṣ-ṣafā, who use, respectively, the expressions "three *umūr*" (ibn al-Muqaffa') and "three *'anāṣir*" (Ikhwān aṣ-ṣafā) ([150], note 9, 539).

Turning back to al-Fārābī's text, let us first consider the singular propositions. The singular propositions may be affirmative, e.g., (1) "Zayd is just," affirmative with a negative predicate, e.g., (2) "Zayd is not-just" or negative, e.g., (3) "Zayd is-not just." The negative predicates are called "indeterminate" ("*ghair muḥaṣṣal*") by al-Fārābī ([14], 85) and *ma'dūl* by Averroes ([26], 106). They do not modify the quality of the proposition, which remains affirmative. Al-Fārābī says that even the metathetic predicates are affirmative in the final analysis, for "not-just" means "unjust", "not-seeing" means "blind", etc. ([14], 86). The same opinion is defended by Avicenna, which shows that for this particular point, the three authors agree with each other and follow Aristotle's opinion, according to which the proposition containing the indefinite (metathetic) predicate is still affirmative.[5]

This analysis of the metathetic predicates indicates the Greek sources that al-Fārābī was using. This point is noted by some scholars and reported in the following quotation:

> "Following Theophrastus (Fortenbaugh et al. 1992, 148–153) he identifies this kind of negation as 'metathetic' (*'udūlī*); he takes the resulting metathetic sentence to be an affirmation, not a denial." ([90], Sect. 8)

In al-Fārābī's frame as well as in Avicenna's and Averroes' ones, the affirmative proposition is considered as having an import, while the real denial of the affirmative proposition, i.e., its contradictory negation, which makes the negative proposition true whenever the affirmative is false, and false whenever the affirmative is true, does not have an import, because it could be true when the subject does not exist. This point is explicit in this part of al-Fārābī's text, given the truth values he gives to the different propositions. He *explicitly* attributes an import to affirmative propositions and denies it from the negative ones, just like Aristotle. Thus, if Zayd does not exist, the proposition:

---

[5]However, the analyses of metathetic terms provided by al-Fārābī and Avicenna are rather different. See, for instance, Paul Thom in his article "al-Fārābī on indefinite and privative names" [142], and Saloua Chatti in ([52], 181).

(3) 'Zayd is-not just'

would be true just because the subject Zayd does not exist. This proposition can be interpreted in the following way:

(4) It is not the case that Zayd is just

this is why it is true when Zayd does not exist, given that its correspondent affirmative is false in that case.

On the contrary, the proposition:

(2) Zayd is not-just

is false if Zayd does not exist, for given that the negation puts on the predicate and not on the copula, the whole proposition is not negative and, being affirmative, it presupposes the existence of its subject. Since its subject does not exist, it is false because it attributes something to a non-existent thing, which (as non-existent) cannot possess any property.

Consequently, the negation which expresses contradiction is external in the three frames. The contradictory negation expresses the denial of the initial proposition. An external negation puts either on the copula when the proposition is singular or on the quantifier when the proposition is quantified. As a matter of fact, as we will see in the following sections, the quantified negative propositions are expressed by means of the particles "*laysa*" or "*lā*", which are put at the beginning of the proposition.

The quantified propositions are expressed as follows:

- Every A is B (universal affirmative: **A**)
- No A is B (universal negative: **E**)
- Some A's are B's (particular affirmative: **I**)
- Not every A is B (particular negative: **O**)[6]

Their truth values matter by matter which the table above exhibits show that the two necessary affirmatives contain predicates which always apply to their subjects, so that they cannot be false, while the two necessary negatives deny the attribution of a necessary predicate to their subjects, which leads to false propositions, e.g., "No man is an animal", or "Not all men are animals". This falsity is due to the fact that "animal" is an essential predicate of the subject "human", a part of its very definition, which determines, among other things, its very nature: no being can be a human if it is not an animal.

As to the impossible matter, things are different for it is the impossible propositions that are always true (e.g., "No man is a stone" is true) and the affirmative ones that are always false (e.g., "Some men are stones" or "All men are stones" are always false).

[6]The vowels **A**, **E**, **I**, and **O** are *not* used by the Arabic logicians, who do not name the quantified propositions. We have introduced them only for convenience.

The possible propositions function in a different way, for in their case, the distinction between the true propositions and the false ones is to be made by taking into account the quantity of the propositions, not their quality. For it is obvious that saying "All men are writers" is false, given that not all men are writers, the predicate "writer" being not essential to the subject "man", while the proposition "Some men are writers" is true, for there are men who are able to write. The negative possible propositions function in the same way for the universal one is false ("No man is a writer" is false), while the particular one is true ("Not all men are writers" is true).

As to the indefinites, both al-Fārābī and Avicenna say that they are or at least ought to be considered as particulars because they may be true together. Averroes says that they are ambiguous: they may be either universal or particular ([26], 92–93). This makes his theory slightly different from those of his two predecessors.

3.2.1 The Oppositions of Categorical Propositions in al-Fārābī's Frame

Al-Fārābī defines the oppositions in several treatises. In *Kitāb al 'Ibāra*, the short treatise corresponding to *De Interpretatione* (which is different from the long commentary of the same *De Interpretatione*, entitled *Sharh al-'Ibāra* ([12]) or *Sharh al-Fārābī li-Kitāb Aristūtālīs fi-l-'Ibāra* ([2])), in *al-Maqūlāt* (the correspondent of the *Categories*) and in *Kitāb al-Qiyās*, the short treatise corresponding to the *Prior Analytics*, he analyzes the oppositions between the quantified propositions as well as the indefinites and the singulars.

He first says that: "Two opposed propositions are either singular or contrary or subcontrary or contradictory or indefinite" ([10], 15) which means that he admits merely three kinds of oppositions, namely, contradiction, contrariety, and subcontrariety. But he separates, in this classification, the singular and the indefinite propositions from the quantified ones. Despite this separation, however, the oppositions between the indefinites and the singulars are still classical and enter into the three kinds above.

The quantified propositions are those that contain a quantifier called "*sûr*" by al-Fārābī. This word "*sûr*" is not in Aristotle's texts. In Arabic, its meaning is not *prima facie* related to logic, for it has a proper meaning, which is "wall", in particular, a wall that surrounds a city, from which a metaphoric meaning can emerge, namely, the idea of encircling something or delimitating it, which the *sūr* achieves. This idea is plausible, since it is the *sūr* (the quantifier) adjoined to the subject of a quantified proposition that indicates that the whole of the subject (in a universal proposition) or only part of it (in a particular proposition) satisfies the predicate. According to Zimmerman who studies this word and some other logical vocabulary in his article "Some Observations on al-Fārābī and Logical Tradition" [150], in the Arabic tradition of logic, the very first author who uses the word *sūr* is ibn al-Muqaffa'. As the table provided by Zimmermann shows, al-Kindī did not use it ([150], 530–531). So it is Al-Fārābī who seems to be the second author to use it, followed by Ikhwān aṣ-ṣafā. However, Zimmermann says first that al-Fārābī did not take it from ibn al-Muqaffa' (who is never cited by al-Fārābī in any of his treatises), since his analysis made him deduce that "ibn al-Muqaffa' is not al-Fārābī's source" ([150], 529), second that "There is no obvious connection between this term and its Greek counterpart *prosdiorismos* which means some additional specification and, in this fixation, a quantifying expression adjoined to an elementary sentence" ([150], 535). This is why he priviledges the hypothesis that the origin of the term may be Syriac, for he says "We will wonder whether *sūr* is a case of a Syriac metaphor transferred into Arabic" ([150], 535). This word seems thus to carry a metaphorical meaning common to its Syriac and Greek counterparts, although as one reviewer suggests it is not an "exact translation" of the Greek word *prosdiorismós*, ("used by Philoponus").

In Ibn al-Muqaffa''s treatise entitled "*al-Manṭiq*" ([97]),[7] we find the following classification of quantifiers: "He said: and the quantifiers (*siwār*), that is, the words (*al-kalām*) that distinguish between the general, i.e. 'every' (*al-kull*), and the particular, i.e. 'some' (*al-ba'ḍ*), are four in number: these are 'every' (*kull*),..., and 'some' (*ba'ḍ*), and also 'no one' (*lā wāḥid*)... and... 'not every' (*lā kull*)" ([97], § 75, 36, my translation). Professor Hodges adds that this sentence "could very easily be an annotated translation of the following passage from Alexander of Aphrodisias" Commentary:

"And again, the subject is that to which the quantitative determination (*posòn diorismós*) of the proposition is annexed ('every' or 'no' or 'some' or 'not every')" ([1], 44.25ff, translated Barnes et al.).

Another reference that Prof. Hodges provided me with comes from the book of Ibn al-Sikkit ([99]) who "reports that *siwār* ...was used to mean 'bracelet'" and that the word was also used by a poet whose verse are quoted by him and where the word *siwār* seems to mean something like "limit".

These references tend to show, as Prof. Hodges notes, that the word *siwār* (whose root is the same as *sūr*) was already used in Arabic with a metaphorical meaning close to the one that we find in the logical treatises, which could explain why the translators chose it to translate a Syriac word carrying the same metaphorical meaning.

Now, what about the oppositions between the different kinds of propositions? These oppositions are defined as follows:

- Two propositions are contradictory if they never share the same truth value, in any matter.
- Two propositions are contrary if they are both false in the possible matter and do not share the same truth value in the necessary and impossible matters.

[7] I owe this reference (and many other ones) to Prof. Wilfrid Hodges. I thank him very much for his help and his fruitful and numerous suggestions and remarks which made me improve significantly my work.

- Two propositions are subcontrary if they are both true in the possible matter and do not share the same truth value in the necessary and impossible matters ([10], 16–17).

Let us consider the definition of contradiction. What al-Fārābī is saying is that in whatever matter we consider (necessary, possible or impossible), when two propositions have the *same* subject and the *same* predicate, but one of them is affirmative while the other one is negative, they are contradictory if they never share the same truth value, i.e., if when the first one is true, the second one is false and vice versa. So when "Every A is B" is true, its contradictory "Not every A is B" will be false, and if the former is false, its contradictory will be true. This difference between the truth values of both sentences can be found in sentences illustrating the three matters. For instance, the following couples of material sentences are contradictory:

(1) Every human is an animal / (2) Not every human is an animal (necessary matter: (1) True, (2) False)
(1′) Every human is a writer / (2′) Not every human is a writer (possible matter: (1′) False, (2′) True)
(1") Every human is a stone / (2″) Not every human is a stone (impossible matter: (1″) False, (2″) True).

So in these examples, the two propositions never have the same truth value in all three matters. The same can be said about **E** and **I** propositions which never share the same truth value, when they have the same subject and the same predicate, in necessary (**E**: False, **I**: True), possible (**E**: False, **I**: True), and impossible (**E**: True, **I**: False) matters.

In the same way, we can explain the definitions of contrariety (and also subcontrariety). Two contrary propositions (**A** and **E**, for instance) are both false in the possible (e.g., "Every human is a writer" (False) and "No human is a writer" (False)), but they don't have the same truth value in the necessary (e.g., ""Every human is an animal" (True) and "No human is an animal" (False)), and they don't have the same truth value in the impossible (e.g., "Every human is a stone" (False) and "No human is a stone" (True)).

The two singular propositions are contradictory when one of them is affirmative and the other negative. In this case, they never share the same truth value. As we said above, the negation must be external, which means that, when the proposition is singular or indefinite, it must put on the copula and not on the predicate. If the negation puts on the predicate, the proposition is not negative; it is affirmative as al-Fārābī says explicitly in the following quotation: "As to the proposition whose predicate is indefinite (*ghair muḥassal*), it is affirmative, not negative" ([8], 147). In this treatise, he provides the following table where he distinguishes between several kinds of opposed singular propositions:

"Zayd is learned (*yūjadu ʿāliman*)	Zayd is not learned (*laysa yūjadu ʿāliman*)
Zayd is not ignorant (*laysa yūjadu jāhilan*)	Zayd is ignorant (*yūjadu jāhilan*)
Zayd is not not-learned (*laysa yūjadu lā-ʿāliman*)	Zayd is not-learned (*yūjadu lā-ʿāliman*)" ([8], 149)

In this table, the propositions of the left side are contradictory to their counterparts (of the same line) in the right side. As to the other propositions, they are opposed in several ways, which al-Fārābī considers in the rest of the text by examining their respective truth values. In examining the truth values of the propositions of the left side, he says first that the proposition

(2) "Zayd is not ignorant"
is more often true than the proposition
(1) "Zayd is learned"
because the latter is true of Zayd if he is adult and savant, while the former is true of Zayd if he is adult and savant and also when he is a child. As to
(3) "Zayd is not not-ignorant"
it has the same truth conditions and the same relations to (1) as (2) ([8], 149).
As to the propositions of the right side, he says that
(4) "Zayd is not learned"
is true if Zayd is a child and is an adult who is not savant, while
(5) "Zayd is ignorant"
is true only when Zayd is an ignorant adult. As to
(6) "Zayd is not-learned"
it is true only of an ignorant adult, so it is less true that (4) and has the same relations with (4) than (5).

The cases of falsity of all these propositions are also stated in the following way:

(1) is false of Zayd if he is adult and not learned, and also if he is a child.
(2) is false of Zayd only if he is an ignorant adult.
(3) is false of Zayd only if he is an ignorant adult.
(4) is false of Zayd if he is an adult who is learned.
(5) is false of Zayd if he is an adult who is learned and also if he is a child.
(6) is false of Zayd if he is an adult who is learned and if he is a child ([8], 149–150).

As a consequence, we have the following truth values and relations:

(1) "Zayd is learned" and (5) "Zayd is ignorant" are possibly false together (if Zayd is a child). But when one of them is true, the other one is false ([8], 150).

They are thus contrary.

(2) "Zayd is not ignorant" and (4) "Zayd is not learned" are both true if Zayd is a child.

But when one of them is false, the other one is true ([8], 150–151). Consequently, we can say that they are subcontrary. In addition, when (4) is false, its contradictory (1) will be true, consequently (5) will be false, being the contrary of (1), and (2) will be true, because it is the contradictory of (5) ([8], 151). Given that when (4) is false, (5) is false, and that there is no case where (5) is true while (4) is false, we can conclude that (4) is the subaltern of (5). The same can be said about (1) and (2). However, although al-Fārābī gives the truth conditions of this relation of subalternation, he does not use any word to qualify it. Only Avicenna introduces it explicitly and calls it "*tadākhul*", as we will see below (Figs. 3.1 and 3.2).

Finally, according to al-Fārābī, (3) and (6) are related to (1) and (4) as (2) and (5) are related to (1) and (4), given that (5) and (6) are equivalent and (2) and (3) are also equivalent ([8], 151). This leads to the following squares:

Given these squares, we can say that the propositions (4) and (6) are not equivalent, although the latter implies the former. This means that the negation must be external to really produce a contradictory proposition. If it is internal, it produces only a contrary proposition.

Almost the same table is provided by al-Fārābī for the indefinite propositions, for he says the following:

"Men are learned (*yūjadu 'āliman*)	Men are not learned (*laysa yūjadu 'āliman*)
Men are not ignorant (*laysa yūjadu jāhilan*)	Men are ignorant (*yūjadu jāhilan*)
Men are not not-learned (*laysa yūjadu lā-'āliman*)	Men are not-learned (*yūjadu lā-'āliman*)" ([8], 151)

However, with regard to the truth values of the propositions, this table is different from the preceding, for the two propositions of the first line, in the left and the right sides are not contradictory; rather they are subcontrary, given that the indefinites are considered as particulars by al-Fārābī ([8], 152). If we apply the same numerals to these propositions, we can also say that (1) "Men are learned" can be true together with (5) "Men are ignorant."

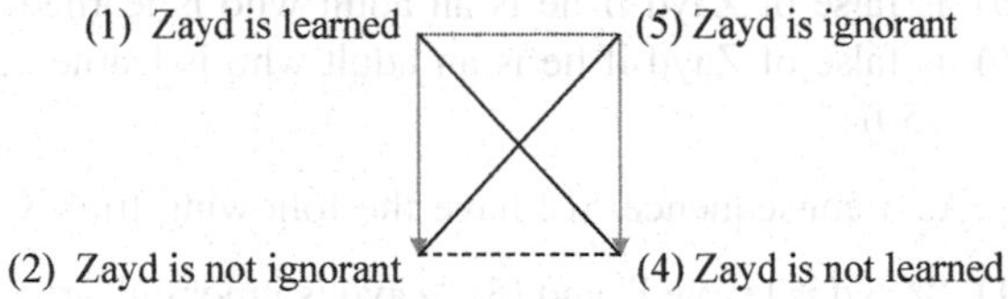
(1) Zayd is learned
(5) Zayd is ignorant
(2) Zayd is not ignorant
(4) Zayd is not learned

Fig. 3.1 The square 1 with singular propositions containing opposed predicates

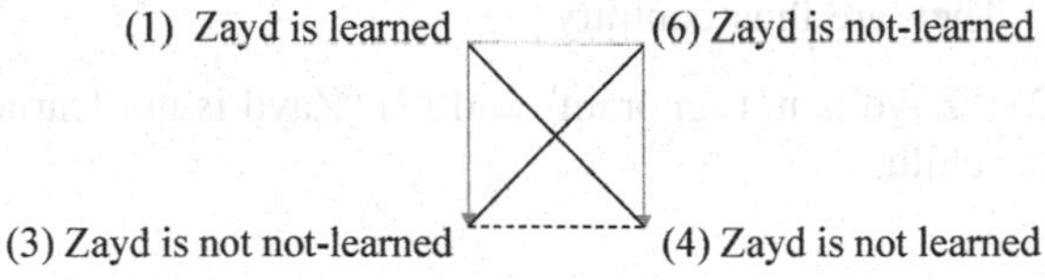
(1) Zayd is learned
(6) Zayd is not-learned
(3) Zayd is not not-learned
(4) Zayd is not learned

Fig. 3.2 The square 2 with singular propositions containing opposed predicates

Now what about the quantified propositions? In *Kitāb al-Qiyās*, al-Fārābī says that the universal affirmative (**A**) and the particular negative (**O**) are contradictory, the universal negative (**E**) and the particular affirmative (**I**) are also contradictory, the two universals are contrary, and the two particulars are subcontrary ([10], 16). However, he does not evoke subalternation in his text, which he does not take into account nor define explicitly, although he does use it implicitly (without naming it) as we saw above. Consequently, the relations he holds may be represented by the following Fig. 3.3:

However,, in *Kitāb al-'Ibāra*, he provides tables for the quantified propositions similar to those containing the singulars and the indefinites. The first table contains the two kinds of particulars (with contrary predicates) and is the following:

"A man is learned	Not every man is learned
Not every man is ignorant	A man is ignorant
Not every man is not-learned	A man is not-learned" ([8], 152)

In this table, the propositions function as the indefinites, for those which are in the left side are not the contradictories of those of the right side. Rather they are subcontrary. In the same way, (1) is the subcontrary of both (5) and (6) for they may be true together. The same can be said about the three negative propositions which may be true together too. However, from top to bottom, both in the left side and in a right side, they are comparable to the singulars and the indefinites for the top propositions imply those of the second and the third lines, in the left side, while in the right side, those of the bottom imply the top propositions.

Another table containing the two kinds of universals is given and commented by al-Fārābī. It is the following:

"Every man is learned (1)	No man is learned (4)
No man is ignorant (2)	Every man is ignorant (5)
No man is not-learned (3)	Every man is not-learned (6)" ([8], 151, numerals added)

In this table, the three propositions of the left side are the contraries of those of the right side provided that they are in the same lines. The contrariety holds also between (1) and (5) and between (1) and (6), for these cannot be true together.

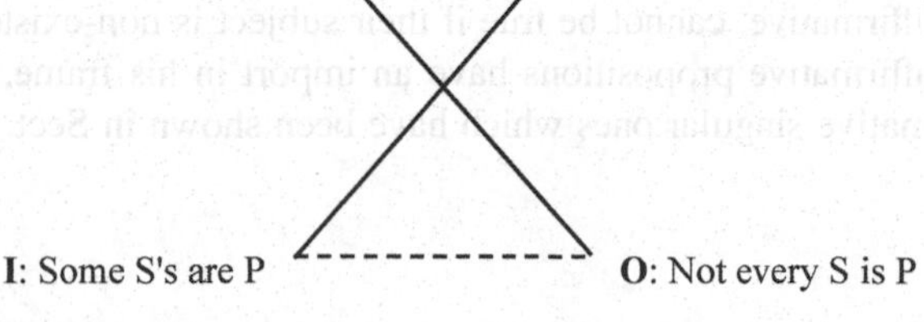
A: Every S is P
E: No S is P
I: Some S's are P
O: Not every S is P

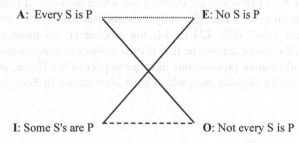

Fig. 3.3 The incomplete square

However, the negatives may be true if "their subjects are non-existent" ([8], p. 152), so all of them are subcontrary rather than contrary.

The two last tables contain contradictory propositions (with contrary predicates): the first one is the following:

"Every man is learned	Not every man is learned
Not every man is ignorant	Every man is ignorant
Not every man is not-learned	Every man is not-learned" ([8],152)

Here the propositions of the left side and the right side are all contradictory provided that they are in the same line. However, the three negatives are subcontrary, because they can be true together when the subject is non-existent, as noted above ([8], 152), while the affirmatives are rather contraries, for "every man is ignorant", "every man is savant", and "every man is not savant" are all false. Regarding the implications, the top propositions of the left side imply those of the bottom in the same side, while the bottom propositions of the right side imply the top propositions of the same side.

The last table is the following:

"A man is learned	No man is learned
No man is ignorant	A man is ignorant
No man is not-learned	A man is not-learned" ([8],153)

This table contains contradictory propositions which do not function exactly like those of the preceding table. For the negatives may be true together, they are thus subcontrary; but the affirmatives are particular, so they are also subcontrary. As to the implications, they are different from those of the other tables for; here, the implication goes from top to bottom in the *right* side and from bottom to top in the *left* side.

What about the import of the quantified propositions?

Al-Fārābī evokes the question of the import in *Kitāb al-Maqūlāt* while considering the truth values of the singular, the indefinite and the quantified propositions, affirmative or negative, in case their subject exists and in case their subject does not exist. He says first that the quantified propositions such as "every fire is hot" and "every fire is cold," or "every man is white" and "some men are black," whose predicates are contrary but which are both affirmative "do not share the same truth value when their subject exists. But when their subject *does not exist*, they are *all false*" ([9], 124.13–14, my emphasis). So these propositions, which are all affirmative, cannot be true if their subject is non-existent. Therefore, the quantified affirmative propositions have an import in his frame, as is the case with the affirmative singular ones which have been shown in Sect. 3.2 to have an import. So he seems to be the *very first author* to *explicitly* attribute an import to affirmative *quantified* propositions and deny it from the negative *quantified* ones.[8]

As to the negative propositions such as "not every man is white," it is *true* if the *subject does not exist*, given that "the affirmative and the negative *do not share* the *same truth value*, whether their subjects exist or not" ([9], 124.14–16, my emphasis) and given that its affirmative *contradictory* "every man is white" is false when its subject is not existent, as we just saw. So this negative quantified proposition is true when its subject does not exist because in that case its contradictory affirmative one is false. As a consequence, the negative quantified propositions do not have an import for when their subject does not exist, they can be true. This idea generalizes the position about the import of singular propositions to that of the quantified ones, and it is important for the analysis of the syllogisms, for instance, whose validity depends sometimes on the import of the propositions. The same can be said about the indefinites which behave in the same way with regard to the import, for the affirmative ones have an import because they can never be true when their subject is non-existent, while the negative ones do not have an import because they can be true when their subject is non-existent.

So we can consider that al-Fārābī may be seen as the very first logician to have generalized Aristotle's position about the import of the affirmative propositions and the absence of import of the negative ones to the *quantified* propositions. Aristotle endorsed this position for the *singular* propositions, but he did not generalize it *explicitly* to the quantified ones. As to Apuleius and Boethius, we cannot really say that they endorsed *explicitly* the absence of import of *all* the negative propositions besides the singular ones, despite what some linguists claim (see, for instance, Horn in [95], 24). For as noted by T. Parsons in his article "The traditional Square of Opposition" (see [121]), the absence of import of the proposition **O** is best shown when this proposition is expressed by "Not every S is P" for only in that case, it can express the possibility for the subjects S to be non-existent. But as appears in the following quotation, Boethius did not write **O** in that way in his comments of Aristotle's text. Here is what Parsons says:

> "In his translation of *De interpretatione*, Boethius preserves Aristotle's wording of the **O** form as "Not every man is white." But when Boethius comments on this text he illustrates Aristotle's doctrine with the now-famous diagram, and he uses the wording 'Some man is not just'. So this must have seemed to him to be a natural equivalent in Latin. It looks odd to us in English, but he wasn't bothered by it." ([121], Sect. 2.3).

Al-Fārābī even evokes what Gilbert Ryle (see [130], Chap. 1) and others now call the "category mistake" case, which he illustrates, for instance, by the following sentences: "every whiteness is odd" (*kull bayāḍun fa-huwa fardun*) and "every whiteness is even" or "every heat is straight (*mustaqīma*)" and "every heat is curved (*munḥaniya*)" ([9], 125.7–8). These sentences are "all false" ([9], 125.8) when their subject is existent. But they do not share the same truth value with their respective

[8]On the problem of existential import in general, see [51] for a full discussion. On that same problem in Aristotle's theory, see [23].

negations, whether the subject is existent or not, for "no whiteness is odd" is true, whether the subject exists or not, while "every whiteness is odd" is false whether the subject exists or not.

3.2.2 *Avicenna's Absolutes and Their Opposites*

Avicenna presents different accounts of the oppositions in his treatises, for in his *al-'Ibāra* (*De Interpretatione*), he presents a complete square with the whole set of oppositions including the subalternation, while in *al-Qiyās*, he presents a different analysis where he introduces time considerations, which makes the theory more complex and modifies the kinds of the opposed propositions. Despite these changes, he holds all kinds of oppositions in all his treatises.[9]

Let us start with *al-'Ibāra* ([35]). In that treatise, Avicenna presents the whole set of logical relations between the quantified propositions, that is, contradiction, contrariety, subcontrariety, *and* subalternation, which he calls "*tadākhul*". He also considers the singular propositions and the indefinite ones. He defines the oppositions by applying the notion of matter modalities as was the case with al-Fārābī. The truth values of the propositions, matter by matter, are thus the same as that of al-Fārābī (see Sect. 3.2.1).

To begin with, Avicenna says that there are three kinds of propositions: the singular ones, the indefinite (unquantified) ones, and the quantified ones. The singulars differ from the two other kinds in that they contain a particular subject; the indefinites (unquantified) contain a universal (or general) subject but they do not contain quantifiers; while the quantified propositions contain a universal (or general) subject and a quantifier. The quantifiers are the words that express the quantity of the proposition; they may be particular or universal. Particular quantifiers are the words "some" and "not all" or "not every", and universal quantifiers are the words "every" (or "all") and "no one" (or "none").

Al-Fārābī and Avicenna say explicitly that there are four quantifiers (*sûr*), not only two. For they distinguish between "every" or "all" (= *kull*), some (= *ba'd*), none (= *lā aḥada* = literally "no one"), and not every (= *laysa kull*), while Averroes distinguishes clearly between the quantifiers (the words expressing the quantity) and the negation, and says that there are two quantifiers to which one could add a negation. He says: "I mean by quantifier the words « all » and « some »" (see [26], 91).

This clarifies things in some respects for the only simple words are the ones used in the affirmative propositions, while the quantified negative ones both contain complex words, i.e., a combination of a quantifier and a negation. However, as one reviewer suggests, and as we will see when analyzing Averroes' theory, this

[9]On the notion of opposition in general in Avicenna's and Averroes' frames, see [50], where the different conceptions of opposition are analyzed and compared with that of Aristotle. See also [33] for Avicenna's specific analysis of opposition in general.

clarification has an inconvenience, namely, the fact that the universal negative quantifier ("No" or "None") can no more be expressed in a clear way. We will examine this idea in Sect. 3.2.3.

As a matter of fact, in the Arabic language, it is easier to separate between the negation and the quantifier, for both "*laysa kull*" (not all) and "*lā 'aḥada*" ("no one") contain a negation *in front of* the quantifier; thus one does not need to put the negation in the middle of the sentence to get the negative proposition, unlike what happens in French, for instance, where **O** is expressed by putting "ne…pas" in the middle of the sentence, and **E** by using words like "nul" or "aucun", which express the negation only when "ne" is added somewhere in the sentence.

Note also that in Arabic, the two negative quantified sentences are comparable in this respect, for the negation in *both* cases puts on the quantifier, so that the so-called problem of the lexicalization of **O** is not raised in the way it is in other languages. This problem arises in some languages because of the fact that **A**, **E**, and **I** are expressed by simple words ("every", "none", and "some" in English, "tout", "aucun", and "quelque" in French), while **O** is not expressed by a single word, but by a complex one ("not all", or "some…not" in English, "quelque…., ne…pas" in French). This absence of lexicalization creates a linguistic (and ultimately logical) problem because of the fact that the fourth vertex of the square is different from the three other ones, leading to some kind of *asymmetry* between the *four* corners (and propositions) and the *three* words provided by the ordinary language. People ask: why *only three* words, while there are *four* propositions? This problem is not raised in the same way in Arabic, just because the Arabic language contains only *two* single words for the quantifiers and adds the negation to these two words to construct *both* **O** *and* **E**. The whole frame is symmetric because we have only *two single* words, that is, *kull* and *ba'ḍ* and *two complex* words, that is, *lā aḥada* (literally "no one") and *laysa kull* (literally "not all") or *ba'ḍ...laysa* (literally "some…not"), given that even **E** is expressed by a complex word so that both negative propositions contain the two quantifiers plus a negation putting on each one (see [58], for a full analysis of this problem and its solution).

Regarding the singular propositions, Avicenna claims in *al-'Ibāra* that:

1. Zayd is just
   and
2. Zayd is-not just

are contradictory, while

3. Zayd is not-just (see [35], 113)

is the contrary of (1), because, in case Zayd is non-existent, both (1) and (3) are false, while (2) is true. This is so because the negation in (2) is external (i.e., in the case of the singular propositions, it puts on the copula), while in (3) it puts on the predicate, which means that Zayd is taken to be existent; this is why the proposition cannot be true. This also means that the affirmative propositions, whether with a simple predicate or with an indefinite predicate, have an import, because they cannot be true in case the subject is non-existent, while the negative propositions do not have an import because they may be true in case the subject is non-existent.

The singular proposition (1) and its denial (2) in this example are contradictory, that is, they can never share the same truth value, in whatever matter, while the two affirmative propositions (with a simple predicate or with an indefinite predicate) are contrary, because they can be false together when the subject is non-existent.

As to the indefinites, Avicenna says that they should be considered as particulars although they do not contain any quantifier. In *al-'Ibāra* he claims: "the indefinite has the force of the particular" ([35], 51). However, he recognizes that this kind of propositions might be considered, in ordinary usage, as universal propositions, or even as singulars in some contexts. Despite this linguistic fact, he insists to interpret the indefinite sentences as particulars and defends a theory which agrees with both Aristotle and al-Fārābī. He even says that an indefinite sentence does not have any contradictory; it only has a subcontrary. It thus functions as a particular proposition. This is what he claims in the following quotation: "the indefinite has no contradictory" ([35], 67) and also "the indefinites [...] are like the particulars, they should be said to be subcontraries" ([35], 66, my translation), given that they are sometimes true together as is the case with the two following sentences: "Men are beautiful" and "Men are not beautiful" ([35], 67).

Note that subcontrariety is defined, as in al-Fārābī's text, as being the relation between the propositions that can be true in the possible matter but do not share the same truth value in the necessary and impossible matters.

What about the quantified propositions?

The quantified propositions, i.e., the universal affirmative **A**, the universal negative **E**, the particular affirmative **I**, and the particular negative **O** are defined in the usual way in *al-'Ibāra*. Their relations are the following:

- **A/O** and **E/I** are contradictory,
- **A/E** are contrary,
- **I/O** are subcontrary, and
- **A/I** and **E/O** are related by the relation of subalternation (*tadākhul*), that is, **I** is the subaltern of **A**, and **O** is the subaltern of **E**.

The first three relations have already been defined, for Avicenna reproduces al-Fārābī's definitions. However, subalternation is defined by Avicenna as follows:

> "As to those that differ in quantity but not in quality, *let us call them* subalterns, we *find* that those which are affirmative are true in the necessary, and that the negative subalterns are true in the impossible, and both do not share the same truth-value in the possible, but the particulars are true in that case, and examine that by yourself" ([35], 48, my emphasis).

In this quotation, Avicenna evokes the fourth kind of opposition, not mentioned by al-Fārābī, which he calls "*tadākhul*" and corresponds to subalternation in the Latin treatises. This name is not new, for according to Zimmermann (see [150]), it can be found in Ibn al-Muqaffa', for instance, who uses the expression "*al-Ikhtilāf al-Mutadākhil*" ([150], 531) to express subalternation. Other logicians use another name for it, for instance, Ikhwān as-Ṣafā (The Brethren of Purity) who call it

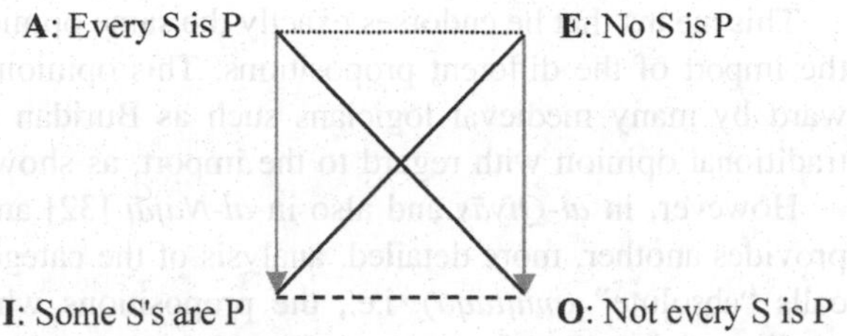
A: Every S is P
E: No S is P
I: Some S's are P
O: Not every S is P

Fig. 3.4 The usual square

"*mutaṭāliyatān*" ([150], 531). According to Zimmermann, subalternation was considered as an opposition by some Greek commentators too and by Paul the Persian ([150], 530, Table). The name "*tadākhul*" used by both Ibn al-Muqaffaʿ and Avicenna (but not, interestingly, by Ikhwān as-Ṣafā[10]), comes from the verb "*dakhala*", that is, "to enter" and from the other verb "*tadākhala*", which means "to enter into each other." Both suggest the idea of inclusion of the part into the whole, since this relation holds between universal propositions and particular ones. The idea then is to say that the particular, whether affirmative or negative, is included inside the universal, or in other words the universals contain the particulars. Therefore, when the universal is true, the particular is true too. The definition of subalternation is, then, the following: "Two propositions are subaltern when the first is false and the second is true in the possible, or both are true in the necessary and false in the impossible, or both are true in the impossible and false in the necessary."

With this definition, Avicenna can be said to hold the following square of oppositions, although, like al-Fārābī, he does not draw any figure at all (Fig. 3.4).

On the other hand, with regard to the import of the different propositions, Avicenna endorses the same opinion as al-Fārābī to the effect that the affirmative ones have an import, i.e., cannot be true if their subject does not exist, while the negative ones do not have an import, since they can be true even when their subjects do not exist. He says that in several treatises, for instance, in *al-Maqūlāt* ([33], 258–259, where he analyzes the sentences whose subjects are empty) and *al-'Ibāra*. For instance, he says that the sentences talking about non-existent objects such as the griffin are all false when they are affirmative and they are true when they are negative. His opinion can even be expressed by means of a principle that Wilfrid Hodges states as follows: "Affirmative Principle. In every true affirmative predicative sentence the subject term is satisfied (i.e. non-empty) (*'Ibāra*, 79. 13).

Negative Principle. A negative predicative sentence is true when its subject term is not satisfied (*'Ibāra*, 81. 3f)." ([84], 120).

[10]The root used by Ikhwān aṣ-ṣafā is "*ṭāla*" (see [150]), which evokes the ideas of length and of superiority. So maybe subalternation is what relates one "superior" proposition (the universal one) to its inferior dependent one (the particular).

This means that he endorses exactly the same opinion as al-Fārābī with regard to the import of the different propositions. This opinion will also be endorsed afterward by many medieval logicians such as Buridan and others. In fact, it is the traditional opinion with regard to the import, as shown by Horn [95].

However, in *al-Qiyās* and also in *al-Najāt* [32] and *al-'Ishārāt* [36], Avicenna provides another, more detailed, analysis of the categorical propositions, which he calls "absolute" (*muṭlaqa*), i.e., the propositions which contain a subject and a predicate (plus sometimes a quantifier and also a negation when they are negative) but no *explicit modal* word ("necessarily" or "possibly"). These propositions are categorical, i.e., either affirmative or negative and they can be either true or false, being declarative, since they report some fact (*khabarī*) ([36], 223). In this respect, they are different from the sentences which express questions (*istifhām*), or requests (*iltimās*), or wishes (*tamannī*), or hopes (*tarajjī*), or astonishment (*ta'ajjub*) ([36], 223). So, one must be able to determine their truth value clearly. But in order to determine this truth value clearly and in a way that can be used in logical arguments, one has to add some conditions to these propositions. For in some cases, one cannot know whether the propositions are true or false if some condition is not added to them. For instance, if one says "Every human is breathing," its natural contradictory is "Not every human is breathing." But breathing is a discontinuous process, since it is inspiring *and* expiring, so that a person who is breathing is inspiring at some times and expiring at some other times. The question is then: can we determine exactly the time of breathing (of inspiring and expiring) for both propositions? For in order to know exactly in which cases the two sentences are contradictory, one has to determine these times precisely, given that when we say "every human is breathing at t_0," and "not every human is breathing at t_1," these two propositions are not contradictory, since the time of breathing is not the same for both propositions. On the other hand, could we say that these propositions are contradictory only when the *exact time* of inspiring (and similarly that of expiring) is always the same for both? According to Avicenna, this cannot be the right way to express the contradictories as he notes in the following passage, when explaining the reason why he added the different conditions to the propositions:

> "It would be appropriate for us to speak warily: some of the things said in the third book (*al-'Ibāra*) were inadequate. Namely, when we say 'Every B is an A' and we want to take into account the time in the sentence 'Not every B is an A', since this is one of [the things that have to satisfy] a condition in order to have a contradictory negation, this makes difficulties for us. When we say: 'Every human breathes' i.e. in the time in which it happens that he breathes, and we say: 'Not every human breathes' i.e. in the time in which it happens that he breathes, so that the time is the same one, then the first sentence is genuinely contradictory to the second. But *this is not how we do [in practice] take [the time] into account when we are using contradictions. Nobody ever demonstrates absurdity this way*" (see [34], book i.5, 38.5–11, translation W. Hodges, my emphasis).

This means that if one says that the universal affirmative sentence is the genuine contradictory of the negative particular one only when we express both in this way: "every human is breathing at t_0" and "not every human is breathing at t_0," one does not determine the contradictories in the usual way; for in this example, nobody can determine exactly the time of inspiring and warrant that this time is exactly the same in the affirmative sentence and in the negative one. Furthermore, this is not what we mean when we say "every human is breathing" and "not every human is breathing" because what we mean is much more general than what is said by the sentence "every man is breathing at t_0" and similarly for the negative one. So, although there is a genuine contradiction between these two **A** and **O** sentences (containing t_0), they do not express exactly what we mean when we say about humans that they are breathing or when we deny that. On the other hand, these propositions ought to have a truth value. Since one of them (the **O** proposition) denies the other one (the **A** proposition), this truth value cannot be the same for both. So, one has to analyze them in a way that makes them always contradictory (not only contradictory at some specific time t_0). This may also be the reason why he says that "some of the things said in *al-'Ibāra* were inadequate" (see quotation above), since in that treatise he did not enter into all the details of the temporal and other conditions that should be added to the propositions in order to determine their truth conditions. The analysis of these conditions takes into account not only what people actually say, but also what they mean by what they are saying. It also accounts for the very nature of the different kinds of propositions, for as will appear in the sequel, there are different kinds of absolute propositions which should not be treated in the same way, depending on the conditions they contain.

However, before proceeding with the analysis of the different kinds of absolutes, we may ask: if the analysis of the propositions in *al-'Ibāra* was so inadequate, according to Avicenna, why did he provide it at all? In other words, why did he present, in that treatise, ideas that he did not believe to be true? A tentative answer can probably be found in the following passage of Avicenna, where he explains his own attitude and some of his reasons:

> "Now since those who are occupied with Philosophy are forcefully asserting their descent from the Peripatetics among the Greeks, we were loath to create schisms and disagree with the majority of the people. We thus joined their ranks and adhered in a Partisan spirit to the Peripatetics, since they were the sect among them most worthy of such an adherence. We perfected what they meant to say but fell short of doing, never reaching their aim in it; and we pretended not to see what they were mistakern about, devising reasons for it and pretexts, while we were conscious of its real nature and aware of its defect. If ever we spoke out openly our disagreement with them, then it concerned matters which it was impossible to tolerate; the greater part of these matters, however, we concealed with the veils of feigned neglect." (*Manṭiq al-Mashriqiyyīn* 3.12–16, trans. Gutas in [78], 38)[11]

In this passage, he says that, because all people who were interested by philosophy used to follow the Peripatetics, he felt obliged to do the same thing, even if he did not always agree with their views. This is so because he was reluctant to create problems by showing radical disagreement about matters on which there was little disagreement. He did manifest his disagreement only when it was impossible not to do so, because what was said was simply wrong or unacceptable. He also says that the Peripatetics' views were "the most worthy of such an adherence,"

[11] I thank Prof. Hodges for bringing my attention to this text and providing me with Gutas' book.

meaning probably that these views deserved to be studied although one might have some disagreement with them. This reflects Avicenna's complex relation with the Greek tradition, since on the one hand, he adhered to it by commenting on all the treatises of the Organon, but on the other hand, he did not hesitate to express his own views, not only in *Manṭiq al-Mashriqiyyīn* which was supposed by some people to be the most personal treatise but also in his much more classical treatise, that is, *al-Qiyās*. So we can say that the disagreement was not total and was not always hidden. Even in *al-Shifā*, it appeared explicitly relatively soon at the level of *al-Qiyās*. There is then no radical difference between this last treatise and *Manṭiq al-Mashriqiyyīn*, as I have shown in a recent article (see [61]) and as will appear in the sequel.

Returning back to Avicenna's analysis of the propositions, let us consider the several ways of interpreting the absolute propositions, depending on the meanings carried by the sentences, which require specific conditions in each case. These conditions can be either related to the time at which the subject S satisfies the predicate P, or they are explicitly related to the subject, i.e., when S is said to satisfy P, only when the individual is described as S, or to the predicate, when we say that S is P at the time where it is described as P. However, as he himself acknowledges in both *al-Najāt* and *al-Qiyās*, several people before him stated some of these conditions. Among these authors, Avicenna evokes in *al-Najāt* Theophrastus and Themistius on the one hand and Alexander of Aphrodisias on the other hand. Here is what he says:

> "There are two opinions about the absolute, Theophrastus's then (*thumma*) Themistius' and others' opinion, and Alexander's and several commentators' opinion. The first one [says] that it is the [proposition] which does not contain any explicit modality (*lam tudhkar fīha jihatun*) to determine whether the judgment is necessary or possible; the judgment is thus totally absolute so that it admits the case where the judgment is necessary and also the one where it is not necessary, i.e. when it holds not permanently. This opinion may very well be that (*laysa yab'udu an takūna*) of the Philosopher (= Aristotle) about the absolute, provided that the Philosopher admits the simultaneous truth of the two absolute universals, affirmative and negative, as when you say 'every horse is sleeping' and 'no horse is sleeping', so that the universal affirmative absolute judgment is carried (*yunqalu*) to the universal negative judgment. Those who endorse this opinion say that this is admitted (*dhālika jā'izun*), but not necessary, because the Philosopher gives sometimes examples of absolutes, where this is not permitted, because these [absolute propositions] are always necessary." ([32], 23.4–12)

In this paragraph, he seems to say that Theophrastus' and Themistius' interpretation of Aristotle allows the absolute universal affirmative proposition (which is the categorical proposition not containing any modal word) to be true together with its correspondent universal negative proposition, in some cases, though not in all cases. The example illustrating that is the simultaneous truth of the two following propositions: "every horse is sleeping" and "no horse is sleeping." According to what Avicenna says in this passage, both Theophrastus and Themistius consider that this case where both universal propositions are true together is admitted by Aristotle. But he adds that they also acknowledge the fact that Aristotle gives other examples where the two universals cannot be true together, because the subject permanently satisfies (or does not satisfy) the predicate. What this passage says is that the absolute proposition according to the opinion endorsed by the two authors mentioned can be necessary or not necessary depending on the kind of sentences. It is thus absolute in that it admits both necessity and absence of necessity, although no specific condition is explicitly added by the two authors cited to indicate when it would be necessary or when it would not be necessary. Avicenna also says that this opinion "may very well be that" of Aristotle himself. Note that the example provided by Avicenna looks like an illustration of what he will call afterwards the "general absolute" propositions [i.e., the propositions containing the condition "at some times"] as we will see below. But he does not say in this passage that this condition is added by Theophrastus or by Themistius.

On the other hand, Alexander's (and his followers') opinion is explained as follows by Avicenna:

> "As to Alexander and several later commentators, they consider that this carrying (*naql*) is necessary (*wājib*) in the absolute, and that the absolute is where the judgment has no necessity in it, except with the four modalities mentioned after the two first ones. It seems then that the absolute, according to these [authors] is where the judgment is true (*mawjūd*) not always as long as the essence of the subject of the predication exists, but at some time, and this time is either as long as the subject is described by which it has been described (*mā dāma al-mawḍū'u mawṣūfan bi-mā wuṣifa bihi*), as when you say: 'every white [thing] (*kull abyaḍ*) has a colour dispersed to the eye' (*dhū lawnun mufriqun li-al-baṣar*), or as long as the predicate is used to make the judgement (*maḥkūman bihi*) or in a specific and determined time (*fī waqtin mu'ayyanin ḍarūrīyyin*) as with the eclipse of the moon and the fact of being in the uterus (*wa al-kawni fī al-raḥim*) for every human or in a necessary but not specific time as is [the time of] breathing for the animal." ([32], 23.13–24.7)

This last opinion is different from the above in that it does not allow the absolute to be necessary. The absolute is thus restricted to the non-necessary. In addition, it is said to contain the following conditions enumerated by Avicenna in this passage and illustrated by examples: "as long as it is S," "as long as it is P," "at a determined time," and "at some (regular but not specified) times." In all cases, the absolute proposition does not contain the idea of permanence, for the conditions above determine in one way or another the time at which the subject satisfies the predicate.

Avicenna adds that he himself does not priviledge one opinion over the other; for he says:

> "We are not concerned by priviledging one opinion over the other [Themistius' opinion or Alexander's one]; rather we consider the two conceptions about the absolute, and this [will] appear to you when we [will] divide (*faṣalnā*) the quantified absolutes. For saying 'every B is A absolutely' means that everything that is described in the intellect or in reality as being B whether it is described as B permanently or described as B at some time, after not having been B, this thing is described as A, we don't know when, whether when it is described as B or at another time or permanently or not permanently as held by Theophrastus. As to the other opinion it is not different from this one with regard to the subject,..., but it differs from it with regard to the predicate because the former considered the judgment carried by the predicate as the most general one, without any specifying condition of permanence or absence of permanence while the latter restricted it with the condition of the absence of permanence, so that saying 'every B is A' means according to them that everything which is described as B, no matter how it is so described, whether with necessity or without

necessity, this thing is described as A not by necessity but at some time, as they said. Likewise when we say 'no B is A' absolutely,…" ([32], 24.9–25.4).

It seems then that his own opinion will take into account both the necessity and the absence of necessity as in Theophrastus' and Themistius' opinion, so it does not restrict the absolute to the absence of necessity, as is the case with Alexander's and his follower's opinion. It thus differs from Alexander's opinion in this respect, but it is not exactly similar to the former either, because it is much more detailed than Theophrastus' and Themistius' opinion, since Avicenna, unlike these two authors, states explicitly all the conditions that can be added to the absolutes and determine the truth values of these propositions. This is the position that he expresses in *al-Qiyās*, when he says what follows:

> "If this is admitted, then we say: there was a disagreement between the ancients about the meaning of the absolute proposition. It was not a real disagreement, but a disagreement about the use of the word. For there is one group who endorsed the opinion according to which what is meant by absoluteness is the very fact that the proposition expresses a judgment, i.e. an assertion or a denial, no matter how (*kayfa kana*), so that this judgment includes all the ways of specifications mentioned (*'āmman li-jamī' wujūh al-takhṣīṣ al-madhkūra*), in which these specifications are not taken into account and where no condition of necessity or of absence of necessity [is stated]. While according to another group, what we mean by absoluteness is the fact that the proposition in so far as it expresses a judgment that is, a denial or an assertion, provided the condition 'as long as it exists' for the individual described by the subject is *not* satisfied, rather another condition [should be satisfied]. Thus the absolute in this sense is more restricted (literally 'more particular': *akhaṣṣ*) [= less general] than the absolute in the first sense." ([34], 26.7–14, emphasis added).

Here too, he evokes without naming their defenders the two opinions already discussed in *al-Najāt* in almost the same words. His position about these opinions appears in the following quotation:

> "The examples that we find in the first teaching show that what is most probably meant is the opinion endorsed by the first group…. For what can be shown from this is that saying: 'every B is A' means that everything that is described and posited as being actually B, permanently or not permanently, this thing is also described as being A, without considering when was that, and whatever specification we consider. Some people said that this is the characterization of the absolute, but they did not define all these specifications; they provided just three conditions: the first one is when B is permanently A, the second is [B is A] as long as it is described as B, and the third is [B is A] as long as it is described as A. So what we say is: 'every B is A' includes (*ya'ummu*) all three divisions (*aqsām*) and is more general than them (*ya'ummuhā kulluhā*). For the generality [of the absolute] is shown either by these three divisions alone or by considering all the specifications that we mentioned above, while the restriction (*al-khuṣūṣ*) [of the absolute] is characterized by two divisions (*bi-ḥassabi qismayni*), so that the absolute in the restricted sense is the one in which the predication is not permanent." ([34], 26.15–27.8).

His position is then close to that of the first group, according to which the absolute in the most general sense can include the necessary as well as what is not necessary, but unlike this first group he does provide the different conditions that can be attached to the absolute propositions. But he also says that in a restricted meaning, the absolute can be seen as what does not include necessity. And this is what he concedes to the second group, who says that the absolute should not include any kind of necessity in all cases. According to Avicenna, the absolute in its very general sense can include necessity as well as absence of necessity, while it does not include necessity only in its restricted meaning. In all cases, its different specifications must be clearly shown by exhibiting the conditions that are attached to it. This is why he states very precisely all these conditions as we show in the sequel. We note, however, that the conditions cited in this quotation are said to be only three conditions ("as long as it exists," "as long as it is S," and "as long as it is P"), while the conditions cited in the above quotation from *al-Najāt* are more numerous, since they also include "at some (regular) times" and "at a determined time."

Note that *al-Qiyās* precedes *al-Najāt*, according to Dimitri Gutas who says in *Avicenna and the Aristotelian Tradition* (see [78]) that the logical sections of the Cure can be dated between 1022 and 1027, while *al-Najāt* has been composed approximately at 1026 or 1027 (see [78], 115). The two other books, namely, *Manṭiq al-Mashriqiyyīn* and *al-Ishārāt wa at-tanbihāt*, were composed, respectively, in 1029 and around 1030–1034, for the latter (see [78], 132 + 155). So the latest work seems to be *al-Ishārāt*.

According to Avicenna, some propositions are true only when one adds the condition "as long as the essence exists"; others are true only when the temporal conditions "at some times" or "at some times but not permanently (= continuously)"[12] are added or when the condition "as long as it is S" is added, for instance, when one says "A moving thing is changing as long as it exists"; this is false, while the sentence "A moving thing changes as long as it moves," i.e., the same sentence with the condition "as long as it is S," is true.

Thus, the main kinds of absolute propositions are the following:

1. The propositions containing the condition: "as long as it exists" (called "*darūrī*" (= necessary) by Avicenna only in *Manṭiq al-Mashriqiyīn*).
2. Those containing: "at some times" (called "general absolutes" by Avicenna).

---

[12]The usual translation of the Arabic adverb "*dā'iman*" is "always". However, when one scrutinizes closely the text, one notes that what is meant by Avicenna when he says "*dā'iman*" is not "always" or "perpetually" (i.e., forever or for a long time), rather it is "permanently as long as the thing exists." This is why one has to be careful when translating the adverb "*dā'iman*" (contained in some propositions) or the adjective "*dā'im*", because 1. Avicenna calls "*dā'im*" the sentences containing "as long as it exists" and not another kind of sentences containing "always", which would be different from them, despite the fact that the text is not always clear (I thank Prof. Hodges for bringing my attention to that particular point), and 2. this condition (as long as it exists) means "continuously" (or "permanently" where the permanence is understood just as the continuous link between the subject and the predicate, not the fact that the subject satisfies the predicate forever or for a very long time), since the idea stressed by Avicenna is not the *length* of the duration but the *duration* itself (even if it is in some cases short as when we talk about some insects which live only a few days and say "Some insects are invertebrate (as long as they exist)" (personal example), i.e., the fact that the link between the subject and the predicate *is continuous* and *not discontinuous* or present only at some specific periods (as for the eclipse of the moon, for instance).

3. Those containing: "at some times but not permanently (= not continuously)" (called "special absolutes" by Avicenna).
4. The determined propositions containing: "at some (determined) time."
5. Those containing a condition involving the subject: "as long as it is S."
6. Those containing a condition involving the predicate: "S is P as long as it is P."

We can also add another kind, which is used to deny kind (5) of propositions and is expressed as follows:

7. Those containing the condition "at some time while S" (see [32], 29, note *).

Now let us explain a bit more what Avicenna means by each of these conditions. In *al-Qiyās*, when distinguishing between the different kinds of propositions, Avicenna claims what follows:

> "And we say 'God is living (*ḥayy*)', that is, eternally, as he has not ceased living and never ceases to be living (*dā'iman lam yazal wa lā yazālu*), and we say: 'Every whiteness is a color' and 'Every man is living', but what we mean is ***not*** that every thing that is white has not ceased and never ceases to be a color (*lam yazal wa lā yazālu*), or that every man has not ceased and never ceases to be living (*lam yazal wa lā yazālu kadhālika*). Rather what we say is: everything that is described as whiteness and said to be whiteness, this thing is a color *as long as its essence is existent*. Likewise for every being that is said to be a human, for it is not the case that this being has not ceased and will not cease [to be] (*lam yazal wa lā yazālu*) an animal; rather [it is an animal] *as long as its essence and its substance exists* (*mā dāma dhātuhu wa jawharurhu mawjūdan*)" ([34], 21.18–22.3, my emphasis).

This quotation explains what Avicenna means by the propositions containing the condition "as long as it exists." Note that "as long as" is the translation of "*mā dāma*", which contains itself the root "*dāma*" (= durates, lasts) from which the adjective "*dā'im*" and the adverb "*dā'iman*" come. So what is "*dā'im*" is what is durable or lasting or in other words, what is continuous (or eventually "permanent", when permanent means simply "continuous" = stable, constant, i.e., which *does not change* and *is not discontinuous*, but *not* if "permanent" is used to signify "lasting forever or for a long time" in which case it would be equivalent to "perpetual"). So what Avicenna means by this condition is that "S is P continuously, i.e. without interruption as long as the individual subject is existent." For these things, there is some kind of permanence (of continuity) which is determined exactly by the time of their existence. So humans, for instance, are animals exactly as long as they are existent, not forever and not perpetually.

The same idea is stressed by Avicenna in the following passage of *al-Qiyās*:

> "An affirmative proposition [stating that its predicate holds] either permanently or at some definite time counts as broad absolute. [Its negation] has to say that "[A] is permanently false of some [B]" where A being permanently false of this individual means being false for *so long as the essence of this individual is satisfied.*" ([34], p. 47.4–6, translation W. Hodges, my emphasis).[13]

[13]I thank Prof. Hodges for bringing my attention to this passage and other ones where the same idea is stressed by Avicenna.

In this passage, the sentence emphasized shows that the permanence is conditioned by the existence of the individual subject, so that the word "*dā'iman*" means simply "as long as the individual exists" or "permanently during the existence of the individual subject."

The same idea is expressed in *Manṭiq al-Mashriqiyyīn,* where he says:

"The contradictory of the sentence 'Some C is a B' in the broader absolute is 'No C is a B' where the intention in this is that every C was not and is not a B for as long as its essence exists *(mā dāma mawjūd al-dhāt)*" (Mantiq al-mashriqiyyin [37]).

So we can say in the light of these passages that the general absolute propositions (containing at some times) are contradicted by the propositions containing "as long as it exists" (= permanently or continuously during the time of its existence), provided one of them is negative while the other one is affirmative and one of them is particular while the other one is universal. It seems then that besides the propositions containing "as long as they exist," there is *no other* kind of propositions which would contain simply the word "always" or "perpetually" as a condition. The condition "at some times" itself means "at some times during the existence of the individual"; this is why the contradictories of the general absolutes are the propositions containing "as long as it exists" (= during the whole time of the existence of the individual). If so, these propositions could not be called "perpetual", because the word "perpetual" usually does not involve any condition, while these "*dā'ima*" propositions are conditioned by the existence of the individual subjects. Since the usual meaning of the word "perpetual" is "lasting forever or for a long time" so that this thing does not cease to be or that its duration is indeed very long, as when one says "the perpetual snows,"[14] it does not correspond exactly to what Avicenna means when he states this *dā'ima* propositions. For as we have seen, by stating the condition "as long as it exists" Avicenna stresses the idea that this condition could not be compared to the one that we find in the sentence "God is eternally living" where there is no condition at all, and where what is meant is really "does not cease and will never cease," i.e., "lasting forever". The word "perpetual" is not exactly equivalent to "eternal" but it is very close to it. It does not correspond exactly to the word "*dā'im*" as it is used by Avicenna, for in Avicenna's text, this word means "lasting the whole time of the existence of the subject," no matter how long is this time, i.e., whether it is long or short. This time has an end, namely, exactly the moment when the individual ceases to be. This being so, maybe it would be better to use the word "permanent" (where "permanent" means simply "continuous") to name this kind of sentences, since permanence can mean "lasting, continuous, constant, stable" and does not necessarily mean "lasting for a long time

[14]Of course, one could object that these snows are also conditioned by the very existence of the mountains. But the reason why they are called "perpetual" is not the existence of the mountains, rather it is the fact that they are always there, whether in winter or in summer. So people who live in these areas have always seen them in all seasons, given that these snows don't disappear in the summer unlike the snow on the ground which is present only in the cold season(s). They thus stress the length of the duration by using the word "perpetual". Note that in French, they are called "neiges éternelles" (i.e., eternal, not only perpetual).

or forever." It is this precise meaning which stresses the idea of continuity, constancy, and stability and the absence of discontinuity that should be retained in this kind of sentences, not the idea of a very long duration, which is carried by the word "perpetual" both in English and in French.

In *al-Najāt*, Avicenna explains even more explicitly this condition, when talking about necessary propositions (containing the expression "by necessity"). He insists that "*dā'iman*" means "permanently as long as the individual subject exists" rather than "perpetually", as we can read in what follows:

> "The second [condition] is that [the predication holds] as long as the essence of the subject exists, i.e. has not perished (*lam tafsud*, literally 'is not decomposed'), as when we say 'Every human is an animal by necessity', that is, every single human is permanently (*dā'iman*) an animal as long as his essence exists, not always (*dā'iman*) without a condition (*dūna shart*) so that he would be an animal [that] has not ceased and never ceases (*lam yazal wa lā yazālu*), before existing and after his perishing (literally 'his decomposition' *fasādihi*)" ([32], 20.5–8)

Here too, he distinguishes between a first kind of permanence that would only apply to the Eternal Being and is not conditioned, while the second kind is conditioned by the existence of these beings (the fact that they have not perished yet). The latter applies to all other kinds of beings and means "as long as they are existent." He adds, nevertheless, that the necessary propositions include the two kinds of necessity, depending on the nature of the subject, for the general meaning of "necessarily as long as he exists" can be "either always if the essence always exists (*tūjadu dā'iman*) or some period of time (*muddatan mā*) if the essence is perishable" ([32], 20.10–11).

Another point worth noticing is the use of the words "*dhāt*" and "*jawhar*" by Avicenna. According to Tony Street, who calls the propositions containing "as long as its essence exists" "substantial" in English and "*dhātī*" in Arabic, "Avicenna is interested in two intrinsic conditions, which came to be called respectively *dhātiyya* and *waṣfiyya*. These translate literally as 'substantial' and 'descriptional', and according to the later logicians writing in Arabic, may condition any temporality or modality" (see [136], 133). So it seems that he applies a distinction made by some post-Avicennian logicians to Avicenna's text. The expression "*dhātī*" by which T. Street qualifies the kind of sentences containing the condition "*mā dāma mawjūda l-dhāt*" (as long as its essence exists) comes from the word "*dhāt*" used by Avicenna in his formulation of the condition. Avicenna's phrasing of the condition is translated as follows by T. Street: "while he exists as a substance" (see [136], 133). However, as W. Hodges has stressed, the English word "substance" does not translate the Arabic word "*dhāt*" but the other Arabic word "*jawhar*" (see [92], 163). So strictly speaking these *dhātī* propositions should not be called "substantial", since "substance" = "*jawhar*", not "*dhāt*". He says "A number of published works refer to Ibn Sīnā's (d) sentences as 'substantial', apparently mistranslating Ibn Sīnā's word *dhāt* 'essence' as 'substance'" ([92], 163).

As a matter of fact, Avicenna does not use the word "*dhātī*" nor does he use the word "*waṣfī*" to qualify these different propositions. Now it is true that in the first quotation provided above ([34], 21–22), the word "*jawhar*" (= substance) is used by Avicenna. But it is adjoined to the word "*dhāt*" (= essence), which is the first one used there and is the most frequently used by Avicenna in his explanations of this kind of propositions. The word used in *al-Najāt*, for instance, is the word "*dhāt*", not the word "*jawhar*", in his explanations of these kinds of propositions (see [32], 23–25, in particular 24.4). In *Manṭiq al-Mashriqiyīn*, Avicenna says what follows about the propositions of this kind: "'B is J as long as its essence exists' (*mā dāma mawjūd al-dhāt*) is [called] the necessary" ([31], 65.6), while in *al-Ishārāt,* Avicenna says what follows: "And the condition is either the continuity (permanence) of the existence of the essence (*dawāmu wujūd al-dhāt*), as when we say "Humans are necessarily speaking bodies (*jismun nāṭiqun*), and we don't mean by this that humans have not ceased and will not cease (*lam yazal wa lā yazālu*) [to be] speaking bodies, for this is false of every individual person. Rather what we mean by this [condition] is "as long as its essence as human exists (*mā dāma mawjūd al-dhāt insānan*), then he is a speaking body" ([36], 265.2–6). So as we can see in all these quotations, the word used by Avicenna is "*dhāt*" rather than "*jawhar*" which is used only in one of them (along with "*dhāt*"). Even in *al-Qiyās*, we find in several other passages the word "*dhāt*", rather than "*jawhar*".

We won't, therefore, use the word "substantial" to qualify this kind of sentences. As we said above, a better word would be "permanent" which involves the idea of continuity, without necessarily involving that of necessity or of perpetuity. Avicenna himself calls them "*ḍarūrī*" in *Manṭiq al-Mashriqiyīn*, for instance ([31], 65). However, in *al-Qiyās* and in *al-Najāt*, he says that the absolute is more general than the necessary because it includes it, as we saw above, when we analyzed the quotations where he was comparing his own view with that of Alexander, Theophrastus, and Themistitus, and where he endorsed the opinion according to which the absolutes can involve both necessity and absence of necessity. They can be divided into several kinds of sentences containing several different conditions. For the condition "as long as it exists" does not always involve necessity in some cases, for instance, in particular propositions, as appears in the following quotation:

> "We say 'every stone is inert', this can be permanently as long as it exists, and it can be at some times. And it is necessary (*lā budda*) that it be at some time, and it can (*yajūzu*) nevertheless be permanently in some of them (*fī ba'ḍihi*), i.e. as long as its essence exists, coïncidentally not necessarily (*ittifāqan lā ḍarūratan*), thus not at every time, but at some time" ([34], 22.14–23.3).

This means that the condition "as long as it exists" does not necessarily mean that the link between the predicate and the subject is necessary, so that the proposition would always be called "necessary"; because of that strong link, this link can be always present but only accidentally or coïncidentally, for instance, when the proposition is a particular affirmative.

In *al-Ishārāt*, he says that the condition "as long as it exists" could be present in some "kinds of absolute propositions which are not necessary (*aṣnāf al-muṭlaq ghair aḍ-ḍarūrī*)" ([36], 268), namely, those where the predicate is permanently (i.e., as long as the essence exists) related to the subject but not necessarily, i.e., not essentially. An example is "'some persons are white, as long as they exist." This proposition contains the condition "as long as it exists" but it is not necessary, because the predicate "white" is not essential for the subject "human" (or person). It simply goes with (*saḥibahu mā dāma mawjūdan*) the individual subject as long as this individual exists, without really being essential to him, i.e., without having the consequence that this individual would not be a human if he did not possess this property. So the absolute propositions could be necessary or not necessary, which means that absoluteness *includes* necessity, it does *not exclude* it, because it is more general than necessity. This could be the reason why Avicenna abandoned the word "necessary" used in *Manṭiq al-Mashriqiyīn*, as W. Hodges notices in his book *Mathematical Background to the Logic of Ibn Sīnā*, where he says "In fact the 'd' in (*d*) stands for *ḍarūrī*. In *Mashriqiyyūn* Ibn Sīnā makes an attempt to set up systematic names for the two-dimensional sentence forms. Most of these names didn't survive to *Ishārāt*, so he must have decided they were not a success" ([92], History 9.1.9, 162).

Other kinds of absolute propositions are those containg the conditions "as long as it is S" or "as long as it is P." These propositions say that the subject individual satisfies the predicate when it is described as S or else when it is described as P. So the individual subject does not permanently satisfy the predicate, and it satisfies it at a certain time, the time at which it is described as S or the time at which it is described as P. According to the second view evoked by Avicenna that we mentioned above, these propositions do not involve necessity, because the subject is not said to be P permanently, rather it is said to be P at the time when it is S, or else at the time when it is P, so the property expressed by the predicate is not essential to it; it is not satisfied all the time of the existence of the individual subject.

The absolute propositions called "general absolutes" are those containing the condition "at some times." But this condition should be understood as meaning "at some times while they exist," i.e., at some (not determined) times in their lives (or their existence). The absolute propositions of this kind do not involve necessity either. Their contradictories are not the propositions containing simply "always", as I used to think; rather they are precisely the propositions containing the condition "as long as it exists," which means "at all times (or 'permanently' or 'continuously') during its existence."[15]

Now let us give some examples of these different kinds of propositions. The examples provided by Avicenna are the following:

The propositions containing the condition "as long as it exists" are illustrated by

1. "Every man is an animal as long as he exists" ([34], 22)

Those containing "as some times" (= at some times during its existence) are illustrated by the following:

[15] W. Hodges provides formalizations of these propositions which he calls two-dimensional propositions in his book *Mathematical Background to the Logic of Ibn Sina* (see [92], 157ff), where he uses time quantifiers and shows that the general absolutes containing "at some times" contradict the propositions containing the condition "as long as they exist," provided they do not have the same quality nor the same quantity.

2. "Every man is not laughing (at some times)" ([34], 82)

This sentence is a universal negative, i.e., an **E** proposition. Avicenna interprets it in the following way: "the predicate 'is laughing' can be negated from every man, at some times" ([34], 82). **E** propositions in Aristotle's and al-Fārābī's frames are convertible, since when "No A is B" is true, "No B is A" is true too. However, according to Avicenna, the **E** propositions which contain the condition "at some times" do not convert because the conversion would lead to the following sentence: "No laughing thing is a man (at some times)," which cannot be true, since "it is impossible to negate the predicate 'man' from what is laughing in effect" ([34], 82) because a laughing thing cannot be said not to be a man, given that if something is laughing, this thing must be a man, while the sentence "every man is not laughing (at some times)" is true, according to Avicenna, since it happens to every one not to laugh at some times. In this case, the permutation of the subject and the predicate leads from a true proposition (every A is not B at some times) to a false one (every B is not A at some times), which invalidates the rule of conversion for this kind of propositions.

Those containing the condition "at some times but not permanently, i.e. not continuously" may be illustrated as follows:

3. "Every human being is eating at some times but not permanently (= not continuously)"

This kind of sentences contains two conditions related by a conjunction. Consequently, the negation of such a sentence contains a disjunction, for it should be expressed in the following way:

"Some human beings are either never eating or permanently (= continuously) eating."

This last sentence is itself equivalent to the following one:

"Some human beings are never eating or some human beings are permanently (= continuously) eating,"

if we take into account the equivalence between "$(\exists x)(Ax \vee Bx)$" and "$(\exists x)Ax \vee (\exists x)Bx$", held by modern and contemporary logicians.

As a matter of fact, Avicenna expresses the negation of a sentence like **1c** above in the following way: "Some C's are permanently (= continuously) B or Some C's are never (= continuously not) B" ([36], 310).

So, his intuitions about the right negation of such a sentence are basically correct, even if he does not state explicitly the De Morgan' law that leads to this formula.

As to 4, Avicenna himself illustrates it by the following example:
4. "The moon has eclipses at some (determined) time"
which he offers in several treatises (e.g., in [34], 23.14, and [31], 68.12).
5 is illustrated as follows:
5. "Every white thing is colored as long as it is white" ([34], 22.10–11).
While 6 may be illustrated by the following example:
6. "Every man is walking as long as he is walking."

Note that the latter (kind 6) does not seem very interesting to Avicenna, who does not use them in the syllogistic and does not even evoke them, for instance, in *Manṭiq al-mashriqiyīn*.[16] As a matter of fact, their logical behavior is very different from that of the other kinds of sentences, for they are always true when they are affirmative, given the truth of their consequents, and they contain a contradiction when they are negative, which makes them always false, according to Avicenna, who says what follows: "[As to] 'Not every B is A as long as it is A', this negative [sentence] is never true" ([34], 41–42.1).

On the contrary, the sentences of kind (1) and of kind (5) are very important, for they are the ones that are used in the syllogisms and that validate the main logical rules such as the conversions.

This analysis modifies the oppositional relations between the assertorics, for the contradictories of the assertorics are now the *dā'ima* propositions, i.e., the propositions containing "permanently [as long as it exists]" or "never".

The contradictories between the general assertorics and the permanent propositions are expressed as follows:

- Every A is B (at some times) (**A**ga) / Some A's are permanently not B's (**O**p).
- Some A's are B's (at some times) (**I**ga) / No A is ever B (**E**p).
- Every A is permanently B (**A**p) / Not every A is B (at some times) (**O**ga).
- Every A is not B (at some times) (**E**ga) / Some A's are permanently B's (**I**p) (see [34], 49, names added),

where p: permanent, ga: general absolute, and "permanently" means "permanently as long as the thing exists" = the whole time of the existence of the thing, whether this period of time is long or relatively short (as for some insects, for instance).

For instance, if one says, "Every man is laughing at some times" (**A**ga) this will be contradicted by "Some men are permanently not laughing" (**O**p). The whole set of permanent propositions and their contradictories is provided in the following table, where "p" stands for "permanent", which means "permanently as long as it exists":

Ap: Every B is C as long as B exists / Not every B is C at some time while B exists.
Ep: No B is C as long as B exists / Some B are C at some time while B exists.
Ip: Some B are C as long as B exist / No B is C at some time while B exists.
Op: Some B are not C as long as B exist / Every B is C at some time while B exists.[17]

Note that the negative propositions, whether general absolute or permanent, do not have an import. So the permanent negative **O**, for instance, which denies **A**ga should say something like "either there is no A or some A are permanently not Bs."

[16] In this treatise, these propositions are abandoned as shown in [60].

[17] The full formalizations of these temporal propositions can be found in W. Hodges ([92], 159ff).

Note also that although Avicenna himself did not draw any figure, his theory gives rise to several such figures and also to more complex ones as has been shown by some authors.[18]

As to the special absolutes and their contradictories, they are the following (where w stands for *wujūdīyya*, which is the word used in *Manṭiq al-Mashriqyyīn* to qualify these propositions):

- **A**w: Every C is sometimes B and sometimes not B / Some Cs are permanently B or Some Cs are never B (*laysa all battata*) ([36], 310) (never = permanently not).
- **I**w: Some C's are sometimes B and sometimes not B / Every C is permanently B or no C is ever B (see [36], 311).
- **E**w: No C is sometimes B and sometimes not B / Some C's are permanently B or some Cs are permanently not B (see [36], 310).
- **O**w: Some C's are not B sometimes but not permanently / Every C is permanently B or no C is ever B (see [36], 311).

However, as has been shown in a previous article (see [55], 54), these contradictories are not all correct. The only correct couple is **A**asp and its contradictory. This correct couple of contradictories shows many interesting things, however, for Avicenna applies intuitively and almost implicitly one of the De Morgan's laws in stating it. This can be shown as follows: the De Morgan's law considered is the first one and it is the following:

- First De Morgan's law: $\sim(P \wedge Q) \equiv (\sim P \vee \sim Q)$

As to Avicenna's couple of contradictories, it is stated as follows:

*Every C is sometimes B **and** sometimes not B/Some C are permanently B **or** Some C are never B.*

If we consider that "sometimes" is contradicted by "never" and "sometimes not" is contradicted by "permanently" (or always), then we will have the following equivalences:

"Not (Every C is sometimes B)" = "Some C are never B."
"Not (Every C is sometimes not B)" = "Some C are permanently (= always) B."

Then the contradictory negation of **A** special absolute would be rendered as follows:

- Not *(Every C is sometimes* B ***and*** *sometimes not B)* ≡ [Not *(Every C is sometimes not B)* **or** Not *(Every C is sometimes B)*].

That is, ~[*Every C is sometimes B* ∧ *Every C is sometimes not B*] ≡ [~(*Every C is sometimes not B*) ∨ ~(*Every C is sometimes B*)].

[18]For instance, Saloua Chatti (see [53], 332–354) deals with modal oppositions and presents a dodecagon containing the whole set of propositions held by Avicenna and their relations. Tony Street (see [137]) gives the whole set of propositions held by Avicenna and their contradictories. All these propositions give rise to several squares of oppositions which are presented in [54], but Avicenna does not draw any figure at all.

If P: Every C is sometimes B and Q: Every C is sometimes not B.
Then the above equivalence is rendered as $\sim (P \wedge Q) \equiv (\sim Q \vee \sim P)$.
That is, by the commutativity of the disjunction: $\sim (P \wedge Q) \equiv (\sim P \vee \sim Q)$,
Which is precisely the first De Morgan's law.

So we can assume that Avicenna is *using* this De Morgan's law, even if he does not *state it* explicitly. This is a major advance in the logic, with which we must credit Avicenna. We will see that Avicenna applies this law implicitly in his modal logic too, when he states the contradictory of the bilateral possible. We will also see below in the chapter on hypothetical logic that he applies the second De Morgan's law [i.e. $\sim(P \vee Q) \equiv (\sim P \wedge \sim Q)$] in his hypothetical logic, when he considers the contradictory of $\mathbf{A}_D$ (= **A** disjunctive hypothetical proposition).

Returning back to the other contradictories stated above, we can say that unfortunately the second one is not correct because Avicenna confuses between "Every C is permanently B or no C is ever B" (explicitly stated in the text) and the real contradictory of **Ispa**, which is rather the following: "Every C is either permanently B or permanently not B." The latter is *not* equivalent to the one provided by Avicenna, since "$(x)Fx \vee (x)Gx$" is *not* equivalent to "$(x)(Fx \vee Gx)$", no more than $(x)[Fx \supset (Gx \vee Hx)]$ is equivalent to "$(x)(Fx \supset Gx) \vee (x)(Fx \supset Hx)$". So the error made by Avicenna is presumably due to the fact that he relies essentially on his intuition and does not have at his disposal a symbolism comparable to the modern symbolism of First-Order Logic, which is very helpful and can considerably clarify these matters. One cannot thus blame him for these errors because they are due to the absence of a formal symbolism and of a rigourous calculus.

As to the two last couples of contradictories, they are also incorrect because in both of them, the negation should put on the whole set of conditions "sometimes and sometimes not" in which case, it is the negative proposition itself that contains a disjunction, not its contradictory.

For instance, if we analyze **Espa** and put the negation in the right place, we get the following:

No C is B (sometimes but not permanently) = $(x)[Cx \supset \sim(x \text{ is sometimes B} \wedge x \text{ is sometimes not B})]$,

which leads by one De Morgan's law to the following:

$(x)[Cx \supset \sim(x \text{ is sometimes B}) \vee \sim(x \text{ is sometimes not B})]$

itself equivalent to the following:

"Every C is either permanently not B or permanently B."

Consequently the negation of this complex proposition will contain a conjunction and not a disjunction as Avicenna says. As a matter of fact, this negation is simply the proposition "Some Cs are sometimes B and sometimes not B," that is,

I special absolute. The same could be said about the last couple of contradictories, which is also incorrect for the very same reason.[19]

The error that Avicenna has made in his analysis of these complex propositions may be due to the fact that *he negated each part* of the conjunction involved in each of the propositions *separately*, so that the result seemed to him equivalent to two separate propositions related by a conjunction, in which case the contradictory would contain a disjunction without any doubt.

His analysis would thus look like the following:

E special assertoric is expressed by: Every S is not P (sometimes but not permanently) = Every S is not P sometimes $\wedge$ Every S is not permanently not P = Every S is sometimes not P $\wedge$ Every S is sometimes P.

Its contradictory is thus: $\sim$(Every S is sometimes $\sim$P $\wedge$ Every S is sometimes P) = $\sim$(every S is sometimes not P) $\vee$ $\sim$(every S is sometimes P) = Some S are not sometimes not P or some S are not sometimes P = Some S's are permanently P or Some S's are permanently not P.

This last proposition is what Avicenna considers as the contradictory of **E** special absolute. His error is then related to the *scope* of the negation, which *should not* put on each *part* of the consequent *separately*, as in Avicenna's analysis, but should on the contrary, put on the *whole* consequent containing the conjunction of both conditions.

As to the propositions containing "as long as it is S" (called *lāzima* in [31], 65), they validate all the relations of the square of opposition, provided that the affirmatives have an import, while the negatives do not, and also that their contradictories are the propositions containing the condition "at *some time* while it is S" (and not "at all times while it is S"). These contradictories are explicitly expressed in *al-Najāt* where it is said that the contradictory of the proposition "Every B is C as long as it is B" is "Not every B is C at a time when it is described as B" ([32], 28–29, note *). The propositions containing the condition "at a time when it is S" (S being the subject of the proposition) are called "*ḥīnīyya*" in that note. They are needed to provide the contradictory negations of the propositions containing "as long as it is S." So we can say that the propositions containing the condition "as long as it is S" and their contradictories are parallel to those containing "as long as it exists" which we called "permanent" above, and their contradictories, as shown by the following table (where we add "s" (for "subject") to express the fact that they contain "as long as it is S"):

"**As**: Every B is C as long as it is B / Not every B is C at some time while it is B
Es: No B is C as long as it is B / Some B are C at some time while they are B
Is: Some B are C as long as they are B / No B is C at some time while it is B
Os: Some B are not C as long as they are B / Every B is C at some time while it is B" (see [61]).

---

[19] I am indebted to Dr. Lorenz Demey for having drawn my attention to that particular feature of the two negative special absolutes, which differ in this respect from the two affirmative ones. I thank him for his enlightening observation.

Those containing "as long as it is P" are tautological when they are affirmative for

- Every S is P as long as it is P,
- Some Ss are Ps as long as they are Ps,

are true since their consequents are always true.

On the contrary, the negatives, that is,

- No S is P as long as it is P,
- Some Ss are not Ps as long as they are P,

are both false if their antecedents are true, given that their consequents are both always false; they would be true if their antecedents are false.

Consequently, the relations of the square do not hold for them if the negatives have an import because in that case, **O** would be always false and the subcontrariety would not hold. If the negatives do not have an import, then **A** and **E** would be possibly true together and the contrariety would not hold.

Finally, we have to note that the analysis provided by Avicenna in *Manṭiq al-Mashriqiyyīn* is not very different from those provided in *al-Qiyās*, *al-Najāt*, and *al-Ishārāt*, except for two things: 1. the propositions containing "as long as it is P" which are abandoned by Avicenna in the former, and 2. the names used in *Manṭiq al-Mashriqiyyīn* are new and different from those used in the other treatises. So we can establish the following correspondences between *Manṭiq al-Mashriqiyyīn* and the other treatises (where the letters inside brackets are used by W. Hodges in [92]):

"*al-Qiyās* (+ *al-Najāt*)	*Mantiq al-Mashriqiyīn*
General absolute (at some times)	*muntashira* (*mutlaqa 'āmma*) [*t*]
Special absolute (at times but not permanently)	*wujūdīya*
Prop containing 'as long as it exists'	*ḍarurīya* [*d*]
Prop containing 'as long as it is S'	*lāzima* [*ℓ*]
Prop containing 'as long as it is P'	abandoned
Prop containing 'at a determined time'	*mafrūḍa*
Prop true at some (future or present) time	*waqtīya* (= *ḥāḍira*) [*z*]
Not evoked (*al-Qiyās* + *al-Najāt*)	*lāzima mashrūṭa*
Prop with 'at some time while S' (*al-Najāt*)	*ṭāri'a* (= *muwāfiqa*) [*m*]" ([61])

So the propositions called "*wujūdīyya*" in *Manṭiq al-Mashriqiyyīn*, for instance, are those containing the condition "at some times but not permanently (while the subject exists)." Those called *ḍarūrīyya* are just the *dā'ima* propositions containing "as long as the essence exists," those called *lāzima* are those containing the condition "as long as it is S," while those called "*ṭāri'a*" or "*muwāfiqa*" are those containing the condition "at some time while it is S," and so on.

We can also note a clear correspondence between these absolute propositions (containing no modal alethic words) and the modal propositions, i.e., those containing the necessity and possibility operators, as shown in the following table:

(1) General absolutes (*muntashira*)	→	One-sided possible propositions
(2) Special absolutes (*wujūdīyya*)	→	Two-sided possible propositions
(3) *Ḍarūrīyya* (as long as it exists)	→	Necessity propositions
(4) *Lāzima* propositions	→	Nec. prop. containing 'as long as it is S'
(5) *Mafrūḍa* propositions	→	Necessity at some determined time
(6) *Ṭāri'a* propositions	→	Possibility at some time while S

These are Avicenna's own kinds of propositions. They are explicitly found in the different texts we have considered. Note that the propositions (6) are listed only because they can be used to deny the kind (4) of propositions. Likewise, the propositions of kind (1) are also used to deny those of kind (3) in the proofs provided. The propositions of kind (5) are not used in the syllogistic although they are listed in all the treatises. We can also add that Avicenna's analysis of the *ṭāri'a* (accidental) proposition is not very clear in *Manṭiq al-Mashriqiyyīn*, for Avicenna gives the universal affirmative *ṭāri'a* "Every S is P at some time while it is S" two different contradictories, which are 1. "Some S is never P as long as it is S" and 2. "Some S is either never P or permanently P, as long as it is S" ([31], 80.12–14). The latter contains a disjunction, and so it should contradict a proposition containing two conditions related by a conjunction. So maybe we could say that, apart from the confusion that this ambiguity creates, he is also considering a proposition containing a condition parallel to the one contained in the special absolute, i.e., "at some times but not permanently." In that case, we would have a seventh kind of propositions which would contain the condition "at some time but not permanently while it is S." This supplementary proposition would correspond to the special absolute which contains the condition "at some times but not permanently while S exists" and it could be added to the lists above since it has its counterpart in modal logic. We could therefore add the following pair of contradictories to the list above:

As_2: Every B is C (at some time but not permanently while B)/Either Some B is permanently C, while B or Some B is permanently not C, while B.[20]

3.2.3 The Oppositions in Averroes' Theory

Averroes' theory of oppositions is different from Avicenna's one and very close to al-Fārābī's and Aristotle's theories, which Averroes follows, by rejecting the temporal conditions added by Avicenna.

Against Avicenna's additions, Averroes' aim is to return back to the fundamentals of syllogistic, so to say, by being as faithful as possible to Aristotle. His theory is also very close to that of al-Fārābī, whom he cites quite often. But his

[20]On the opinions defended by post-Avicennian logicians such as Abhari and Katibi on the analysis of propositions and terms, see [141] and [146].

analysis of the different propositions and their oppositional relations are nevertheless slightly different from al-Fārābī's one, especially with regard to the indefinites, which are ambiguous in his frame, while they are treated as particulars in al-Fārābī's one.

According to Averroes, there are exactly six oppositions, namely, the following: (1) The opposition between the Singular propositions, (2) the opposition between the Indefinites, (3) the first couple of quantified Contradictories (**A/O**), (4) the second couple of quantified Contradictories (**E/I**), (5) the Contraries (**A/E**), and (6) the Subcontraries (**I/O**) (see [26], 91–93).

The opposition between the singular propositions is a contradiction, for when one of them is true, the other is false, and vice versa ([26], 92). He adds that "it is not possible [for them] to be true together or false together. For instance, when you say 'Zayd went out' and 'Zayd did not go out'" ([26], 92). According to this opinion, the real negation in a singular proposition is external, for only in that case, the negative proposition is the contradictory of the affirmative one.

Unlike al-Fārābī who distinguishes as we have seen above (Sect. 3.2.1), between the singular negatives, such as "Zayd is-not learned," and those containing a metathetic predicate, such as "Zayd is not-learned" and says that the first is contradictory to the affirmative, while the second one is contrary to it, Averroes says that the metathetic particle (*lā*, in the metathetic expression "*lā-'ādilun*" [not-just]) has "the strength of the negative particle (*ḥarf*)" [= *laysa*] in "the singular proposition, when their subject is taken to be *existent*" ([26], 107), for in that case, the two propositions "Socrates is not-just" (*lā-'adilun*, i.e., with the metathetic predicate) and "Socrates is-not just" (*laysa bi 'ādilin*, i.e., with an external negation) have the same truth value and contradict the affirmative one, when Socrates exists. So if "Socrates is just" is true, then both "Socrates is not-just" and "Socrates is-not just" would be false, when Socrates exists. And vice versa, if "'Socrates is just" is false, both "Socrates is not-just" and "Socrates is-not just" would be true.

If Socrates does not exist, the affirmative would be false, and the negative true, but what about the metathetic one? Would it be true like the negative, or false like the affirmative? In his *al-Maqūlāt* (The *Categories*), Averroes says that both propositions "Socrates is sick" and "Socrates is healthy" do not share the same truth value if Socrates exists, but "are *both false* if Socrates does not exist or if Socrates is a fœtus" (see [26], 66), while the two propositions "Socrates is sick" and "Socrates is not sick" never share the same truth value, "whether Socrates exists or not" (see [26], 66), for in both cases if one of them is true, the other is false. So the propositions containing contrary predicates (sick and healthy, in this example) do not have the same truth values as the real contradictory propositions.

Now, what would be the truth value of the proposition containing a *metathetic* predicate in case the subject "Socrates" is a fetus? In al-Fārābī's analysis, in this case, this proposition would be false in the first sense considered by him, which makes the metathetic predicate comparable to the privative one, such as "sick" or "ignorant".[21] But in the second and the third senses considered by al-Fārābī, the metathetic predicate is closer to the negative one, which makes the singular proposition containing it true, in case the subject is an infant (or a fœtus), and when the subject does not exist. These distinctions are not made in the same way by Averroes, who treats the metathetic predicate as a negative one. So we would say that his theory is different from that of his predecessor, because it would make the sentence containing the metathetic predicate true when the subject is a fetus. It is thus close to the second and the third meanings of the metathetic predicates considered by al-Fārābī, although maybe not exactly equivalent, for we don't know if it has an existential import in his frame (i.e., if the existence of the subject is required, in order for the proposition to be true), while al-Fārābī says explicitly that the subject of a proposition containing a metathetic predicate has to exist, unlike the one of a real negative proposition. As Thom says: "If *z* is non-*f* in the *third sense*, all that is required (*in addition to x's existence*) is that it not be the case that *z* is *f*, as shown in Fig. 3.3" (see [143], 202, my emphasis).

Regarding the indefinite propositions (called *muhmala*), Averroes' theory is different from both al-Fārābī's and Avicenna's theories, which follow Aristotle in that they both consider the indefinites as particulars. For he says explicitly that these propositions should not always be interpreted as particulars. Sometimes, they are particulars but they may be interpreted as universals. For the two sentences "Men are white" and "Men are not white," which are possible with regard to their matter, may be interpreted as particular. But the indefinites may also be interpreted as universals. He says in *Talkhīs Kitāb al-'Ibāra* what follows:

> "As to the indefinites (*al-muhmalāt*), they may be both true in the possible matter, and they may be considered as contrary. The reason for that is that the propositions starting by the article "*'al*" (*al 'alif wa-l-lām*) and what replaces them in all the languages, sometimes signify what the universal quantifier signifies, and sometimes signify what the particular quantifiers signify. If they signify what the universal quantifier signifies, they are [literally: "have the power of": *kānat quwwatuhā quwwata al mutaḍādda*] contrary propositions, and when they signify what the particular quantifier signifies, they are subcontrary [*kānat quwwatuhā quwwata mā taḥta al-mutaḍadda*], that is because the two propositions "Men are white" and "Men are not white" may be true together, when the article "al" signifies the same as what the quantifier "some" signifies, and they could be false together if the article "*al*" [*al 'alif wa-l-lām*] signifies what the universal quantifier signifies." ([26], 92.26–93.6, my translation)

Note that to illustrate the case where the indefinites may be viewed as universals, and consequently as contrary propositions, Averroes does not give an example in this part of the text. But his analysis shows that it is the *possible* indefinites that may *sometimes* be interpreted as universals and consequently as contraries, in which case they would be *both false*, given that contrariety means that both propositions may be false together. While the propositions whose matter is either necessary or impossible never share the same truth value, one of them being true, while the other

[21]See, for instance, al-Fārābī ([8], 147), and the analysis provided by Pr. Paul Thom in ([143], 193–209). On the different readings of the subject term, see [138]. On al-Farabi's opinion about the categories, see [145].

one is false. Averroes' opinion thus seems to apply to the *possible* indefinites, not to the necessary or impossible ones. This is corroborated by the very definition of contrariety that he provides in the same treatise, which is the following:

> "As to the contrary propositions, they never share the same truth value in the necessary and the impossible, but are both false in the *possible*, and they cannot be true together, for when one of them is true, the other one is false." ([26], 92.18–19, emphasis added)

This being so, his opinion is different from that of his predecessors, since Averroes endorses explicitly the position that the indefinites are ambiguous, while Aristotle, al-Fārābī, and Avicenna all agree in saying that it should rather be a particular proposition, even if it seems in some ordinary usages, as ambiguous.

Now, in some parts of his text, Averroes does indeed give an example of indefinite propositions which behave like universal ones, which is the following: the contrary of the "conviction that [so-and-so] is good" is "the conviction that [so-and-so] is bad," which is, according to him (and to his interpretation of Aristotle, whom he is commenting on in this part of the text), comparable to "the conviction that all [things] are good" is contrary to "the conviction that nothing is good" (see [26], 131.12–15).

As to the quantified propositions, Averroes defends a theory which is closer to Aristotle's and al-Fārābī's theories. As we saw above, the contradictories are **A/O** on the one hand and **E/I** on the other. These contradictory propositions are said to "never share the same truth value in any matter" (*taqtasimu al-ṣidqa wa-l-kadhiba fī jamī'i al-mawāddi*) ([26], 92), for instance, the following two couples of propositions: "Every man is white" versus "Not every man is white" or "Some men are not white" ([26], 92.3–4), and "Some men are white" versus "No man is white" ([26], 92.6–7).

The contrary propositions are the two universal propositions, affirmative and negative, such as the two propositions "Every man is white" and "No man is white" ([26], 92.1). They are defined as follows: "As to the contrary propositions, they never share the same truth value in the necessary and the impossible, and are both false in the possible; and they are not possibly true together, for *when one of them is true, the other one is false*." ([26], 92.18–20, emphasis added).

The subcontrary propositions are those that contain "the particular quantifier" ([26], 92.7), such as "Some men are white" and "Some men are not white" ([26], 92.8). These subcontrary (called *mā taḥta al-mutaḍādda*) propositions are defined as being "never true nor false together in the necessary and the impossible, but true together in the possible; and *when one of them is false, the other one has to be true* (*ṣadaqat al-ukhrā ḍarūratan*)" ([26], 92.21–22, emphasis added).

It may seem that these definitions do not add anything new to those that have already been given by Aristotle, al-Fārābī, and Avicenna. But we can nevertheless note some distinctive features that are present in Averroes' definitions which are slightly different from what we find in al-Fārābī's and Avicenna's texts.

First, there are conditions that Averroes adds in his definitions of both contrariety and subcontrariety that we don't find in Avicenna's text, for instance, although they are in some way present in al-Fārābī's one, namely, the following conditions "and when one of them is true, then the other one is false" in the definition of contrariety, and "if one of them is false, then the other one *has to* be true" in the definition of subcontrariety. Avicenna does not add this supplementary condition, although he does give the whole truth conditions of all the necessary, impossible and possible propositions. However, al-Fārābī says in his short treatise *al-'Ibāra*, while analyzing the propositions with metathetic predicates and comparing them with those containing simple predicates, that when one of the propositions is true, the other one has to be false (see [14], 101.5–10) [because they are contrary]. But he does not include this condition in the very definitions of contrariety that he provides in his treatises, for these definitions are the following:

> "As to the contrary [propositions], they do not share the same truth values in the necessary and the impossible, and are false together in the possible" ([15], 122.3–4)
> and
>
> "…the so-called contraries, such as "Every man is an animal" and "No man is an animal", they do not share the same truth value in some cases, that is, in the matter necessity and the matter impossibility, e.g. when we say … "Every man flies" and "No man flies", and they are sometimes false, that is, in the possible matter, as when we say "Every man is white" and "No man is white"" ([16], 158.11–14)

So we can say that the additional conditions provided by Averroes are not so clearly stated in his predecessors' writings. These additional conditions introduce more precisions by specifying not only what *are* the truth conditions of such (contrary and subcontrary) propositions, but also what they *should be*.

Another clarification provided by Averroes, which we don't find in any of his two predecessors in the Arabic tradition (and not in Aristotle either), concerns the definitions of the quantifiers, which are strictly separated from the negation in his frame, while they are mixed with the negation in al-Fārābī's and Avicenna's ones. For Averroes says explicitly that there are only *two* quantifiers, that is, "every" (*kull*) and "some" (*ba'd*) ([26], 91.11), which determine the quantity of the propositions. While both his predecessors consider that there are four quantifiers, which are "every", "some", "not every", and "none", thus mixing between the quantifiers per se and the negation. For instance, al-Fārābī says that the quantifiers are the following: "All, None (*lā wāḥid*), Some and Not all" ([15], 118.10), while Avicenna mentions the same classification in *al-'Ibāra* (see [35], 54.11). This may be seen as a further clarification by Averroes, because he clearly separates between the negation as such and the quantifiers, i.e., the particles expressing the quantity. This is what modern and contemporary logicians do when they symbolize the quantified propositions. However, there are major differences between his quantifiers and the modern ones. For in modern logic, for instance, in FOL, the existential quantifier and the universal quantifier are both expressed by symbols ($\exists$ and $\forall$, respectively), and these symbols have precise meanings which are, respectively: $\exists$: "there is an x or more (…)" and $\forall$: "for all x (…)." While Averroes does not have a formal symbolism by which he can express these quantifiers, rather he uses the ordinary Arabic language, where the only words available are "*ba'd*" (for the particular quantifier) and "*kull*" (for the universal one). When one adds the negation to these, one gets "*laysa kull*" and "*laysa ba'ḍ*". The problem is that both these expressions render the particular negative quantifier (= the one used in **O**), and none of them renders the universal negative one, which is usually expressed by "*lā aḥada*" or "*lā waḥida*" (= no one). So what seemed at first sight a clarification (which it is in some way) has the unwelcome consequence that the universal negative cannot be rightly expressed by Averroes, with his separation between the quantifier per se and the negation, as noted by one referee, who says "by his definition, 'No Human is a horse' will be unquantified."

The separation between the quantifiers and the negation is indeed important in logic, for it makes it possible to define the quantifiers and to relate them with the negation in all possible ways without confusing both kinds of particles. For the duality laws, for instance, [that is, $\forall\sim\ \equiv\ \sim\exists$ and $\exists\sim\ \equiv\ \sim\forall$ therefore $\forall \equiv \sim\exists\sim$ and $\exists \equiv \sim\forall\sim$] may be stated because the negation may be put either in front of the quantifier or after it or both. But one can use only one quantifier, as in Frege's theory, where only the universal quantifier is used, and where the difference between the universal negative proposition and the particular negative one is rendered by means of the scope of the negation, since the quantification used in **E** is rendered by "$\forall\sim$", while that used in **O** is rendered by "$\sim\forall$".

However, despite these clarifications and precisions, we find some confusion in his text too. For instance, when talking about the particular negative proposition (**O**), he does not distinguish between two—somewhat different—ways of expressing it, which are "Not every S is P" and "Some Ss are not P," as we saw above. Although these ways are generally considered as almost equivalent, they are not quite so, despite the validity of the duality laws. For the difference between the first sentence and the second one is related to the problem of existential import, which is not explicitly raised in Averroes' theory, nor in fact in his predecessors' ones. Now, as some contemporary logicians showed, such as Terence Parsons (see [121]), it is when **O** is expressed by "Not every S is P" that it does not have an import, while it does have an import if it is expressed by "Some Ss are not P." Since Averroes identifies between both sentences, he does not realize the difference between them, which is important, because it may affect the truth value of the proposition.

On the other hand, Averroes did not provide the whole set of oppositions, for unlike Avicenna, who evokes and defines subalternation, which he calls *tadākhul*, Averroes did not define nor even evoke this relation in his analysis of the oppositions, despite his knowledge of Avicenna's texts. This may be due to the fact that he follows Aristotle, who does not talk about subalternation explicitly in *De Interpretatione*, and presumably also because he did not have any idea about the square of oppositions, given that the logicians of the Arabic tradition, including Avicenna and al-Fārābī very probably did not know Apuleius (125–180 AD), who was the *very first* logician to draw the square, which we find several centuries later in Boethius' writings (6th Century). Let us now turn to the syllogistics presented in the three frames.

3.3 The Syllogistic

3.3.1 Al-Fārābī's Syllogistic

Al-Fārābī presents his syllogistics in *Al-Qiyās* and *Al-Qiyās al-Ṣaghīr*. In the latter, he summarizes his views and gives concrete examples, while the former contains the formal and general proofs of all the syllogistic moods. He provides also definitions of the syllogism as such in other treatises without developing the whole syllogistic, i.e., without stating and proving the different valid moods in each figure. We will consider these definitions and compare them with the ones provided in *Kitāb al-Qiyās* and *al-Qiyās al-Ṣaghīr* (see [11] and [16]) in order to evaluate the theory as a whole.

The (deductive) syllogisms (considered in this book) are basically of two kinds:

1. The predicative syllogisms (*qiyāsāt ḥamlīya*), which are the classical Aristotelian assertoric syllogisms.
2. The hypothetical syllogisms (*qiyāsāt sharṭīya*), which correspond to the Stoic kind of syllogisms.

(2) is subdivided into

2a. The connected syllogisms (*qiyāsāt muttaṣila*).
2b. The separated (or disjunctive) syllogisms (*qiyāsāt munfaṣila*).

Some indirect syllogisms called "*Qiyās al-Khalf*" correspond to the *reductio ad absurdum*.

In this section, we will examine the assertoric syllogisms, i.e., those that involve the assertoric propositions which are the affirmative and negative categorical (i.e., subject–predicate) propositions that do not contain modal words, and can be either true or false.

Let us first see how al-Fārābī defines the syllogism. Then we will examine the moods held in the system and the proofs provided by al-Fārābī.

In *al-Alfāẓ al-musta'mala fī al-Manṭiq*, al-Fārābī defines the syllogism as follows:

> "The syllogisms in general (*bi-l-jumla*) are things (*ashyā'*) arranged in the mind (*turattabu fī al-dhihni*) in a certain order, so that when they are so arranged, they let the mind see (*ashrafa 'alā*) inevitably (*lā maḥāla*) some other thing that he did not know before but comes now to know, forcing him to admit (*yaḥṣulu inqiyād*un) what he just looked at and to [consider it] just as what he knows." ([4], 100.3–5).

This definition seems to be al-Fārābī's personal view about the syllogism, for it does not appeal explicitly to Aristotle's or the Greek commentators' texts. The ideas that al-Fārābī stresses in this definition are the following:

1. The syllogism is something that the *mind* conceives (in Aristotle's definition, the mind is not evoked).
2. The syllogism is composed of things [= premises] that are arranged in such a way as to lead to a given conclusion *lā maḥāla*, i.e., *inevitably* (or necessarily).

3. There is an order in this specific *arrangement* of the [premises].
4. It is this specific ordered arrangement that makes the mind *admit,* i.e., hold true the conclusion.
5. The conclusion that the syllogism yields to *was unknown* before being admitted by the mind and becomes known once the mind comes to see it.
6. The syllogism is thus productive of a *new* knowledge that the mind did not have before deducing it from the premises.
7. Finally, it seems that the syllogism is mainly composed of the premises, the conclusion being what the premises yield or entail.

By using the word "things" to refer to the premises (and also the conclusion), al-Fārābī seems to stress the idea that the syllogism involves what the mind thinks (i.e., thoughts or intellected meanings) not only what he says (i.e., linguistic expressions).

In his explanation of this definition, al-Fārābī adds that these "things" of which the syllogism is made (i.e., the premises) are *not* words (*alfāẓ*) ([4], 100.7); rather they are meanings or thoughts that the mind can conceive and understand, for "it is not possible (*laysa yajūzu*) that the arrangement of meaningless words alone could be thought to lead to a meaningful [conclusion] which would follow from them by necessity" ([4], 100.12–14). So the syllogism has to do with the meanings of words or of sentences; it is not concerned by language alone or by these words alone. This idea is stressed in all the succedent passage where al-Fārābī says "…So it is clear that the things that are arranged (composed) in the mind are not words but they are rather intelligible meanings (*ma'ānin ma'qūla*)" ([4], 101.5–6).

The definition above seems to refer indirectly to the famous debate between the grammarian al-Sirāfī and the logician Mattā Ibn Yūnus, by stressing the idea that logic and the logical arguments do not deal with a particular tongue or language but with thoughts, which are meanings that can be expressed in any language. This is confirmed by the following passage where he stresses the idea that the things learnt are "the same for everybody (*wāḥida 'inda al-jamī'*), while the signifying words (*al-alfāẓ al-dālla*) are not the same for everybody (*laysat wāḥida bi-a'yāniha 'inda al-jamī'*)" ([4], 101.7). He also says that the things conceived by the mind are conceived "naturally and necessarily (*bi-at-tab'i wa al-ḍarūrati*)" while the words are constructed by using "conventions (*bi-isṭilāḥin*)" ([4], 101.11–12). This, according to him, is completely in accordance with Aristotle himself who says in *Posterior Analytics* that the demonstrations and the syllogisms are not "constituted by external speaches (discourses: *nuṭq*) but internal discourses" ([4], 102.8–9), that is, they are not what people say, but what people think.

By stressing the importance of meanings and minimizing that of words, he also seems to refer to the criticisms that have been addressed to the very word "*manṭiq*" the Arabic counterpart of "logic", whose root is the verb "*naṭaqa*" (= to pronounce, to say) and which means "speech" in Arabic. As noted by D. Gutas (1988):

> "The Arabic word mantiq means "speech" and was for this reason selected by the translators "to serve as a literal and artificial translation of the technical meaning of Greek *logos*" ("speech" / "reason" among other meanings). The selection proved particularly

discomfiting to later apologists of logic in Islam because it provided ammunition to the arguments of its opponents: if the science of logic, which, in addition to its reprehensibility for being a foreign – or in this case, Greek – speech, then it is worthless to a speaker of Arabic who needs only the grammar of his native tongue. In the debate between the grammarian Sirafi and the logician Abu Bishr Matta, this was precisely Sirafi's point at one stage of his argument" (see [77], 271–272).

These criticisms are made by the grammarians who say that if logic, as its name indicates, studies Greek speech then it is not needed by any Arabic speaker whose language is Arabic, not Greek. So we could say that Al-Fārābī's focus on meanings, which are universal, whereas languages are particular, and natural whereas languages are conventional, makes a point against the grammarians—though indirectly in this particular text—by stressing the difference between what logic ("*manṭiq*") studies and what grammar studies.

However, as noted by W. Hodges, Avicenna did give a great importance to the expressions and symbols, and that made his logic close to modern and contemporary systems as we can read in the following quotation:

"In short, for Al-Fārābī, ordering the expressions can't help for ordering the meanings. The expressions can't even be ordered in the mind, since expressions are spoken. (Ibn Sīnā precisely answers this in (28) with his observation that reasoners talk to themselves in their minds.) Al-Fārābī goes on to say that if expressions could help to order meanings in the mind, then so could pictures and gestures, and this would be 'laughable and beneath contempt'. In this exchange, Al-Fārābī presents views based on some a priori notion of what is proper, and Ibn Sīnā replies with sound empirical observations about human cognition. Compared with Al-Fārābī's view, Ibn Sīnā's position marks a distinct advance towards a realistic notion of how logic rests on symbol-processing and is open to methods of computation." ([87], Sect. 7).

According to Avicenna's opinion which is reported in this quotation, it is not by stressing the importance of meanings that one should define logic and distinguish it from grammar, since logic itself uses its own expressions and symbolism. One must then distinguish the two disciplines by their specific subject matter, their aims, and their methods which are all different from those of grammar.

But we have to relativize Avicenna's criticism of al-Fārābī as it is reported in the above quotation, for al-Fārābī does not deny the importance of language and of symbols and he also talks about the *internal speech* that Avicenna is evoking above, for instance, when he analyzes the word *manṭiq* (logic), which comes from *nuṭq* (speech), a word that does not seem at first sight to mean the study of reasoning or of valid arguments. *Manṭiq* and *nuṭq* are compared to the Greek *logos* (from which "logic" comes too) and this link, together with the very choice of this word by the translators are justified as follows by al-Fārābī in his essay "*Iḥṣā al-'ulūm*":

"Logic (*al-Manṭiq*), in so far as it provides the rules governing the expressions, provides rules that are common to the expressions of [several] nations (*tashtariku fīhā alfāẓ al-umam*), and are taken in so far as they are common (*min ḥaythu hīa mushtaraka*), and it does not study anything that is peculiar to the expressions of some nation…This science, in so far as it provides rules about external speech and rules about *internal speech* (*al-nuṭq al-dākhil*), and by these rules provided in both these things, rectifies (*yuqawwimu*) the third [meaning of] speech [i.e. the power to speek and to think], which is innate in humans, and guide it (*yusaddiduhu*) so that in both things it only makes the rightest, the most complete and the best things, [this science] has been called by a word coming from '*nuṭq*' which is used in the three meanings [i.e. as external speech, internal speech and the innate power of the soul to produce speeches and to think]" ([3], 78.15–79.4, my emphasis)

So al-Fārābī *does* talk and admit what could be called an *internal speech*, i.e., thoughts that are expressed by language inside the human mind. He does consider that the thoughts are expressed by languages and function like external speeches and that "humans talk to themselves in their minds" as stressed by Avicenna. The idea that he stresses to distinguish between logic and grammar is rather the idea of the universality of logic *which contrasts with* the particularity of grammar: logic is expressed by something like a universal language, i.e., a language that could be common to all particular tongues. In this respect, it is not specifically related to one specific tongue (to Greek, for instance) as the grammarians tend to think. It is much more general and can be useful to all nations and all people, whatever tongue they use.

This universal character of logic is also expressed explicitly in other treatises, as witnessed by the following passage taken from the short essay "*al-Risālah*" where al-Fārābī distinguishes in the same way between logic and grammar:

This art, since it gives rules to the rational faculty for the interior speech which is the intelligibles, and rules shared in common by all languages for the exterior speech which is the expressions, and directs the rational faculty in both matters at once towards what is right and protects it from error in both of them together, is called *manṭiq*, logic. Grammar shares with to some extent and differs from it also, because grammar gives rules only for the expressions which are peculiar to a particular nation and to people who use the language, whereas logic gives rules for the expressions which are *common to all languages*." (al-Fārābī, *Risālah* ([6] cited in [77], 272).

Returning back to the definition of the syllogism, we find in *Kitāb al-Qiyās* the following definition, which is much closer to the Aristotelian one, for it is expressed as follows:

"The syllogism is a discourse where more than one thing is stated (*tūḍa'u fīhi*), which when composed (*ullifat*), something other than [these things] follows from them by themselves, not by accident but necessarily. And what follows from the syllogism is called the "conclusion" or, alternatively, "what follows"" ([10], 19.7–8, my translation).

This definition is almost the same as Aristotle's one which is the following:

"The syllogism is a discourse in which, if more than one thing is stated (*wuḍi'at*), something other than what is stated follows by necessity, by means of their being so (*li-wujūdi tilka al-ashyā'i bi-dhātihā*)" ([40], 142, my translation).

As we can see, the main difference between Aristotle's and al-Fārābī's present definition is the mention "not by accident," which is added by al-Fārābī in his definition. Another feature noted by Lameer is the expression "from which" used in this definition (instead of "in which") which suggests, according to him (and to Thom, to whom Lameer refers in this part of the text), that the syllogism is "composed of premises only" (see [108], 92), which means that the conclusion is not a part of it. According to this interpretation, the assertoric syllogism is then a pair of premises in al-Fārābī's view. This interpretation seems to be corroborated by the following quotation: "and what is implied *by* the *syllogism* is called the conclusion…" (*wa al-lāzimu 'an al-Qiyāsi yusammā al-natīja…*) (see [10], 19.8), which suggests that the conclusion *follows from* something called the syllogism; therefore, it is *not a part* of it, but can be deduced from it.

Now we have to search in other treatises to check whether a syllogism is just a pair of premises or a pair of premises and a conclusion, or as Rescher expresses it in *Galen and the Syllogism* "…the distinction between an Aristotelian two-premisses-cum-question and a galenian two-premisses-cum-conclusion conception of the syllogism…" (see [126], 17). According to N. Rescher, this distinction is important because it determines the admission or the rejection of the fourth figure. If one defines the syllogism as a pair of premises, his definition forces him to reject the fourth figure because this fourth figure would be in his account just a variant of the first one, since the real difference between the fourth figure and the first one does not appear at the level of the premises; it is rather shown by the subject and the predicate inside the conclusion. This would explain why al-Fārābī does not talk about the fourth figure in his extant writings, although according to Rescher, he might have talked about it and rejected it in his long commentary of the *Prior Analytics*, which is now lost ([126], 13).

So let us see what al-Fārābī says about the syllogism in his other treatises. In *al-Risālah*, al-Fārābī says what follows:

"And by this order (*bi-hādhā al-tartīb*), the number of premises of which the syllogism is composed (*allatī minhā yalta'imu al-qiyās*) [is determined]" (see [6], 58)

This suggests that the syllogism is composed of premises only, although the idea of ordered composition is also present.

Elsewhere, al-Fārābī says what follows:

"And the syllogistic discourses (*al-aqāwīl al-qiyāsīyya*) are composed of the simple discourses (*al-aqāwīl al-basīṭa*) so that they come to be complex discourses (*aqāwīl murakkaba*). And the smallest complex discourse is the one containing two simple discourses, while the biggest [one] is not determined. So in every syllogistic discourse, the great parts (*ajzā'uhu al-'uḍmā*) are the simple discourses while its small parts, that is, the parts of its parts, are the single intelligibles (*al-mufradāt min al-ma'qūlāt*) and the expressions that signify them." ([3], 86)

Here too, he expresses the idea that the syllogism is composed of premises, but he adds that these premises are in turn composed of simple discourses (the terms), which are thus the smallest components of the syllogism, being the "parts of its parts," or the components of its components, whether these are linguistic expressions or what these expressions signify, i.e., simple intelligibles. Despite this precise division of the respective parts of the syllogism, the conclusion is not evoked and so it does not seem to be considered as one part of the syllogism.

In *al-Qiyās aṣ-ṣaghīr*, he says what follows:

"The syllogism is a discourse composed of premises that are posited and which when composed, something else than [these premises] follows from them by themselves not by other things, necessarily. And what comes to be known by the syllogism is called the conclusion or what follows (*al-radf*)" ([16], 160.2–4)

This definition is very close to that provided in *al-Qiyās*, and it says explicitly that the conclusion follows from the syllogism; it is therefore not a part of it.

However, in another essay, he seems to add a further idea, which is expressed as follows:

"The syllogism is composed of (*murakkabun min*) two things: the first one is the premises of which the syllogism is made, the second one is the figure by which the syllogism is structured (*alladhī bihi yatashakkalu al-qiyās*)" ([18], 9)

Here he evokes explicitly the figure that appears to be part of the syllogism just as the premises. Does this mean that he takes the conclusion to be also part of the syllogism? Not if we consider that the figure in the Aristotelian account evoked by Rescher above, which is opposed to the Galenian one, can be determined only by the premises and the places of the middle term inside them, in which case only three figures are considered: 1. The one where the middle is subject or predicate in either premise, 2. the one where it is predicate in both premises, and 3. the one where it is subject in both premises. So the addition of this particular point related to the figure does not seem to change radically his position to the effect that the syllogism is composed of premises and that the conclusion is not part of it. This being so, it seems that Lameer's interpretation with regard to the composition of the syllogism in al-Fārābī's logic is corroborated by several texts and seems to be the right one.

In *al-Burhān*, the definition contains some different new features, which are expressed as follows:

"The syllogisms which are composed of premises that are certain and necessary are divided into three kinds: The first one yields by itself (*yufīdu bi-dhātihi*) the knowledge of the thing only; the second one yields by itself the knowledge of the cause only; while the third one yields by itself the knowledge of both things." ([17], 272)

Even in this definition, which is different from the above ones in that it considers only some kinds of syllogisms—the demonstrative ones—where the premises are certain and necessary, it seems that the syllogism is defined as premises which lead to the answer of the question raised by the goal or objective (*maṭlūb*). This answer can lead to the knowledge of the thing or of the cause or of both, the latter being the most perfect one, according to al-Fārābī. So here too, the definition is not significantly changed despite the special character of the demonstrative syllogism.

However, in all cases, the syllogism does not seem to be an implication, despite what Lameer's analysis suggests (since he is following Łukasiewitcz in that specific point as we will see more precisely below). This is so because the syllogism in al-Fārābī's view cannot, in any case, be interpreted as a complex proposition containing an implication, which could be either an axiom or a theorem, depending on the figure to which it belongs, as in Łukasiewicz's interpretation of the Aristotelian syllogism. The reason is that al-Fārābī uses the expression "it follows that" (*yantuju*) (*al-Qiyās*, in almost all the syllogisms) or "it follows from it" (*yalzamu 'anhu*) ([11], 76) and occasionally the word "therefore" (*fa-idhan*) (for instance, [11], 78.13), indicating clearly that the conclusion is *deduced* from the premises, which is not always the case in Aristotle's texts because Aristotle sometimes expresses the syllogism by means of "if…then", for instance, in the following formulation of *Barbara*: "If A is predicated of every B, and B of every C, A must be predicated of every C" ([24], Book I, 25b32–26a2), where the comma replaces the word "then", and the word "must" (or sometimes in other syllogisms its equivalent "it is necessary") indicates the necessity of the link between the premises (antecedent of the implication) and the conclusion (consequent of the implication). But this feature may not necessarily be a clear departure of al-Fārābī from Aristotle's way of expressing the syllogism, since his use of the word "therefore" or its equivalents by al-Fārābī may simply be due to the fact that he is following "some post-Neoplatonic teaching texts" as one reviewer suggests. According to him, this may be so because the word "therefore" is used by Paul the Persian, for instance, whenever the conclusion really follows from the premises.

The premises are predicative propositions (*qaḍāyā ḥamlīya*) containing a subject and a predicate. The subject and the predicate are called "terms" (*hadd*) ([15], 125, [10], 20.8). The terms of a syllogism are the two extremes (*tarafāni*), that is, the major and the minor, and the middle term (*al-awsaṭ*). The major (*hadd akbar*) is defined as "the predicate of the conclusion (*maṭlūb*)" ([15], 126, [10], 21.2) (in *al-Qiyās al-Ṣaghīr*, the conclusion is called *natīja*). The minor is defined as the "subject of the conclusion (*maṭlūb*)" ([15], 126, [10], 21.2–3). The middle is defined as "the common part of the two premises conjoined" ([15], 126, [10], 20.16–21.1). It is also called the "cause" (*'illa*), for instance, in *Kitāb al-Qiyās* ([10], 24.10). The major premise is the one containing the major term, while the minor premise is the one containing the minor term.

Note, here, that the definitions of the two extremes (the major and the minor) are not exactly equivalent to those provided by Aristotle for al-Fārābī, following Alexander of Aphrodisias, defines these terms by means of their places in the conclusion, thus by appealing directly to the structure of the syllogism. This definition, however, raises some problems, since the syllogism as we saw above is composed of the two premises only, so how can we decide what term is the major one and what term is the minor one in the valid moods? In other words, how can we distinguish between the valid moods and the invalid ones, since the validity of a mood depends also on the place of the subject and of the predicate in the conclusion? Consider, for instance, the following premise pair:

Every A is B
Every B is C

In this example, the middle term is B, but among the two other terms A and C, which one should be the subject of the conclusion and which one should be its predicate in order for this premise pair to be productive? Should the right (i.e., the true) conclusion be "Every A is C" or "Every C is A"? How can we choose between these two conclusions, if we don't know in advance the structure of the valid mood and if we use letters rather than concrete terms? If we don't include the conclusion inside the syllogism, so that the places of all the terms in the premises and in the conclusion are already fixed, it is hard to distinguish between the productive premise pairs and the non-productive ones, since the only difference between the following combinations (where one is valid while the other one is invalid) lies in their *conclusions*:

(1) *Valid* combination (*Barbara*)	(2) *Invalid* combination
Every A is B	Every A is B
Every B is C	Every B is C
Therefore *Every A is C*	Therefore *Every C is A*

So, here, something must be added in order to show in which case a premise pair is really productive, i.e., in which case it yields to a *true* conclusion, whenever the premises are themselves true. This problem is not raised by al-Fārābī, but Avicenna raises it and solves it by fixing the order of the premises and by defining the major term as being the one of the major premise, i.e., the second premise in his frame.[22] So in this example, the major term should be C, just because C is the term of the major premise which is *not the middle* term (the middle term being the one which is present in *both* premises).

However, in the case of al-Fārābī, we could note that he presents the syllogistic moods not only by using letters as term variables but also by using concrete terms. For instance, in *al-Qiyās aṣ-ṣaghīr*, almost all the syllogistic moods are expressed with concrete terms. And this use of concrete terms makes it easier to find the right subject and the right predicate in the conclusion, although it is not a good method in logic, since in deductive logic, it is the form of the argument that makes it valid, not its matter. But the fact remains that when one uses concrete terms, the choice of the subject of the conclusion and of its predicate is easier, since one sees immediately which conclusion is true. For instance, in the combination above, we can replace A by "Greek", B by "human", and C by "mortal", giving rise to the following two combinations:

(1) Every Greek is a human	(2) Every Greek is a human
Every human is mortal	Every human is mortal
Therefore *Every Greek is mortal*	Therefore *Every mortal is Greek*

Thus presented, the valid combination is obviously (1), since the conclusion of (2) appears immediately as false. So maybe he did not raise the problem because he relied on these concrete presentations of the syllogistic moods and because *he did not, for this reason, see* that there was a problem. As a matter of fact, concrete examples are helpful in particular when one searches for counterexamples (as was the case for Aristotle himself). And even if it is true that no concrete example can

[22] I owe this remark to Prof. Wilfrid Hodges whom I thank for his comments and suggestions.

ever "prove" the productivity of a premise pair, given that we can find many invalid combinations where the conclusion is true, they can nevertheless have a pedagogical utility because people understand things much better when they are illustrated by examples. And we know that al-Fārābī was concerned by pedagogy and by teaching (*ta'līm*) from the fact that he talked about teaching in many of his treatises (see, for instance, *Iḥṣā al-'Ulūm*).

This definition is different from Aristotle's one, which is stated as follows: "By extremes, I mean both that term which is itself in another and that in which another is contained." ([24], Book I, 4, 25b32–26a2, [40],[23] 148.2–3). However, Aristotle's definition applies only to the extremes that can be used in the first figure (we can even say only to *Barbara* in the first figure, as will appear below), for it uses the notion of inclusion of a term into another, which is not satisfied by the terms in the moods of the second and third figures. For instance, in the following *Barbara* syllogism: "If Every human is mortal and Every Greek is a human, then Every Greek is mortal," the term "Greek" is contained into the term "Human", which is itself contained into the term "Mortal". So it is in the moods of that figure that the extension of the minor is the smallest one, while the biggest one is the extension of the major, and the medium one is the extension of the middle. While in a syllogism of the second figure or of the third figure, this inclusion is not really present and the extensions of the terms are not always comparable to the ones described above. For instance, in the following *Darapti* syllogism: "If Every table is a piece of furniture, and Every table is wooden, then some wooden things are pieces of furniture," the middle term "table" does not necessarily contain the minor "wooden thing," whose extension is larger than that of "table", while the extension of the major "piece of furniture" is not larger than that of the middle, for we could have many wooden things which are not necessarily pieces of furniture, for instance, doors or windows, etc.

While al-Fārābī's definition (and Alexander of Aphrodisias' one), which does not appeal to the notion of inclusion at all, but to the places of the major and the minor in the conclusion, applies to all syllogistic moods without exception, and is therefore more general, more formal, and more adequate than Aristotle's one.

The place of the middle in the two premises determines the figure (called *shakl*). Al-Fārābī, like Aristotle, admits three figures. The first figure is the one "where the middle is predicate in one of the premises, subject in the other one" ([15], 127, [10], 21.9). Although he does not specify in which premise the middle is subject or predicate in this definition, it is clear from the syllogisms held that the middle is

[23]Abderrahman Badawi, *Manṭiq Arisṭu*, volume 1, Dar al Kalam, Beirut 1980. The two volumes of this book contain the old Arabic translations of the main Aristotle's treatises. Volume 1 contains 1. *Al-Maqūlāt* (*Categories*, translation Isḥāq ibn Ḥunayn, pp. 33–98), 2. *Kitāb al-'Ibāra* (*De Interpretatione*, translation Isḥāq ibn Ḥunayn, pp. 99–136), 3. *Kitāb al-Taḥlīlāt al-'Ūlā* (*Prior Analytics*, translation Tayadurus, corrected by Ḥunayn ibn Isḥāq, pp. 137–316), while the volume 2 contains 1. *al-Taḥlīlāt al -Thāniya* or *Al-Burhān* (*Posterior Analytics*, translation Ishaq ibn Hunayn from Greek to Syriac and Abu Bishr Mattā from Syriac to Arabic, pp. 329–485), 2. *Kitāb al-Tupīqā* (*Topics*, translation Abu Uthmān al-Dimishqi, pp. 489–595).

subject in the major premise and predicate in the minor. The second figure is the one "where the middle is predicate in the two premises" ([15], 127, [10], p. 21.6) and the third figure is the one "where the middle is subject in the two premises" ([15], 127, [10], 21.6).

However, given that the syllogism is composed only of the two premises, should we consider the non-productive premise pairs such as (2) above as syllogisms pertaining to some figure? Should the non-productive premise pairs be allowed to have a figure? This problem too is not clearly raised by al-Fārābī, but as we have seen above, one of his definitions of the syllogism says that it is a premise pair that pertains to a figure. So maybe he presupposes from the start that only the productive premise pairs have a figure and not any other premise pair. This would mean that he is not really searching for new valid moods and consequently that he considers that the available syllogistic system he is presenting is not in need of important revisions or improvements.

Al-Fārābī counts 36 possible combinations of premises in each figure if one considers the universal, the particular, and the indefinite, each being either affirmative or negative. He rules out first 21 combinations in each figure, such as "the two negatives, the two particulars, the two indefinites, the couples major particular/minor indefinite and major indefinite/minor particular" ([15], 127, [10], 22.2–3) by just enumerating the premise pairs that could not be productive given the rules admitted and given that the indefinites are considered as particulars. By "ruling out" what I mean is that he does not count these premise pairs as possibly productive, since they cannot be so, if we apply some obvious rules like "from two particulars nothing follows" or "from two negatives nothing follows." So these rejections seem quite natural and do not raise problems, since they follow from these obvious rules.

From the remaining 15 combinations, he rules out other ones different from figure to figure, for in the first figure, he rules out the ones where "the minor premise is negative," and the ones where "the major premise is either particular or indefinite" ([15], 127, [10], 22.4–6), while in the second figure, he rules out the ones "where the two premises are affirmative" and the ones "where the major premise is particular or indefinite" ([15], 127, [10], 22.6–7), and in the third figure, he rules out the ones "where the minor premise is negative" ([15], 128, [10], 22.7). In this case, the rules are less obvious and should in principle be justified with more precision and clarity. But al-Fārābī does not really present justifications as if he were considering these rules as in no need of further clarifications. This could also mean that he is relying on some already constituted system that he admits without feeling the need to justify its "rules". As a matter of fact, as noted by W. Hodges, these rules were first provided by Philoponus and it seems that al-Farabi is using them, because they were commonly used in his intellectual milieu. We will return to this point below.

He thus gets four valid moods in the first figure, four valid moods in the second figure, and six valid moods in the third figure, given that the indefinites are treated as particulars, so that there is no further added combination which would be different from that containing the particulars. Although he does not *explicitly say* in that treatise *why* these combinations are ruled out, *we* will show below that he was right in doing so, by providing counterexamples of the pairs rejected that appear to be obviously invalid. These counterexamples clarify the reason *why* all the moods of the first figure where "the minor premise is negative," and the ones where "the major premise is either particular or indefinite" ([15], 127, [10], 22.5) are ruled out.

If we take into account that the indefinite is a particular, the non-conclusive moods of the first figure contain the following premises:

(1) major universal affirmative, minor universal negative;
(2) major universal affirmative, minor particular negative;
(3) major particular affirmative, minor universal affirmative;
(4) major particular affirmative, minor universal negative;
(5) major particular negative, minor universal affirmative.

(1) is illustrated by the following concrete example (where I have placed the major premise in the first place as in the *usual western tradition*, so that the mood appears clearly to all readers):

- Every *square* is a figure (**A**)
- No triangle is a *square* (**E**)

Therefore - No triangle is a figure (**E**)

which is obviously not valid. So there is no valid mood of structure **AEE** in the first figure. As a matter of fact, the only valid mood with a negative universal and an affirmative universal in the first figure is *Celarent* (that is, **EAE**, ***not*** **AEE**)

The moods corresponding to 2–5 are *not* valid either as shown below.

(2) is illustrated by the following:

- Every human is mortal (**A**)
- Some animals are not humans (**O**)

Therefore - Some animals are not mortal (**O**)

(3) is illustrated by the following:

- Some figures are squares (**I**)
- Every triangle is a figure (**A**)

Therefore - Some triangles are squares (**I**)

(4) is illustrated by the following:

- Some humans are animals (**I**)
- No duck is human (**E**)

Therefore - Some ducks are not animals (**O**)

(5) is illustrated by the following:

- Some mortals are not animals (**O**)

- Every human is mortal (**A**)

Therefore - Some humans are not animals (**O**)

As we can see, these counterexamples prove the invalidity of all the combinations ruled out by al-Fārābī. But al-Fārābī himself does not provide them and it seems that he privileges the "syllogistic rules" which were used by some commentators (for instance, Philoponus, as shown by some historians) over the counterexamples. This departs from Aristotle who justifies his rejection of syllogisms of kind (1) above (that is, **AEE** in the first figure) by saying the following:

> "But if the first term belongs to all the middle, but the middle to none of the last term, there will be no deduction in respect of the extremes; for nothing necessary follows from the terms being so related; [...]. As an example of a universal affirmative relation between the extremes we may take the terms animal, man, horse; of a universal negative relation, the terms animal, man, stone." ([24], I, 4, 26a3–26a12, [40], 148.15–16)

Thus the following combination:

Every human is an animal

No horse is human

Therefore No horse is an animal

is invalid, given the truth of its premises and the falsity of its conclusion.

But no example of this kind is provided by al-Fārābī in *al-Qiyās,* nor in any other treatise, apart perhaps in the *now lost* long commentary on Aristotle's *Prior Analytics* which we could not check, given that the available parts of it are not sufficient to check this particular point. In *al-Qiyās aṣ-ṣaghīr*, al-Fārābī does not proceed to the calculations that we find in *al-Qiyās* and presents directly the valid moods. These valid syllogistic moods themselves are presented with concrete terms, although the second and third figure moods are proved in the same way as in the present treatise, i.e., by using the same kind of proofs.

Now, Tae-Soo Lee suggests in "*Die Griechische Tradition der Aristotelischen Syllogistik in der Spätantike*" ([110]) that the Roman Empire logicians such as Philoponus replaced the counterexamples by some "syllogistic rules" which helped check the productivity of the premise pairs. These rules are evoked in the following quotation by W. Hodges:

> "Philoponus [10] 70.1–21 collected together a group of rules, taken from this and other places in Aristotle, that determine whether or not any given formal premise-pair is productive. From al-Fārābī (10th century) onwards there was a growing tendency in Arabic logic to give lists of 'conditions of productivity' based on the Philoponus rules, rather than

using case-by-case arguments as Aristotle did. If the rules were justified at all, it was by informal set-theoretic arguments. These arguments tended not to be rigorous, and for modal logic they were totally inadequate." ([89], 10)

This shows that al-Fārābī in that particular treatise is borrowing the rules from Philoponus (see [122]), although he did know Aristotle's method, since he commented on Aristotle's treatises, including *Prior Analytics*. So it seems that al-Fārābī was using already admitted rules which were probably known in his intellectual environment.

In addition, as W. Hodges notes, some logicians after Philoponus still used the counterexample method to check the non-productive premise pairs. Among them, he evokes Paul the Persian and in the Arabic tradition, Ibn al-Muqaffa‘ and Ibn Zur‘a. In the sequel, we will see that Averroes uses also the counterexample method. So it seems that there were two methods for checking non productivity: Aristotle's method of counterexamples and a rival method which relies on "syllogistic rules."

On the other hand, al-Fārābī does not follow (and indeed never mentions) the Arabic authors who preceded him like Ibn al-Muqaffa‘ who did use Aristotle's method of counterexamples, as W. Hodges notes. The question is then the following: why did he use these "conditions of productivity" instead of Aristotle's method? Was he aware of the "dubious" character of the rules on which they are based? Before answering this question, we have to note first that the results at which these "syllogistic rules" arrive are exactly the same as those at which we just arrived by using the counterexamples. So in categorical logic, these "syllogistic rules" arrive at the same results as Aristotle's more reliable method, which means that they are not really misleading. Second, nothing indicates that al-Fārābī uses the same method in modal logic, where the "syllogistic rules" are indeed misleading. And we don't know if he used them or not in that context, since the relevant passages of his commentary on *Prior Analytics* are lost. Third, from his treatises, especially the short ones, we know that al-Fārābī very often proceeds by dividing things and classifying them in a quite systematic way. So maybe we can say that al-Fārābī has preferred a method which he considered as more general, more systematic, and more pedagogical than Aristotle's method, given his general tendency of using classifications.

Returning back to the text, we find him claiming that the rules of the second figure lead to the rejection of the combinations where "the two premises are affirmative" (p. 22.6) and those where "the major is particular" (p. 22.6) (the indefinite being considered as a particular).

So the combinations to be ruled out would be the following:

(1) Major universal affirmative, minor universal affirmative;
(2) Major universal affirmative, minor particular affirmative;
(3) Major particular affirmative, minor universal affirmative;
(4) Major particular negative, minor universal affirmative;

given that the combinations with two particulars (indefinites) and with two negatives are invalid.

The examples illustrating the invalidity of these combinations could be, for instance, the following (where the major is the first premise and the minor is the second one, as is now usual):

(1) - Every human is mortal (**A**)
- Every plant is mortal (**A**)
Therefore - Every plant is human (**A**)

(2) - Every plant is mortal (**A**)
- Some animals are mortal (**I**)
Therefore - Some animals are plants (**I**)

(3) - Some humans are animals (**I**)
- Every cat is an animal (**A**)
Therefore - Some cats are human (**I**)

(4) - Some living beings are not animals (**O**)
- Every human is an animal (**A**)
Therefore - Some humans are not living beings (**O**)

In the third figure, the combinations ruled out are those "where the minor premise is negative" ([15], 128, [10], 22.7). To illustrate these combinations, we can consider the following list:

(1) Major universal affirmative, minor universal negative;
(2) Major universal affirmative, minor particular negative;
(3) Major particular affirmative, minor universal negative.

The first combination can be exemplified by the following syllogism, whose invalidity is obvious:

Every human is an animal (**A**)
No human is a bird (**E**)
Therefore No bird is an animal (**E**)

The second combination can be illustrated by the following concrete syllogism:

Every human is an animal (**A**)
Some humans are not birds (**O**)
Therefore Some birds are not animals (**O**)

which is also obviously invalid.

As to the third combination, it is also invalid as the following shows:

Some humans are animals (**I**)
No human is a bird (**E**)
Therefore Some birds are not animals (**O**)

So as we can see, the counterexample method confirms the results obtained by applying the rules. But the fact that al-Fārābī is using them rather than searching for eventually new moods means that he considers the number of valid syllogistic moods as already settled and that there is no need for searching for new ones. He is thus accepting the theory as it is and does not seem to be willing to improve it, even if as we will see in what follows he does improve it in some ways.

Now what about the valid moods in the different figures? How does al-Fārābī express them?

The four valid moods (called *ḍarb*) of the first figure are presented in *al-Qiyās* ([15], 128–129, [10], 23–24) in three different ways which are the following:

(1) A way where the predication is expressed, by *"mawjūdun fī"*, which corresponds in Aristotle's texts to "belongs to", in which the predicate is put at the beginning of the sentence, while the subject is in the end.
(2) The second way used is what Lameer calls the "class 3 predication" ([108], 98), which is the usual one, that is, the one where the subject is at the beginning and the predicate at the end of the sentence.
(3) A third way, which is called "class 2 predication" by Lameer, is expressed by "is predicated of" (*qīla 'alā* or *maqūlun 'alā*) ([108], 97) and is used at the very end of the chapter ([10], 24.5–10), where all the syllogisms of the first figure are stated once again in that way, after having been stated in the first way, and in the third way ([10], 23.3–24.4). But this way of stating the syllogisms functions more or less like the first class predication, for in both cases, the sentence starts by the predicate.

Since "class 1" and "class 2" function in almost the same way, we will state only one of them, together with class 3, which is different, and will be privileged by al-Fārābī in other treatises, for instance, *al-Qiyās al-Ṣaghīr* ([11], 76–82).

In the Lebanese edition ([10]), these moods are stated as follows[24]:

Mood 1: *Barbara* (A: major, B: middle, C: minor)

(1) ('class 1' in Lameer)	(2) ('class 3' in Lameer)
- A belongs to every B (*mawjūdun fī*)	- Every C is B
- B belongs to every C	- Every B is A
Therefore - A belongs to every C (p. 23.3–4)	Therefore - Every C is A (p. 23.11–12)

Mood 2: *Darii*

(1) (class 1)	(2) (class 3)
- A belongs to every B	- Some C's are B's
- B belongs to some C	- Every B is A
Therefore - A belongs to some C (p. 23.5–6)	Therefore - Some C's are A's (p. 23)

Mood 3: *Celarent*

(1) (class 1)	(2) (class 3)
- A belongs to no B	- Every C is B
- B belongs to every C	- No B is A
Therefore - A belongs to no C (p. 23.5–6)	Therefore - No C is A (p. 23.13–14)

Mood 4: *Ferio*

(1) (class 1)	(2) (class 3)
- A belongs to no B	- Some C's are B's
- B belongs to some C	- No B is A
Therefore - A is not in some C (p. 23.7–8)	Therefore - Some C is not A (or Not every C is A) (p. 23)

As we can see, in all the moods expressed in (2), the premises have been permuted, but the terms are all in their right places. One might ask about the pertinence of this permutation, which is used by al-Fārābī in the first figure syllogisms. The answer might be that these syllogisms are more obvious when the major premise is the second one and the minor premise is the first one, in case the syllogism is stated with propositions where the subject is the first term in the sentence (i.e., in class 3 predications). While if the syllogisms are stated in the manner of Aristotle, they are more obvious if the major is the first premise and the minor is the second premise. This reason is also given by Lameer in his book, for instance, when he says, commenting on Aristotle's order of the premises: "The evidence of the conclusion following from the premises in *Barbara* for example, is brought about by first stating the premise containing the major term: 'If A is predicated of every B, and B of every C, A must be predicated of every C' (I.4, 25b

[24]The name *Barbara* as well as the other names of the syllogistic moods is *not* in al-Fārābī, since their origin is, as is well known, Latin. Note that neither of the logicians of the Arabic tradition gives names to the moods, they only enumerate them, by saying: the first mood of the first figure, the second mood of the first figure, etc. We use the usual names here only for convenience.

37–39…), this is in contradistinction to the obviously less evident sequence 'If B is predicated of every C and A of every B, A must be predicated of every C'" ([108], 101). Here, the first ordering used by Aristotle is obvious because B is the link between A and C. Thus, the major A includes the middle B which in turn includes the minor C, given that the extension of the major is the biggest one, that of the middle is medium, and that of the minor is the smallest one. In other words, the syllogism taken as example, that is, *Barbara* expresses the *transitivity* of the relation of containment, which is the reverse of inclusion, for if the major (A) includes (or as Leibniz says "contains" (see [62], 20–21)) the middle (B), and the middle (B) includes the minor (C), and then the major (A) includes the minor (C). Some diagrams using circles represent these inclusions clearly too, when the circle representing the minor is inside the one representing the middle, which in turn is also inside the one representing the major as are, for instance, the diagrams used by Euler and also Leibniz (see [62], 21, Fig. 1).

However, if one uses the class 3 predication, these containments (and consequently the inclusions) are not so obvious when the first premise is the major one and the second premise is the minor one. For the correspondences may be stated as follows:

'A is predicated of every B' corresponds to 'Every B is A'
'B is predicated of every C' corresponds to 'Every C is B'
'A is predicated of every C' corresponds to 'Every C is A'

In this case, the obvious inclusion relations are better rendered when one starts with the second premise, for then, we have the following syllogism:

Every C is B
Every B is A
Therefore Every C is A

which expresses really the *transitivity of inclusion* as it is usually formalized, that is, the following formula: "$[(C \subset B) \,\&\, (B \subset A)] \rightarrow (C \subset A)$" = If C is included in B and B is included in A, then C is included in A. Here the relation is between classes rather than concepts; thus it is extensional since it involves the extensions of the terms.

Similarly, the same syllogism expressed in the way of class 3 predication may be interpreted in terms of implications, in which case it expresses also the *transitivity of implication*. For one can say: If C implies B, and B implies A, then C implies A, that is, when formalized: $[(C \supset B) \,\&\, (B \supset A)] \rightarrow (C \supset A)$. In both cases, it is more evident to start with the minor premise, although the order of the premises whatever it is, does not alter the validity of the syllogism, the conjunction being commutative.

As to the class 1 predication, expressed by the verb "belongs to", it seems to appeal rather to the notion of comprehension. For "belongs to" means "is part of the comprehension of" the term. As Robert Blanché observes it in his book *La logique et son histoire, d'Aristote à Russell*: "Saying that the predicate A (ὑπάρχει) belongs to the subject B is obviously expressing things intensionnally, for taken in extension it is on the contrary B, namely, the species, that belongs to A, that is, to the genus, as being included in it"[25] ([44], 36, my translation). In that case, this way of predicating would not appeal to the extension of the terms, that is, to the notion of class (the species being a subclass of a genus) but to their comprehension, that is, to the notion of attribute. Thus saying, for instance, "Mortal belongs to every Man and Man belongs to every Greek, therefore, Mortal belongs to every Greek" means not only that the class of the Greeks is a subset of the class of Men which is itself a subset of the class of mortals, but rather that the comprehension of the term "man" contains the attribute "mortal", and the comprehension of the term "Greek" contains the attribute "man" and consequently the attribute "mortal" too. So *Barbara* thus formulated is valid from the point of view of comprehension too. This is why this particular syllogism is the most obvious one. As noted by Couturat in his book on Leibniz's logic, "Presumably, *Barbara*'s scheme will always be the same, whether the inclusion of concepts is interpreted in terms of extension or in terms of comprehension"[26] ([62], 21, my translation).

This being so, we may say that al-Fārābī's acceptance of all kinds of predications indicates that he does not privilege one approach over the other, for instance, the extensional one over the comprehensional (or intensional) one, unless there are other evidence that would prove his adoption of one perspective rather than the other.

In this respect, it would be interesting to see what approach he finally endorses and to compare it with that of Aristotle and the other logicians of the Greek or the Arabic traditions. Now we find in the literature some commentators who claim that he endorses the extensionalist view. For instance, Joep Lameer says what follows:

> "al- Fārābī's understanding of the significance of the major premise is indicative of the fact that he must have taken the concepts major, middle and minor term as corresponding to successive degrees of relative extension, with the major containing the middle and the middle encompassing the minor. For it is only from such a perspective, that one might say that the major premise necessitates the conclusion, because the major term *truly sets a limit* to what the conclusion can assert. In the following, we shall refer to this view as the "extensionalist" view." ([108], 140, my emphasis)

He justifies this view by referring to *Kitāb al-Qiyās*, ([10], 40.7–8), that is, to the quotation provided above, which says the following:

> "... and among the two premises of the syllogism, the minor is clear and *the major premise, which is the one which ought to always be universal in order to justify the necessary entailment of the conclusion*, is not so clearly universal... [...*fa takūnu ṣoghrā muqaddamatay al-Qiyāsi bayyinatan wa kubrāhumā wa hia allatī sabīluhā 'an takūna abadan kullīyatan li-tafīda ḍarurata luzūmi al-natījati ghairu bayyinin 'annahā kulliyatun* ...]" ([10], 40.7–8, emphasis added).

[25] "Dire que le prédicat A (ὑπάρχει) appartient au sujet B, c'est évidemment s'exprimer intensivement, car en extension c'est au contraire B c'est-à-dire l'espèce, qui appartient à A, c'est-à-dire au genre, comme y étant incluse".

[26] "Sans doute le schème de *Barbara* sera toujours le même, que l'on interprète l'inclusion des concepts en extension ou en compréhension".

In this quotation, the idea stressed by Lameer is expressed by the sentence emphasized, namely, the fact that it is the *major* premise that *justifies* the necessity of the link between the premises and the conclusion, or in other words the validity of the syllogism. Because the conclusion is "necessitated by the major premise" ([108], 139), especially in the first figure as appears in the last part of the quotation above, that is, in the following:

> .... For it is *only from such a perspective*, that one might say that the major premise *necessitates the conclusion*, because the major term truly sets a limit to what the conclusion can assert. In the following we shall refer to this view as the "extensionalist" view ([108], 140)

However, as we saw above, unlike Aristotle, al-Fārābī does not define the major and the minor terms by appealing explicitly to the extensions of these terms, for the definitions in terms of the relative extensions of the minor and the major do not apply anyway to the syllogisms of the second and the third figures. Furthermore, the perspective attributed here to al-Fārābī applies only to the first figure syllogisms; it is not generalized to the other figures as he himself recognizes in the following: "with regard to the deductions belonging to the second and the third figures, on the other hand, he sees their conclusions as sometimes being necessitated by the major premise, and at other times by the minor (*Kha.*: 91.10–12 (L) = 45.11–13 (S))" ([108], 139) and he talks a few words later of "al-Fārābī's ambivalence" ([108], 139) with regard to the role of the first premise in the second and the third figures, to finally say that this ambivalence does not in any case reject the fundamental role of this first premise ([108], 140).

Consequently, it seems that Lameer's interpretation, according to which the importance of the first premise indicates that the extension of the major includes that of the middle, which in turn includes that of the minor can be challenged. First, the predominance of the first premise and its role in necessitating the conclusion does not necessarily mean that the major should always include the middle which itself includes the minor. These inclusions, as we saw above, apply only to *Barbara* for we can show that they do not apply even to the three last moods of the first figure (*Darii*, *Celarent* and *Ferio*).[27] And we can notice that he defines *in fact* the major and the minor terms not by means of their relative extensions, but by means of their places in the conclusion, even if by doing so, he is just following his sources and ultimately Alexander of Aphrodisias. The view he himself endorses, even if it is not really personal does not indicate *by itself* that the extensions of the terms have anything to do with the predominance of the major premise. Rather,

[27] As a matter of fact, although al-Fārābī does not say it and presumably is not aware of it, they apply only to *Barbara*, for even *Darii* may contain a minor whose extension is larger than its major, since its second premise is particular, for instance, in the following example: "every human is an animal, some living beings are human; therefore some living beings are animals," where the extension of the minor "living being," which includes plants apart from animals is larger than the extension of the major (animal) and of the middle (human), which both are parts of it. As to *Celarent* and *Ferio*, where some propositions are negative, the middle is not included in the major, as shown in the example illustrating *Celarent* below.

al-Fārābī's opinion seems to be that the major premise necessitates the conclusion *mainly* because it is *universal* in all the syllogisms of the first figure, and because without a universal premise, nothing follows. This universal character of the major premise is stressed in al-Fārābī's quotation above, where he says that the major premise "*ought to always be universal*" ([10], 40.7). Because it is the *one* premise in all syllogisms of the first figure that is *always* universal, its role in the deduction of the conclusion is fundamental. Now, it is the place of the middle term in the first figure, in particular, in *Barbara* that makes its extension intermediary between the extensions of the minor and the major. Therefore, the medium status of the middle term does not have much to do with the universal character of the first premise, but rather to its place in the two premises. In the other figures, the first premise could be particular (for instance, in *Bocardo* and *Disamis*). This is perhaps the reason why al-Fārābī considers that the second premise necessitates the conclusion in some cases, for in *Bocardo* and *Disamis*, it is the *second premise* that is *universal,* not the first one. And neither in *Bocardo* nor in *Disamis* we can say that the major contains the middle which in turn contains the minor.

Even in the syllogisms of the first figure which contain negative premises, such as *Celarent* or *Ferio* (or particular affirmative ones, such as *Darii*), it is not obvious that the major contains the middle which in turn contains the minor. Take the following *Celarent* syllogism:

No animal is a stone
Every human is an animal
Therefore No human is a stone

Could we say here that the major (stone) contains the middle (animal)? No, of course, for the class of animals is not a subclass of the class of stones. The only containment that we find here is the inclusion of the minor into the middle.

As to the concept of limit, used by Lameer, it could as well apply to the comprehensions of the terms, for a term, whatever comprehension it may have, cannot satisfy all attributes. There must be a limit in this satisfaction too. For instance, one cannot say about a man that he is a stone, or that he is a plant. So the notion of limit or limitation does not only apply to the extensions of the terms, it could also concern their intensions or comprehensions.

To conclude, the arguments given by Lameer in favor of the so-called extensionalist view or perspective of al-Fārābī's syllogistic do not conform to the evidence provided by the texts. On the contrary, the definitions given by al-Fārābī seem to show that he does not endorse this view, since he does not follow Aristotle in his characterizations of the minor and the major terms. Furthermore, he seems to endorse the position that the major premise in the first figure syllogisms is fundamental, not because of the extensions of the terms, but because of its universality.

As to Lameer's criticism of this so-called "extensionalist" view, it relies on an example given by Łukasiewicz (see [112]), who shows that *Barbara* is valid whatever extension the terms may have.

As it is reported by Lameer, the example is the following:

"I.	II.
If all crows are birds	If all animals are crows
and all birds are animals	and all crows are birds
then all crows are animals	then all animals are birds" ([108], p. 141)

While the syllogism I respects the hierarchy of extensions of the terms in the first figure, the syllogism II does not. But both of them are valid, according to Łukasiewicz and Lameer. And they are indeed formally (and *semantically*) valid because of their general structure which, as we saw above, exemplifies the transitivity of implication. However, in example II, the first premise and the conclusion are false, which is coherent with Łukasiewicz's interpretation of Aristotle's theory, since he treats all the Aristotelian syllogisms as implications having as antecedent the two premises and as consequent the conclusion. In this perspective, one could tolerate the falsity of the premise(s) and of the conclusion, because the main implication, expressed by "then", is always true.

In fact, Łukasiewicz's example shows two things: 1. There are syllogisms having the structure of *Barbara*, hence valid, but where the major term does not contain the middle term and the middle term does not contain the minor term; 2. There are syllogisms having the structure of *Barbara*, hence valid, but where the conclusion is false. By this example, it is shown that the validity of a syllogism does not require that the major term contains the middle term which in turn contains the minor term. It is also shown that the conclusion of a valid syllogism is not necessarily true. In other words, the validity of *Barbara* is not so much due to the transitivity of *inclusion*; rather it is much more due to the transitivity of *implication*. This feature is shown by Łukasiewicz's example where the minor is included neither inside the middle nor inside the major. We can even go further by taking completely independent terms, where there is no inclusion at all, for instance, the terms "table","house", and "cat". Using these terms, we get the following *Barbara* stated as an implication: "If every table is a cat and every cat is a house, then every table is a house." This implication is indeed valid since it is always true, even if the three propositions are all false and no term is included in any other one. So *Barbara* remains valid in all cases, i.e., whether the premises are true or not and even when there is no inclusion at all. However, this is not the way traditional logicians, whether Aristotle or al-Fārābī, illustrate the valid syllogisms, just because Łukasiewicz's example and the example above do not seem intuitively plausible. In particular, the idea that an implication is always true when its antecedent is false seems very strange to many people, even nowadays. In the case of al-Fārābī, the syllogism is not presented as an implication; rather it is presented as an inference, i.e., as a pair of premises from which the conclusion follows. These premises are asserted, hence considered as true, and in that particular case (i.e., when the premises are true), the conclusion ought to be true and the terms can be related by the relation of inclusion.

Now let us see what al-Fārābī says about the inclusion of terms in *al-Burhān*, for instance, where the premises of the syllogisms considered are evidently true. In that treatise, he provides several classifications of the *Barbara* syllogisms using Porphyry predicables. For instance, he provides a first classification containing eight combinations among which we find the following:

"2. A is the genus of B (*A jinsun li-B*) and B is the genus of J, 3. A is a difference to B (*faṣlun li-B*) and B is a difference to J, 4. A has B as definition (*A ḥadduhu B*) and B has J as definition" (*al-Burhān*, p. 280.7). Kind (2) is illustrated as follows: "Every human is an animal, and every animal is a body, therefore every human is a body" ([17], 280.18) where the class of humans is included into the class of animals, which in turn is included into the class of bodies. As to kind (3), it is illustrated by the following: "Every human is speaking (*nāṭiqun*), and every speaking (being) is conscious (*mudrikun*), therefore every human is conscious." Here, we cannot say that the class of humans, that of speaking beings and that of conscious beings are really different in their extensions, so that the first one is included into the second and the second into the third. Rather the three classes are co-extensive, since speaking is specific to humans and so is consciousness. So the *Barbara* syllogisms do not always involve inclusion strictly speaking, but in this case they do involve the use of intensional properties specific to some kinds of individuals which are the matters of the terms and it is because of these matters or the meanings of the terms that the premises and the conclusion are true. Considered from this perspective, the *Barbara* mood appears as justified by the matters of its term, not only by its formal structure, since the basis on which such syllogisms are constructed is not really the structure of the mood but the matters of the terms and their respective relations to each other.

These examples rely on the predicables of Porphyry, such as the species, the genus, *differentia* (difference), and so on and construct the syllogistic moods only on that basis. So they tend to show that the validity of the *Barbara* syllogism can also be justified and based on the matters of its terms, which are either genus and a species, or differences, and so on. Given this reliance on the nature of terms and their relation to each other in terms of inclusion or by means of their respective meanings, the deductive character of the *Barbara* syllogism is not really emphasized in these examples, even if the formal validity of *Barbara* is already known and obvious by itself, in al-Fārābī's frame as in all other ones.

Finally, we have to say that in *al-Qiyās al-Ṣaghīr*, al-Fārābī uses only the class 3 syllogisms, which are the usual ones. He even does not use variables and expresses many syllogisms with the same concrete terms. For instance, *Barbara* is expressed as follows:

"Every body (*jismun*) is composed (*mu'allafun*)
and Every composed (thing) is created (*muḥdathun*)
It follows necessarily (*lā maḥāla*) that Every body is created" ([11], 76.5–6)

In the same way, *Celarent* is expressed as follows:

"Every body is composed
and No composed (thing) is eternal
It follows from that: No [single] body is eternal" ([11], 76.12–13)

As we can see, here too, the two premises are permuted, for the first premise is the minor, while the second one is the major.

The two other syllogisms of the first figure are expressed by means of the terms "composed", "created", "existing thing", and "eternal", depending on the mood.

As to the first mood of the second figure (*Cesare*), it is expressed as follows:

"Every body is composed (minor)
and No eternal (being) is composed (major)
It follows that No [single] body is eternal" ([11], 77.9–10)

Here too, the minor is the first premise, while the major is the second one, given the subject and the predicate of the conclusion. So, the same order of the premises is kept in the second figure too, at least in this particular treatise. Note that this order will be common in the whole Arabic tradition.

Let us now see how al-Fārābī demonstrates the syllogisms of the second and the third figures. These syllogisms are proved by using, among other things, the conversions which are the following rules:

E-Conversion: "No A is B" is equivalent to "No B is A."
I-Conversion: "Some A's are B's" is equivalent to "Some B's are A's."
A-Conversion: "Every A is B" entails "Some B's are A's."

How does al-Fārābī prove these conversions? Is he aware of the problem raised by **A**-conversion (which is not valid when one formalizes the propositions in a modern way)?

Al-Fārābī provides a proof of **E**-Conversion both in *Kitāb al-Qiyās* and in *al-Qiyās al-Ṣaghīr*. In the former treatise, the proof is the following:

"The universal negative converts [as a negative universal] because if it is true, its two parts (*juz'āhā*) are completely separated from one another (*muftariqāni ghāyata al iftirāq*), so that they do not join (*lā yajtami'āni*) in anything at all (*fī 'amrin aṣlan*) nor at any time (*wa lā fī waqtin mina l-awqāti*). So if either of its parts is satisfied by something (*wujida fi 'amrin mā*), the other one would not be [satisfied by] this thing (*lam yakun 'an yūjada fīhi al-ākharu*), for if they are joined in some thing, then what satisfies the subject (*mā yūjadu fīhi mawḍū'uhā*) would satisfy the predicate (*yūjadu fīhi maḥmūluhā*), and this is impossible (*muḥālun*), because it is the contradictory of what was laid down at the beginning, to the effect that its predicate is not satisfied by anything which satisfies the subject (*mahmūluhā lā yūjadu wa lā fī shay'in mimmā yūjadu fīhi mawḍū'uhā*)" ([10], 18.7–11).

In this quotation, the proof is presented in an informal way. It relies on the complete *separation* of the two terms, which means that these terms can never apply to the same thing. What it says is that if two terms A and B are completely foreign so that they can never be satisfied by the same thing, then if one truly says

"No A is B" this would mean that no "thing" could at the same time be A and B, that is, if something is A, it cannot be B, and if it is B, it cannot be A (since al-Fārābī says "if either of its parts is satisfied by something, the other would not [be satisfied] by this thing"). Now let us call this thing (or "*'amr*") C. Then, we could express his claims by the following sentence: "If C is A, then C is not B and if C is B, then C is not A." If "p" stands for "C is A" and "q" stands for "C is B", the whole sentence says the following: "$(p \supset \sim q) \wedge (q \supset \sim p)$". So if something (that is, C) is at the same time A and B, this would be formalized by "C is A and C is B," i.e., "$p \wedge q$" when formalized. But then, if "$p \wedge q$" is true, $(p \supset \sim q)$ would be false [for we would have $\sim(p \supset \sim q)$]. But this is impossible since $(p \supset \sim q)$ was assumed to be true at the beginning. Therefore, if "No A is B" (which generalizes "If C is A, then C is not B") is true, then "No B is A" must be true too.

However, when analyzing this proof, Lameer says that "this is not a demonstration at all" ([108], 113). His negative judgment is due to the way he formalizes the proof, which is the following:

> "BêA → AêB. This is so because, if BêA, then *not*-(∃C)(CâB & CâA) [note 3: (∃C): There is at least one C, for which….]. Now if (∃C)(CâB & CâA) [and therefore BîA] were true, this would be the contradictory opposite of what was laid down in the beginning, viz. *not*-(∃C)(CâB & CâA) (which is true given the truth of BêA), reason for which BêA → AêB. If we replace BêA by the propositional variable p and AêB and *not*-(∃C)(CâB & CâA) by the variables q and *not*-r respectively, al-Fārābī's statement boils down to the following: p q, because p is the case and p → *not*-r, which does not admit of *not*-r being not the case - without providing any further information that would enable us to relate p → q to p → *not*-r." ([108], 113–114).

He thus formalizes al-Fārābī's argument, but his formalizations seem bizarre, for he treats "*not*-(∃C)(CâB & CâA)," "AêB" and "BêA" as if they were independent; this is why, he has the two implications "p → q" and "p → *not*-r" instead of "p → *not*-q". Nevertheless, the point he stresses is that al-Fārābī's explanations are insufficient to prove the rule. As a matter of fact, al-Fārābī does not evoke the commutativity of conjunction by which modern logicians justify very easily the permutation of the two elements of a conjunction.

Now what about the proof given by al-Fārābī in *al-Qiyās al-Ṣaghīr*? In that treatise, al-Fārābī says what follows:

> "For if "No man flies" [is true], then it cannot be [the case] that "any of the things that fly are men", for if "some flying thing is a man", this thing would be a man who flies; therefore the sentence "No man flies" cannot be true, for "there is a man who flies" (*idh kāna fī al-ṭā'iri insānun*) [that is, "some men fly"]. So if we want the sentence "No man flies" to be true, then it should not be the case that "some flying things are men". Therefore if we assert the first [proposition], we have to assert the other one." ([11], 78.4–8)

When commenting on this proof, Lameer rewrites it in the following way: "BêA → AêB. If not, then BêA & AîB (the contradictory of AêB). But AîB → BîA. However, we assumed that BêA, therefore BêA & BîA, which is impossible. Conclusion: BêA → AêB" ([108], 112). He observes that this proof uses **I**-conversion, which will be demonstrated afterward by means of **E**-conversion, by saying: "It is only further on in that treatise, that al-Fārābī provides such a proof, and then it is established on the basis of the rule for the conversion of *e*-propositions demonstrated here" ([108], 112). This means that there is some kind of circularity in both proofs. But he does not so far say *where exactly* al-Fārābī gives his proof of **I**-conversion, since he does not refer to any particular passage or section in the book, where such a proof is given.

As a matter of fact, al-Fārābī evokes **I**-conversion just after his demonstration of **E**-conversion in *Kitāb al-Qiyās*, for instance, by saying the following: "As to the affirmative particular, its two elements (*juz'ayhā*) are not separated at all in anything of that part (*ba'ḍ*) which is in both of them, for that part is inside them both (*dhālika al-ba'ḍu huwa ba'ḍun lahumā jamī'an*), and this is why their truth is preserved when the [elements] are converted, always and in every matter." ([10], 18.11–13).

In this passage, we don't find any recourse to **E**-conversion at all, for this rule is not evoked, nor even suggested implicitly in the justification of **I**-conversion. This justification is not itself a real proof, for it does not deduce the validity of the rule by means of some other rules. It only relies on the *absence of separation* between the two elements of the particular propositions, which one observes in both "Some A is B" and its converse "Some B is A," which is the reason why the truth of the first is preserved in the second. The idea involved seems thus the following: A and B in the particular affirmative have *something in common*, and this something in common is the reason why the conversion holds, because *it is still present*, whether A is the subject and B is the predicate, or the other way round (B is the subject and A the predicate). Thus, it seems that **I**-conversion is considered by al-Fārābī as more obvious than **E**-conversion; this is presumably why he does not really demonstrate it. Since he does not provide a demonstration of **I**-conversion by means of **E**-conversion, there is no circularity as Lameer claims.

Now if we return back to the second treatise, which J. Lameer is evoking, that is, *al-Qiyās al-Saghīr*, we *do* find something like a "demonstration" of **I**-conversion in the proof of one of the syllogisms of the third figure, namely, the 11th syllogism in al-Fārābī's classification (= *Datisi* in the usual classification). In that proof, al-Fārābī says what follows:

> "If "Some body is active" is true (*in saḥḥa lanā 'anna "jisman mā fā'ilun"*), it follows necessarily that "Some actives are bodies" (*lazima 'an yakūna shay'un min al-fā'ilīna jisman*). This is because if there are no bodies in the actives (*lam yakun fī al-fā'ilīna mā huwa jismun*) in effect (*ḥaṣala*) and "no single active is a body" (*wa lā fā'ila wāḥidan jismun*), and this is the universal negative, which contains (*wa yanṭawī fīhā*) "no single body is active" (*lā jisma wāḥidun fā'ilun*), then it is not true that "Some body is active" (*lā yaṣuḥḥu 'an yakūna jismun mā fā'ilan*), therefore if "Some body is active" is true, then "Some active is a body" is true too…." ([11], 81.2–5)

Here, he uses indeed the **E** propositions converted. The verb used (*yanṭawī fīhā*) means "to contain" or "to include", thus evoking the notion of inclusion.

What about **A**-conversion? This is the most problematic rule from a modern point of view. But al-Fārābī considers it as obvious for he says: "As to the universal affirmative, its conversion is clear [*wa ammā al-mūjiba al-kullīya, fa 'amru* *in'ikasihā bayyinun*]" ([10], 18.13–14). He illustrates this conversion by the following universal affirmative "every man is an animal" which converts to "some animal is a man." This conversion is justified by the fact that the entailment cannot hold if the second proposition were universal, since "every animal is a man" is false. So the conversion can only lead to a particular proposition given the truth values of the two propositions, and given the *inclusion* of the subject into the predicate in such a proposition.

Now from a modern perspective, **A**-conversion is valid only if the universal proposition has an import, for a universal proposition is a conditional which could be true even if its antecedent is false (that is, if its subject is non-existent), in which case it cannot lead to a particular proposition, which is formalized as a conjunction. Thus, if the antecedent of the universal proposition $(x)(Ax \supset Bx)$ is false, it will also be false in the conjunction $(\exists x)(Ax \land Bx)$ representing the particular proposition which would invalidate the entailment. Only in case an extra premise is added, the conversion holds.

So the admission of **A**-conversion is a further reason to consider that the affirmative universal has an import in al-Fārābī's theory. As a matter of fact in Sect. 3.2.1, we have already noted that al-Fārābī assumes the affirmative propositions to have an import, given the truth conditions he attributes to these propositions when their subjects exist and when their subjects do not exist. This being so, **A**-conversion is valid in his frame and consequently the moods proved by means of it, such as *Darapti* and *Felapton*, are also valid. We will return to these moods and their proofs in the sequel.

Now how does al-Fārābī demonstrate the syllogisms of the second and the third figures?

The first mood of the second figure, that is, *Cesare*, is proved by reducing it to *Celarent*. This reduction is made by applying **E**-conversion to the major **E**-premise. This reduction proves its validity ([15], 130, [10], 25.2–3).

Thus *Cesare* is expressed in the two following ways:

Mood 1: *Cesare*

	(1) class 1	(2) class 3
	- B belongs to no A	- No A is B
	- B belongs to every C	- Every C is B
Therefore	– A belongs to no C (p. 130) ;	therefore - No C is A

Where the first premise is obtained by **E**-conversion, which leads from "A belongs to no B" to "B belongs to no A" in class 1, and from "No B is A" to "No A is B" in class 3. Note here, that the premises are not permuted, unlike those of the first figure, for *Cesare* is expressed in its usual way, since the major premise is the universal negative proposition.

As to the second mood of the second figure, it is expressed as follows:

Mood 2: *Camestres*

	(1) class 1	(2) class 3
	- B belongs to every A	- Every A is B
	- B belongs to no C	- No C is B
Therefore	- A belongs to no C	therefore - No C is A

This mood is also reduced to *Celarent* by applying E-conversion to both the second premise of *Celarent* and its conclusion ([15], 131, [10], 25.5–11).

We have to note that al-Fārābī has permuted C and A in his reduction, for *Celarent*, in his frame, is the following syllogism:

- Every C is B
- No B is A
Therefore - No C is A

So the conversion of the two negative universals in *Celarent* should lead to

- Every C is B
- No A is B
Therefore - No A is C

This inversion may be explained by the fact that, in his frame, A is the variable corresponding to the major, B to the middle, and C to the minor. By presenting Camestres, he thus replaced all the variables in their "right" places.

The third valid mood in the second figure is presented as follows:

Mood 3: *Festino*

	(1) class 1	(2) class 3
	B belongs to no A	- No A is B
	B belongs to some C	- Some C's are B's
Therefore	– A is not in some C	therefore - Some C's are not A's

This uses the reduction to *Ferio* by converting the universal premise. Here too, the premises of *Ferio* have been transposed in the class 3 predication, since *Festino* starts with the universal premise, while *Ferio*, as it was presented in the class 3 predication, started by the particular premise. In both cases, however, the major is universal and the particular is the minor and all the terms are in their right places.

The fourth mood is presented as follows:

Mood 4: *Baroco*

(1) class 1	(2) class 3
- B belongs to every A	- Every A is B
- B is not in some C	- Some C's are not B's
Therefore - A is not in some C	therefore - Some C's are not A's

Al-Fārābī's proof is the following: Since "B is not in some C" (premise 2), then B is negated of "all this part" ([15], 131, [10], 25.15–17), then he continues saying:

> "Suppose that this part is designated on its own (*mufradun 'alā ḥiālihi*) and let us call it D, then we obtain 'B belongs to every A; and B belongs to no D', this leads to the second mood of the same figure [i.e. *Camestres*], which as has been shown, is reduced to [*Celarent*] in the first figure by the conversion of the universal which leads to 'D belongs to no B'. Since we had 'B belongs to every A', we deduce 'D belongs to no A', then by conversion we obtain 'A belongs to no D'; given that 'D is some C', we can deduce 'A is not in some C'" (slightly modified, [15], 131, [10], 25.15–26.4).

This proof is different from Aristotle's one (which is made by *reductio ad absurdum*). It is a proof by *ekthesis*. According to Lameer, it is similar to that of Alexander of Aphrodisias ([108], 131). It can be expressed as follows:

What we have to prove is that from "B belongs to every A" and "B is not in some C," it follows that "A is not in some C".

1. B is not in some C

therefore B is "negated from all this part", so if this part is called D, this means that:

{ 2. B belongs to no D
But we have also
3. B belongs to every A (premise 1) } the two premises of *Camestres*

[*Note that from 2 and 3, we may deduce 4 below, but al-Fārābī does not provide 4.*
4. A belongs to no D (by *Camestres*)

Rather, he appeals to the conversion of 2 to obtain:

5. D belongs to no B

And deduce, from 5 and 3:

6. D belongs to no A (by *Celarent*)

Then by conversion, he deduces from 6:

7. A belongs to no D

But he had also:

8. D is some C (assumption)

From which he deduces:

9. A is not in some C (from 7 and 8, by *Ferio*) ([10], 25.15–26.4)

This conclusion follows by *Ferio*, for "D is part of C" is understood as "D belongs to some C" and combining the universal "A belongs to no D" with "D is some C" leads by *Ferio* to "A does not belong to some C" ([108], 132). According to Lameer, the use of *Ferio* in this proof is specific to al-Fārābī ([108], 132). It is due to the fact that D is a *universal* in al-Fārābī's proof, while it is an *individual* in Alexander's one, as he notes in the following passage: "In this case too, D is a universal concept, which is clear from al-Fārābī's use of the term 'man' in his example of the proof in class-three predication with concrete terms a little further in the text (*MQ* 27.10)" ([108], 131), while according to him, "Alexander sees *ekthesis* as relating to the individual thing. And as a consequence of this, we had to object to his suggestion that the proof of *Baroco*, if *ekthesis* is thus conceived, can be accommodated within the system of regular Aristotelian syllogistic" ([108], 130). For given that D would be a singular individual, the proposition having D as a subject would be itself singular and cannot therefore be part of any mood of Aristotelian syllogistic, which contains only universal and particular propositions. As a consequence, the proof by *ekthesis* in this case, even if it can be valid, would not be syllogistic.

Now, as we have shown above, al-Fārābī could have deduced "A is not in some C" by *Ferio* just from step 4, which is the conclusion of *Camestres*, if he had provided this conclusion. He did not therefore really need to use *Celarent*, given that he already had the first premise of *Ferio*, to which he would have added the assumption (8) and have obtained the conclusion (9) above.

As to the proof provided with class 3 predication, it is the following:

1. Every A is B (*al-Qiyās* 1986d, p. 27.17–18)
2. No D is B (assumption, idem, p. 27.18)

} "This is the combination of the second mood of that same figure" ([10], 27.17) (= *Camestres*)

3. No B is D (by the conversion of 2, [10], 27.19)

From which it follows that:

4. No A is D (1, 3, by *Celarent*, ([10], 27.19)
5. No D is A (by the conversion of 4, p. 27.20)
6. Some C is D (assumption)

From which it follows:

7. Some C is not A (by *Ferio* from 5, 6) ([10], 27.21–28.1)

Note that, here too, he could have appealed only to *Celarent* via the conversion of (2), which gives the premise 3, from which, when combining it with (1), he can deduce (4). Or he could also have appealed directly to *Camestres* without evoking *Celarent* at all, in order to get (5), which is what he needs to apply *Ferio*. So here too, there is a superfluous step in the deduction, maybe justified by the fact that *Celarent*, a mood of the *first* figure, is more *obvious* than *Camestres*, which is itself deduced, consequently not so obvious.

Now with regard to the concrete terms illustrating these syllogisms, which show, according to Lameer, that D is a universal, we find the following terms: A: Horse, B: to whinny, C: animal, D: man, for al-Fārābī says:

> If we consider that the animals from which we have denied the whinnying, are men, for example, we have then "every horse is whinnying" and "No man is whinnying". It follows "No man is a horse" as we showed above. And "men are some animals", therefore "Some animals are not horses" ([10], 27.10–12).

However, the use of *ekthesis* in this proof of *Baroco* gives rise to another objection raised by Wilfrid Hodges ([87], 18),[28] which is that the assumption "Some C is D" is not warranted by and does not follow from the premises of *Baroco*. For since **O** does not have an import in al-Fārābī's frame, the second premise of *Baroco*, that is, "B is not in some C" (or "Some C is not B") could be true even if there is no C at all. Consequently *ekthesis*, which relies on the fact that D is part of C and B is negated from *the whole* of D (i.e., that part of C), presupposes that the C's exist (i.e., that there are individuals satisfying C). This presupposition is not logically justified (since **O** does not have an import), but al-Fārābī seems to consider it as justified, given the material example that he gives, where C is the term "animal". As a consequence, the premise "Some C is D" can be false in case there are no Cs, but al-Fārābī does not seem aware of this possibility. This is why al-Fārābī seems to rely in some of his proofs on real world justifications, without really taking into account the logical structures of the propositions. For this reason, his proof above is not pertinent from a logical point of view, since he relies on some propositions whose truth is not warranted. His proof is then based on empirical facts rather than on logical forms or on purely deductive rules. This is why we could express some doubts about the formal character of some of his proofs and consequently about his general conception of logic, which could appear as not really formal.

What about the moods of the third figure? How does al-Fārābī prove them?

The first mood of the third figure is *Darapti*; it is presented as follows:

Mood 1: *Darapti*

(1) class 1	(2) class 3
- A belongs to every B	- Every B is A
- C belongs to every B	- Every B is C
Therefore - A belongs to some C	therefore - Some C's are A's

Its proof uses the conversion of the minor premise, by which al-Fārābī reduces *Darapti* to *Darii*, that is, to

[28] Professor Hodges is talking about Avicenna's *ekthetic* proof in his article and he justifies Avicenna's use of it by saying what follows: "Ibn Sīnā violates this picture by introducing a *step* where the conclusion of the *step* is not a logical consequence of the premises of the *step*, but it doesn't matter because the conclusion of the proof as a whole is a logical consequence of the premises of the proof" ([87], 17–18). However, according to him, this justification does not apply to al-Fārābī's proof, which does not rely on the same conception of the logical deduction.

- A belongs to every B
- B belongs to some C

Therefore - A belongs to some C

Thus, the proof relies heavily on **A**-conversion and assumes it to be valid without any doubt, as is the case with Aristotle's proof. But if one rejects **A**-conversion, *Darapti* would not be valid.

But although he never raises any problem about **A**-conversion or *Darapti*, he not only presupposes but considers explicitly that the affirmative propositions have an import, given that they can never be true when their subject does not exist as he says in *al-Maqūlāt* (see Sect. 3.2.1). Since the import of **A** validates **A**-conversion and consequently *Darapti* (and *Felapton*), this explains the fact that he did not raise any problem in his proof of *Darapti*.

As to *Felapton*, it is presented as follows:

Mood 2: *Felapton*

(1) class 1	(2) class 3
- A belongs to no B	- No B is A
- C belongs to every B	- Every B is C
Therefore - A is not in some C	Therefore - Some C is not A ([10], 28.4–8).

Here too, it is by the conversion of the minor that *Felapton* is reduced to *Ferio*, that is, to the following:

- A belongs to no B
- B belongs to some C

Therefore - A is not in some C

The same problem evoked above is raised for *Felapton*, whose proof is also made by using **A**-conversion, which is valid only when the subject of **A** is satisfied. But since A-conversion is valid in his frame because of the import of **A**, there is no reason to raise any problem about it. As a matter of fact, al-Fārābī considers thc proof as simple, brief, and obvious.

The third syllogism, that is, *Datisi*, is proved by converting the particular premise, and reducing it to *Darii*, as follows:

Mood 3: *Datisi*

(1) class 1	(2) class 3
- A belongs to every B	- Every B is A
- C belongs to some B	- Some B's are C's
Therefore - A belongs to some C	Therefore - Some C's are A's

This is reduced to *Darii*, which is the following:

- A belongs to every B
- B belongs to some C

Therefore - A belongs to some C

Since **I**-conversion raises no problem, the proof is correct.

The fourth mood is *Disamis* and is presented as follows:

Mood 4: *Disamis*

(1) class 1	(2) class 3
- A belongs to some B	- Some B's are A's
- C belongs to every B	- Every B is C
Therefore - A belongs to some C	Therefore - Some C's are A's

This is reduced to *Darii* by the conversion of the major (particular), the transposition of the premises, and then the conversion of the conclusion. Neither of these steps raises any problem.

As to the fifth mood, which is *Ferison*, it is presented as follows:

Mood 5: *Ferison*

(1) class 1	(2) class 3
- A belongs to no B	- No B is A
- C belongs to some B	- Some B's are C's
Therefore - A is not in some C	Therefore - Some C's are not A's

It is proved by reducing it to *Ferio* by converting its second premise, which is particular. Thus, the proof does not raise any problem, given the validity of **I**-conversion.

Finally, the last mood, that is, *Bocardo*, is presented as follows:

Mood 6: *Bocardo*

(1) class 1	(2) class 3
- A is not in some B	- Some B's are not A's
- C belongs to every B	- Every B is C
Therefore - A is not in some C	Therefore - Some C's are not A's

It is proved by *ekthesis* as appears in the following quotation:

> We reduce it to the first figure, not by conversion but because A is negated from some B. Let 'this part of B' from which A is negated be D, if C belongs to every B, then C belongs to every D; we have then A belongs to no D and C belongs to every D, which by reduction leads to the second mood of this figure [i.e. *Felapton*] ([10], 28.16–29.1)

The reduction leads to the following:

- A belongs to no D
- C belongs to every D

Therefore - A is not in some C ([10],28.19–29.1)

Barbara is also part of the proof for the two premises

"C belongs to every B" (premise 2)
and
"B belongs to every D" (assumption)
lead to
"C belongs to every B".

Now, as Lameer points it, Aristotle "effects the proof of *Bocardo per impossibile*" ([108], 134), but he suggests the possibility of proving it by *ekthesis* too, when he says "Proof is possible also without reduction, if one of the Ss be taken to which P does not belong" ([24], I, 6, 28b20–28b21). However, he notes that Aristotle's [potential] proof of *Bocardo* by *ekthesis*, as it is interpreted according to him by Wieland, is different from the proof presented by al-Fārābī, for Aristotle would have used *Datisi*, then *Ferison* in his proof ([108], 134–135), given that if we state *Bocardo* as Aristotle states it, i.e., as follows:

If Every S is R ("R belongs to every S")
and Some S is not P ("P does not belong to some S")

It is necessary that Some R is not P ("P does not belong to some R") ([24], I, 6, 28b16–28b21)

Then the proof by *ekthesis* would have been the following:

Let D = The Ss that are not P, then we have

1. No D is P (assumption)

But we had 2. Every S is R (major premise)

and 3. Some Ss are D (assumption)

So 4. Some Ds are R (from 2, 3 by *Datisi*)

Therefore 5. Some Rs are not P (from 1, 4 by *Ferison*)

This proof is different from al-Fārābī's one which, as we saw above, uses *Barbara* and *Felapton* instead.

The above proofs show a remarkable feature, which is that al-Fārābī does *not use reductio ad absurdum* in his demonstrations, unlike Aristotle who uses it in the proofs of *Baroco* and *Bocardo*. This may be due to the fact that al-Fārābī privileges direct proofs over *reductio ad absurdum*, even when these proofs are not evident at first sight, as is the case with the proof of *Baroco*. Although al-Fārābī did not explain explicitly why he avoided the proof by *reductio ad absurdum* (*qiyās al-khalf*), his choice may be justified by the *indirect* character of such a proof, as one of the meanings of the word "*khalf*" in Arabic is "behind". In this sense, *reductio ad absurdum* is opposed to direct proofs which are called "*mustaqīm*" ([11], 86), i.e., straight. So maybe he thought that straight proofs are clearer and more efficient than indirect ones, and that it is always better to prove the moods by following clear and direct rules rather than "from behind" and by means of "doubtful" ([11], 86) assumptions, and for that reason, that *reductio ad absurdum* should be used only when there is no possible direct (*mustaqīm*) proof. This attitude could justify and clarify his departure from Aristotle with regard to the proofs of *Baroco* and *Bocardo*. However, in his proof of E-conversion, we find some kind of reduction to the impossible for "No S is P" must imply "No P is S" because S and P in that case are "completely separated," which means that no thing satisfies both S and P. So when "No S is P" is true so must be "No P is S". Now if the rule is invalid, then "No S is P" is true, while "No P is S" is false. But if "No P is S" is false, this means that "Some P is S," i.e., there is a thing which satisfies both S and P. But this is *impossible* because "it is *contradictory* of what was laid down at the beginning, to the effect that its predicate is not satisfied by anything which satisfies the subject" ([11], 18.7–11). Although this proof is informal, it relies on some kind of reduction to the impossible.

In addition, as we will see below, he provides a short analysis of the proof by *reductio ad absurdum* in a separate chapter of *al-Qiyās*, just after his discussion of the hypothetical syllogisms.

Now, what about the fourth figure?

Some people like Ibn al-Ṣalāh (cited in [108], 137) say that al-Fārābī rejects the fourth figure but Lameer says that "there is no evidence in support of his alleged rejection of the fourth figure" ([108], 139) for nothing in the texts left shows that al-Fārābī has any knowledge of the fourth figure. Nicholas Rescher goes even further by saying that the fourth figure could not have been admitted by al-Fārābī as an independent figure, given his very definition of the syllogism, which is considered as a pair of premises.

Anyway, we do not find in the remaining texts any discussion of the fourth figure, while Ibn Sīnā, for instance, does discuss (and reject) the fourth figure as we will see in the next section. Let us now turn to Avicenna's syllogistic.

3.3.2 *Avicenna's Syllogistic*

As we have seen above, Avicenna's analysis of the assertorics is different from that of both Aristotle and al-Fārābī because he classifies the assertorics into several kinds, whose logical behavior is rather different. As a consequence, his syllogistic is more complex and slightly different from that of his predecessors. We will then examine the novelties that he introduces inside the syllogistic and compare it with those of his predecessors.

The first thing to note is that Avicenna does not use the standard categorical propositions such as "Every S is P" or "No S is P" which do not contain conditions, in his *first* presentation of the syllogistics and his proofs of the moods. Instead, he uses the propositions containing the conditions that we considered above when he first presents his categorical syllogistic, as we will see in the sequel. In particular, he uses the propositions that he considers as convertible, that is, mainly those that contain the condition "as long as it is S" or those containing the condition "as long as it exists." These conditions are strongly presupposed, even if not always explicitly formulated in the propositions composing the syllogisms.

However, we must add that he sometimes uses in different places the standard categorical propositions (i.e., without additional conditions), for instance, in his treatise called *Dāneshnāmeh* as noted by Tony Street in his article "An outline of Avicenna's syllogistic" ([136]), where he says that Avicenna "'does give the standard account of the Aristotelian assertoric syllogistic in the introductory *Dānishnāmeh*'" ([136], 132, note 7). We also find the standard Aristotelian quantified propositions in *al-Qiyās* too, in particular in book 9, Sects. 3 and 6, where Avicenna analyzes the compound syllogisms containing several premises stated successively, and where the propositions used are all of the usual kind. For instance, he considers the following compound syllogism "'Every J is B, and every B is D, therefore every J is D. And every J is D and every D is H, therefore every J is H' ([34], 437), where all propositions are stated in a usual way, i.e., without any additional condition (see also [38] where the usual categorical propositions are evoked). So we can say that he does not reject the standard account of the categorical propositions and moods. But in his own syllogistic, he introduces the additional conditions.

Now, in order to analyze his syllogistic, let us first see how Avicenna defines the syllogism. His definition is the following:

> The syllogism is some discourse (*qawlun mā*) in which if more than one thing is posited, something other than these things follows from them by their very nature (*bi dhātihā*), not accidentally but by necessity ([34], 54.6–7)

This definition is Aristotelian and close to al-Fārābī's definition (Sect. 3.3.1), but it does not include the last sentence in al-Fārābī's definition where it is said that "what follows *from the* syllogism is called the conclusion" (idem). So it seems that according to Avicenna, the syllogism is not only a pair of premises. We will verify this interpretation in what follows.

The whole chapter is devoted to an analysis of the Aristotelian definition and what follows from it. In this analysis, Avicenna stresses the following points:

1. The syllogism contains *more* than *one* premise ([34], 58.8–13), which means that it could have at least two premises and sometimes more, when it is complex (*murakkabun*) ([34], 58.13).
2. It involves not only what is literally said (or heard) *(al masmū')* but what is meant *(al ma'qūl)* too ([34], 55.4–6). This point is important because the conditions added to the assertoric propositions are often meant by the speaker, more than explicitly added.
3. It involves necessity *between* the premises and the conclusion, *not* in the conclusion *itself*. This necessity is internal and due to the nature of the premises together with the structure of the syllogism, i.e., the combination of the

premises. Therefore, the syllogism is formal,[29] for it is its structure (*ta'līf*) that makes it conclusive. So the conclusion follows from the premises, not only because the premises are propositions, but because they are combined in such a way that they must lead to that conclusion, for he says "because what follows [does not follow] from these premises that are the matter of the combination, whatever it is, but rather from them *and* from their combination in that [particular] discourse" ([34], 59.10–12, emphasis added). The necessity itself is indicated by the fact that the conclusion *always* ensues from the premises, whatever matter they have (*wa qawluhu bi-l-iḍṭirāri 'ay dā'iman, laysa fi māddatin dūna maddatin*) ([34], 64.8–9). To confirm this idea, he gives an example which shows that when the syllogism depends exclusively on the matter of the propositions, it is not a syllogism at all, because it is not valid, even if its conclusion is true. Thus, the syllogism must be *formally valid*, that is, valid by means of its structure ([34], 64.9–13).

4. The syllogism does not depend on some particular ordinary language, for it is conclusive in *any language (ayata lughatin kānat)* ([34], 55). Its validity does not depend on the grammatical rules of a particular tongue.
5. The premises are posited (*sullimat*) that is *assumed* to be *true*. They are not necessarily true by themselves (*bi nafsihā musallama*) ([34], 55.12).
6. Not only the premises are *not* necessarily *analytically true*, they are not necessarily *simply true by themselves* too. This feature is supposed to include and account for several kinds of syllogisms, that is, the demonstrative (*burhānī*) syllogism, the dialectical (*jadalī*) one, the poetic one (*shi'rī*), and the *reductio ad absurdum* (*qiyās al-khalf*), among others ([34], 55). This is corroborated by the following quotation:

> The dialectical syllogism does not necessitate the truth (*lā yūjibu al-ḥaqqa*) for (*ḥaythu*) it does not necessitate (*lā yūjibu*), because *its premises are not true by themselves*; nevertheless if they are assumed to be true (*idhā sullimat*), what follows from them, really follows (*yalzamu 'anhā mā yalzamu*) ([34], 55.14–16, emphasis added).

In this quotation, as the sentence emphasized shows, he talks about the premises of a syllogism as being *just a part* of it (for he says "its premises," that is, the premises *of the* syllogism). Thus, they are not the whole syllogism as was the case with al-Fārābī's syllogism in Lameer's interpretation above.

However, the conclusion in a syllogism must be different from the premises, which is expressed in the definition above by the word "other"; it cannot be "one of the things that we have posited (*iḥdā mā sullima*)" ([34], 64.6)

Avicenna includes the hypothetical syllogism in his definition, for he says: "Like the categorical syllogism, the hypothetical syllogism too posits [premises]" ([34], 58. 5).

---

[29]On the notion of form, see [62].

He excludes from the syllogism, the induction (*istiqrā*), the analogy (*mithāl*), and the arguments based on signs (*'alāma*), which are not truth-preserving ([34], 60) and not conclusive ([34], 64).

The deductive syllogism is either *categorical* or *hypothetical* or *modal*. Among the deductive syllogisms, Avicenna evokes the following which contains the *equality relation*: "- C equals B
- B equals D
Therefore - C equals D" ([34], 59)

This is *conclusive* because of the following principles:

1. "C is equal to what equals D"
2. "All equal to what equals a thing is equal to it" (*musāwiyāt al-musāwiyāt musāwiya*) ([34], 59)

As a matter of fact, this expresses the transitivity of equality. In this respect, it is not foreign to *Barbara*, which, as we saw above, expresses also the transitivity of inclusion and of implication.

Now, what about the categorical syllogism as such? How does Avicenna present it and analyze it? How does he demonstrate the moods of the second and the third figures? Why doesn't he admit the fourth figure?

Avicenna presents his categorical syllogistic in *al-Qiyās*, Sect. 2, Chap. 4 ([34], 106–122) after having defined and proved the conversions, and having specified in which conditions the different propositions are convertible.

As we saw above, the **E** propositions do not convert when they contain "at some times," for then, the first propositions are true while the second one is false. So there is no entailment when the subject and the predicate are permuted (See Sect. 3.2.2). So what are the **E** propositions that convert? And how does Avicenna prove E-conversion?

According to Avicenna, the first kind of **E** propositions to be convertible are the ones containing the condition "as long as it is S," for he says that the following sentence "No white thing is black, as long as it is white" converts to "No black thing is white, as long as it is black" ([34], 77.1–2), given that if the first one is true, the second will be true too.

He provides a proof by *ekthesis* of E-conversion, which is the following:

> "For if "No C is B", then "No B is C", otherwise "Some B is C". Let us identify (*fa-l-nu'ayyin*) this something and call it D, then D is described as B and as C, so that it is both B and C (*fa yajtami'u fīhi 'annahu* [B] *wa 'annahu* [C]). So it is a unique thing satisfying both C and B (*fa yakūnu shay'un wāḥidun yajtami'u fīhi 'annahu* [C] *wa 'annahu* [B]). But we said that: No [thing among] C is described as B (*lā shay'a min* [C] *yūṣafu bi 'annahu* [B]), that is, as (*ma'a mā yakūnu*) something which is both C and D, although "C is B" (*ma'a 'anna* [C] *huwa* [B]), and this is a contradiction." ([34], 76.12–16)

What it says is the following: If "No C is B" then "No B is C." If this entailment does not hold then "Some B is C" will be true, for in that case "No B is C" will be false. Now if "Some B is C" is true, then there is something which satisfies both B and C. Let us call this thing D. So there is a thing D which is C and which is B; therefore "Some C is B." But we said above that "No C is B," consequently we have "No C is B" together with "Some C is B," which is a contradiction.

However, he notes that this proof, as it is presented by some people, uses **I**-conversion, which is not proved yet, and whose proof uses **E**-conversion, which leads to a circle ([34], 77.5–9). One could avoid the circle only if "a thing that one identifies either by the senses or by the intellect (*bi-l-ḥissi aw bi-l-'aqli*) as being in the same time C and B, [this thing] if it is described as C will be B, and if it is described as B, will be C" ([34], 78.1–2). So here, the proof does not appeal to the conversion itself (that is, to the rule "Some C is B" therefore "Some B is C"), given that the two propositions "Some (or this) C is B" and "Some (or this) B is C" are both evident and immediately true by themselves, so that none is *deduced* from the other. Since **I**-conversion as a rule is not used, there is no circle.

Anyway, later on in his text, he also provides a proof of **E**-conversion attributed to an author whom he calls "The eminent later scholar (*al-fāḍhil min al-muta'akhkhirīn*)"[30] ([34], 81.1), that is, al-Fārābī as shown by Tony Street. This proof is made by *reductio ad absurdum* and runs as follows: "No C is B" leads to "No B is C," "otherwise "Some B is J," but we said "No C is B." And this is a complete syllogism, known to be conclusive…For what follows from this is "Some B is not B," which is a contradiction" ([34], 81.2–4). When formalized, this proof is the following:

If - No C is B $\nrightarrow$ No B is C, then:
1. No C is B
2. ~(No B is C)

Therefore 3. Some B is C (from 2, by definition)
But then - No C is B (1)
- Some B is C (3)
Therefore - Some B is not B (from 1 and 3 by *Ferio*, C = middle)

Since we arrive at a contradiction when we suppose that "No C is B" is true while "No B is C" is false, it follows that "No C is B" must lead to "No B is C."

This proof uses the definitions of **E** and **I**, together with the mood *Ferio*, which is an obvious mood (because it is a first figure mood) and proceeds by *reductio ad absurdum*, for it deduces the validity of the rule by proving that the supposition of its falsity leads to a contradiction.

However, **E**-conversion is said to hold also when the propositions contain the condition "as long at S exists," for Avicenna says the following: "And just as we

[30] This author has been considered by some people as being Alexander of Aphrodisias, but Tony Street has shown in an article entitled "'The eminent later scholar' in Avicenna's *Book of the Syllogism*", *Arabic Sciences and Philosophy* ([135], 205–218) that he is in fact al-Fārābī, by providing convincing arguments.

say 'No stone is an animal,' i.e., permanently and as long as it exists, in the same way 'No animal is a stone, as long as it exists'. So if the first judgment [is true], so will be the second" ([34], 77.3–4). Note that the condition "as long as it exists" does not mean that the proposition has an import, since that condition is part of the antecedent of the conditional expressing the whole proposition, so that the conditional can be true even when that part of its antecedent is false.[31] The whole proposition does not hence require the existence of its subject to be true, even when that condition is present. It does not therefore have an import.

Now, what about **I**-conversion, and **A**-conversion? In what conditions do they hold? Are they proved or obvious?

A-conversion is valid and partial for **A** leads to an **I** proposition, when the subject and the predicate are permuted. The proof provided uses **E**-conversion, for it is the following:

> "If we say "Every C is B" it follows that "Some B is C". Usually, this is proved by saying: if "Some B is C" is not true, then "No B is C" will be true. Since this [proposition] is convertible, "No C is B" will be true too. But we said "Every C is B"; this is absurd (*wa hādhā khalf*)" ([34], 88.9–12).

This proof is considered as valid provided **E** is really the contradictory of **I**, that is, if it is expressed with the condition "as long as it is S" or with the condition "as long as it exists," in which case the link between the subject and the predicate is continuous. Otherwise, **E** and **I** could be true together and would not be contradictory, as, for example, the two sentences "No man is laughing at some times" and "Some men are laughing at some times."

Avicenna provides two more proofs to **A**-conversion: the first is by *ekthesis*, and the second is by *reductio ad absurdum*, but different from the usual one. The first one runs as follows: "If 'Every C is B', then let us consider one of the things described as C and call it D. This D is C and is also B. So what is described as B, which is D, is also described as C" ([34], 90.8–9).

As to the proof by *reductio ad absurdum*, it is the following: "If 'Some B is C' is not true, then 'No B is C as long as it is B' is true, but we said 'Every C is B', it follows, according to an obvious syllogism [i.e. *Celarent*] 'No C is C', which is contradictory" ([34], 90.13).[32]

In this proof, the **E** proposition that contradicts "Some B is C" contains the condition "as long as it is B." Since the contradictory of "No B is C (as long as it is B)" is "Some B's are C (at some time while B)," it seems that the initial **I** proposition contains the condition "at some time while S." Avicenna's argument runs thus as follows:

---

[31]For more details about the formalizations of these propositions, see W. Hodges ([92], 159).

[32]There is an error here, for what is written in the text is "No C is D", but this could not be so, first because Avicenna does not use in that second proof the letter D at all, second because "No C is D" is not deducible by *Celarent* from the two premises given, and it is not itself contradictory.

If 'Some B's are C at some time while S' is false, then:

No B is C (as long as it is B)

But Every C is B (as long as it is C)

This conclusion is contradictory, which shows that the conversion should hold, i.e., that "Every C is B (as long as it is C)" entails "Some B's are C (at some times while B)" by conversion. So **A**-conversion leads to a proposition containing "at some time while S" starting from a proposition containing the condition "as long as it is S."

According to Avicenna, **A** general absolute, which contains the condition "at some times" converts as an **I** proposition which contains the same condition, for if we say "Every writer is awake at some times" this does not lead to "Some awaken [beings] are writers as long as they exist or as long as they are awake" given that the latter is false while the former is true. But it can lead to "Some awaken beings are writers at some times." Likewise, **I** general absolute propositions containing "at some times" convert symmetrically as **I** propositions containing the same condition. So **A**ga converts to **I**ga and **I**ga converts to **I**ga, where "ga" stands for general absolute. We will see that these conversions are parallel to the conversions of **A** possible and **I** possible. These conversions have been stated by Wilfrid Hodges too in his book *Mathematical Background to the Logic of Ibn Sīnā*, where he says that "The 2D sentences that convert symmetrically (cf. Definition 3.3.9) are those of the forms (9.3.7) (e–d), (e–l), (e–z), (i–m), (i–t), (i–z)" ([92], 170) and some lines later he lists the "sentences that convert but not symmetrically" among which we find "(a-t) (B, A) has converse (i-t) (A, B)" ([92], 171). If we recall that in his frame (a-t) means **A** general absolute (i.e., **A** with the condition "at some times"), and (i-t) is **I** general absolute, this means that **A** general absolute leads to **I** general absolute but not conversely and that **I** general absolute leads to **I** general absolute and vice versa. We will have to use these conversions too in the proofs provided by Avicenna, in particular, in his modal logic.

The proposition **A** that contains the condition "as long as it exists" is convertible, just as the one containing "as long as it is S," Avicenna says. However, Avicenna's example, which is the following: from "Every man is an animal permanently and as long as he exists," one can deduce "Some animals are men *as long as they exist*" ([34], 91.2–3, emphasis added), does not seem to be correct. For although this claim is intuitively plausible, it raises some problems related to the *condition contained in* ***I***, given that if **A** is expressed as "Every C is B (as long as it exists)," the condition contained in **I** could not be "as long as it exists." This is so because if we take into account the proof above and what Avicenna says about **E**-conversion, the **I**-proposition to which the conversion leads to should be the contradictory of the following **E**-proposition: "No B is C (as long as it exists)." Now the contradictory of that **E**-proposition is not (1) "Some B's are C (as long as they exist)"; rather it is (2) "Some B's are C (at some times)." If we say that the **I** proposition is (1), then its contradictory (which is evoked and used in the proof) is "No B is C (at some times)"; but "No B is C (at some times)" does *not convert* as Avicenna stresses in several places. Since the proof uses the *conversion* of **E**, the **E** proposition must be convertible, which means that this **E** proposition must be either (1') "No B is C (as long as it exists)" or (2') "No B is C (as long as it is S)." But (2') is *not* the contradictory of (2), as we saw above; rather the contradictory of (2) is (1'). So the **I** proposition should rather be (2), i.e., "Some B's are C (at some times)." What we can deduce is that "Every C is B (as long as it exists)" converts to "Some B's are C (at some times)." As to "Every C is B (as long as it is C)," it converts to "Some B's are C (at some times while B)," as we saw in the above proof.

In addition, conversion in general, whatever **A** proposition is considered, is valid only when **A** has an import, and this is crucial to validate some syllogisms like *Darapti* and *Felapton* as we will show below.

As to **I**-conversion, it seems to be clear and evident, as he says in *al-Qiyās* (p. 78.1–3), since whatever thing is at the same time B and C, satisfies both predicates at once, so that if it is described as B, then this B is also C, and if it is described as C, then this C is also B. However, here too, we can add, as one reviewer suggests, that when **I** contains the condition "as long as it exists," i.e., when it is, according to Avicenna, a *ḍarūrīyya* (= necessary) proposition, then its converse should not contain the condition "as long as it exists"; rather it should contain the condition "at some times." This is so because the proof uses **E**-conversion like the one above, and because the contradictory of the **I** proposition containing "at long as it exists" is the **E** proposition containing "at some times," which does not convert. Consequently, "Some C's are B (as long as they exist)" converts to "Some B's are C (at some times)."

To summarize, **A**-conversion is expressed thus, depending on the conditions contained in the respective **A** propositions:

A-conversion 1[33]: Every S is P (as long as it exists) ⊃ Some Ps are S (at some times while it exists).[34]
A-conversion 2: Every S is P (as long as it is S) ⊃ Some Ps are S (at some times while S).
A-conversion (with the condition "at some times"): Every S is P (at some times) converts to "Some P are S (at some times)."

Likewise:

A-conversion (with "at some times while S"): Every S are P (at some times while S) converts to "Some P are S (at some times while P)."

---

[33]As we will see below (Sect. 4.3.1.1), the two **A** propositions are named, respectively, Ap and As, the only difference between them being the conditions they contain. The same remark applies to **I**-conversion, which is expressed in two ways too, depending on the conditions contained in the initial **I** propositions.

[34]According to W. Hodges ([92], 171), (a–d) (which I have called here **Ap**) converts to (i-m), i.e., to an **I** proposition containing the condition "at some times while S" (S being the subject); he also says that (a–d)(BA) implies both (i–m)(AB) and (i–t)(AB) ([92], 171).

As to **I**-conversion, it is expressed thus, depending on the conditions contained in the initial **I** propositions:
I-conversion 1: Some Ss are P (as long as they exist) ⊃ Some Ps are S (at some time while they exist).
I-conversion 2: Some Ss are P (as long as they are S) ⊃ Some Ps are S (at some times while P).
I-conversion (with "at some times"): Some S are P (at some times) converts to Some P are S (at some times).

Likewise:
I-conversion (with "at some times while S"): "Some S are P (at some times while S)" converts to "Some P are S (at some times while P)."
E-conversion: "No S is P (as long as it is S)" converts to "No P is S (as long as it is P)."
E-conversion (with the condition "as long as it exists): No S is P (as long as it exists)" converts to "No P is S (as long as it exists)."

The remaining **E** propositions (= those with the conditions "at some times" and "at some times while S") do not convert. Likewise, no **O** proposition converts.

Now what about the syllogistic? It is in section II, chapter 4, that Avicenna presents the Aristotelian moods and proves them. The categorical syllogisms are called conjunctive (*iqtirānī*) and predicative (*ḥamlī*), since all their propositions are predicative. Every syllogism contains two premises and three terms (*ḥadd*), among which one term, the middle, is either subject in one premise and predicate in the other or subject in both premises or predicate in both premises (*al-Qiyās*, p. 106). In the first case, the figure could be either the first one, when "it is predicated of the subject of the conclusion (*maṭlūb*) and subject of the predicate of the conclusion" ([34], 106.15–107.1) or the fourth figure when "it is predicated of the predicate of the conclusion and subject of the conclusion's subject" ([34], 107.2–3). This fourth figure is, however, rejected (*'ulghā*), because "it is not natural, not acceptable and not compatible with what is usually conceived and 'admitted' (*ghair mulā'imun li 'ādati al-naḍari wa-l-ru'yati*), so that one does not strongly need it (*mustaghnā 'anhu bi quwwatin*)" ([34], 107.11–13). So the unnaturalness of the fourth figure and its lack of clarity are the main reasons why it is rejected. Note that Avicenna evokes, here, Galen whom he calls *fadhil al-aṭibbā* (the great physician), who talks about the fourth figure too, but "not in the same way" as he himself does ([34], 107.11).

As to the major and the minor, they are defined in the same way as in al-Fārābī's frame, for he says that "the minor is the subject of the conclusion" ([34], 107.16), while the major is the predicate of the conclusion ([34], 108.1). The major premise is the one which contains the major, while the minor premise is the one which contains the minor term. The combination of the premises is called *qarīna* ([34], 108.3), while the syllogism is a combination of premises from which a conclusion follows by its very nature (*li dhātihā*) ([34], 108.3). The figures are called *shakl*, as in al-Fārābī. Note that the conclusion is called *maṭlūb*, i.e., what is "searched for," when it is not deduced yet ("*mā dāma lam yalzim ba'du, bal yusāqu 'ilayhi al-qiyāsu*" ([32], 33.1–2) and *natīja* (that is, *mā yantuju* = what follows), when it is already deduced ("*fa idhā lazima yusammā natījatan*") ([32], p. 33.2).

The first figure is considered as perfectly clear, first because "its syllogisms are perfect (*kāmila*)," second because their conclusions can be of "all kinds," while in the second figure, "only the negatives are deductible, and in the third figure only the particulars are deductible," finally because it also "deduces the best conclusion, i.e. the universal affirmative" ([34], 108.5–8). By "perfect" syllogisms, Avicenna means the syllogisms that do not need to be proved, because in these syllogistic moods the conclusion is deduced immediately and clearly without any need of additional rules. This conclusion follows from the premises in an obvious and clear way; this is why it is deduced directly.

Now before presenting the syllogistic moods and instead of calculating the whole number of possible combinations of premises in order to rule out the non-conclusive ones as al-Fārābī does, he presents some general rules that hold for all figures. These rules are the following ([34], 108.8–11):

- *From two negatives, nothing follows* ([34], 108.8).
- *From two particulars, nothing follows* ([34], 108.8).

These are obvious and are known to al-Fārābī too, who uses them to rule out the non-conclusive moods.

Then he states the following:

- *From a minor negative premise and a major particular, nothing follows, unless the negative is possible* ([34], 108.8–9).

This is less obvious and evokes the modal syllogisms as well, which gives it a general connotation. As we saw above, al-Fārābī too rules out the moods where the minor is negative and the major particular but only in the first figure, and he does not evoke the modal syllogisms at all (see Sect. 3.3.1). While the rule stated here applies to all figures, since Avicenna does not say that it only applies to the first figure moods. As a matter of fact, no valid syllogism of the second and the third figures contain **I** as a major together with **E** as a minor (**O** being excluded anyway in that case).

The fourth rule is the following:

- "*The conclusion (natīja) follows the least (akḥass) premise, with regard to quantity and to quality, but not with regard to modality*" ([34], 108.9–10, [32], 33.6–7)[35]

This rule states explicitly what the Aristotlian syllogisms already show, for it means that the conclusion should be particular if one premise is particular, and should be negative if one premise is negative. Of course, it should also be particular *and* negative, if one premise is particular *and* negative, as in *Baroco* or *Bocardo*, for instance. However, this condition does not apply to the modalities, which seems to

[35]Here too, there is an error in *al-Qiyās*, for what is written in that treatise is the word *aḥsin* (= the best) instead of *akhass* (= the least). This is why we also cite *al-Najāt*, where there is no error, and the right word is used.

suggest that if one premise is assertoric, while the other one is necessary, for instance, the conclusion should not necessarily be assertoric; it thus could be necessary. It also suggests that when one premise is possible, the conclusion is not always possible. This idea is close to Aristotle's view with regard to the moods containing one necessary premise and one assertoric premise as witnessed by the following passage:

"It happens sometimes also that when *one* proposition is necessary the deduction is necessary, *not however when either* is necessary, but only when the one related to the major is, e.g. if *A* is taken as necessarily belonging or not belonging to *B*, but *B* is taken as simply belonging to *C*" ([24], I, 9, 30a20–23, italics added).

However, Aristotle adds the precision that the conclusion is necessary only when the necessary premise is "the one related to the major," thus allowing only for the validity of the **LXL** (necessary-assertoric-necessary) moods, while Avicenna's rule is more general in that it does not add the precision that the conclusion could be necessary only when the *major* premise is necessary. This general character could make the reader think that when *whatever* premise is necessary, the conclusion could be necessary too. We will return to these points later in Sect. 3.4.

He also states some other rules specifically related to each figure and explains why these rules should be admitted by taking into account the role of the middle in each figure. So he says what we already noted above when talking about the different extensions of the terms in the first figure in al-Fārābī's theory (see Sect. 3.3.1, note 24), namely, that the extension of the minor term is smaller than that of the middle (or "enters into that of the middle") only when the minor premise is affirmative *and* universal, for "if [the middle] is not universal, then the minor could be larger than it (*yumkinu 'an yafūtahu al-'aṣgharu*), for it could be the case (*yajūzu*) that it is not itself that some (thing) about which the judgment is made (*al-ba'd alladhī 'alayhi al-ḥukmu*), whether it is necessary or possible" ([34], 108.16–109.1). So, for instance, as we saw above, if we consider the following *Darii*:

Every human is an animal
Some living beings are human
Therefore Some living beings are animals

The extension of the middle (human) is smaller than that of the minor (living being), so its extension is included inside the minor's one.

However, in the first figure moods which contain a negative premise, the minor is included inside the middle, as it is with an affirmative universal minor premise as we can read in the following quotation:

> "If there is in the major a universal affirmation about all what is said about the middle, or a universal negation from all what is said about the major in whatever way it is said, then the minor enters into the middle [*fa idhā kāna fi al kubrā ījābun kullīyun 'alā kull mā yuqālu 'alayhi al-'awsaṭu, aw salbun kullīyun 'an kull mā yuqālu 'alayhi al-'awsatu kayfa qīla, dakhala fīhi al 'aṣgharu*]" ([34], 108.14–16)

As a matter of fact, as we saw above, in the first figure moods where one premise is negative, the minor is indeed included inside the middle, but the middle and the major can be completely foreign to each other, so that there is no containment at all between them. For instance, in the following *Celarent*:

No animal is a stone
Every human is an animal
Therefore No human is a stone

Here, the minor (human) is included into the middle (animal), so its extension is smaller than that of the middle, but the extension of the major is not bigger than that of the middle. In fact, they are entirely independent, so that we cannot say if the extension of the major is bigger or smaller or equal to that of the middle. What is important, however, is that in this mood as well as in *Darii* and *Ferio*, the hierarchy of extensions does not matter for the validity of the mood. In fact, as we saw above, this hierarchy is present only in *Barbara*, and even in *Barbara*, it is not what matters the most for the validity of the mood.

Like al-Fārābī, he says that the indefinites should be treated as particulars, but he also adds that the *singular* should be treated as a *universal*, which al-Fārābī does not say. As an illustration of a syllogism with singular propositions, he provides the following:

"Zayd is the father of Abdullah
The father of Abdullah is the brother of 'Amr
Therefore Zayd is the brother of 'Amr" ([34],109.13–14).

These singular propositions are comparable to universals because the syllogism provided functions like a *Barbara* syllogism. However, when the premises are singular the conclusion should be singular too ([34], 109. 14–15). Note that the singular propositions are also treated as universals in many western traditional syllogistics, such as Leibniz's theory[36] or in the syllogistic defended by the Port-Royal logicians.[37]

Now how does Avicenna present the valid moods?

[36] See for instance, ([62], 4), where it is said: "...But he (is right in) blaming Hospinianus for having identified the singular propositions with the particulars, and he shows that they are, on the contrary, equivalent to the universals, given that the subject, in these propositions, is taken in the totality of its extension" (my translation).

[37] Arnault, Antoine & Nicole, Pierre ([25], 158) say: "although the singular proposition is different from the universal one in that its subject is not general, it is nevertheless closer to the universal than to the particular; because its subject, being individual, is for this very reason, necessarily taken in all its extension, which is the essence of a universal proposition and distinguishes it from the particular" ([25], my translation).

First, we can say that, unlike al-Fārābī, he starts by the subject term in all the propositions and not in two ways like al-Fārābī. Second, in all the moods he presents, the major premise is the second one and the minor is the first one. This will become usual in all Arabic logical writings. Third, the moods are written either by using the expression "if…then" or by using "therefore" or else "it follows."

All the syllogisms of the first figure are expressed as follows, in *al-Qiyās*:

Barbara: "If every C is B and every B is A, it is clear that every C is A" (*al-Qiyās*, p. 109.16).
Celarent: "If every C is B and no B is A, then it is clear that no C is A" (*al-Qiyās*, p. 110.1).
Darii: "If some C is B and every B is A, then it is clear that some C is A" (*al-Qiyās*, p. 110.2).
Ferio: "If some C is B and no B is A, then it is clear that not every C is A" (*al-Qiyās*, p. 110.3).

While in *al-Najāt*, *Barbara*, for instance, is expressed thus: "From two universal affirmatives, what follows is a universal affirmative. For instance, Every C is B and Every B is A lead by a perfect syllogism to Every C is A" ([32], 33.11–12), and *Ferio* is expressed in the following way:

Ferio: "From 'Some C is B' and 'No B is A', it follows that 'Not every C is A'" (*al-Najāt*, p. 34.1).

In these cases and the other first premise moods, the conclusion is deduced from the two premises, a commentary is made before stating the mood, which is expressed by using term variables and illustrated by concrete examples.

So, the syllogism is expressed both as a deductive argument and as an implication. But even when he uses "if…then", he deduces the conclusion from the two premises, for he says "it is clear that…".

In addition, his very definition of the syllogism shows that he clearly considers it as an inference which leads to a conclusion called "*natīja*", that is, what "follows from" the premises. So the syllogism is not only an implication, it is an inference, which leads necessarily to a conclusion.

Now how does Avicenna prove the moods of the second figure?

In *al-Qiyās*, he starts by stating some further rules related to the second mood, and by justifying them, for instance, the following:

- *No second figure mood contains two affirmative premises* ([34], 111.10)

He explains that by saying that in this figure, the middle is always predicate in the two premises, therefore it can be predicated of all kinds of subjects, for instance, the term "body" can be predicated of two opposed subjects such as "human" and "stone", as well as of two compatible ones such as "human" and "laughing" (*al-Qiyās*, p. 111.11). For this reason, when the two premises are affirmative, it is not conclusive.

Let us develop the example given to clarify this point. If the middle is "body", the major and the minor are "man" and "laughing", we have these two possibilities, when the minor is universal:[38]

(1) Every laughing (thing) is a body	(2) Every stone is a body
Every human is a body	Every human is a body
Therefore Every laughing (thing) is a human	Therefore Every stone is a human

Thus the invalidity of this "mood" appears clearly with example (2), where the conclusion is clearly false while the premises are true. This shows that the structure of that "mood" is not correct and must be rejected, even if this invalidity does not appear clearly in example (1). The same may be said if the minor is particular in (2), for instance, for "some stone is a body" is true but "some stone is a man" is false.

The second rule states that

- *No second figure mood contains two negative premises*

For instance, from "No talking being (*nāṭiqun*) is a stone" and "No man is a stone," nothing follows ([34], 111.12), for the "conclusion" "No talking being is a man," is obviously false.

While the third rule states that

- *No second figure mood contains two particular premises* ([34], 111.13).

And the fourth rule states that

- *In no second figure mood, the major is particular and the minor universal* ([34], 112.1).

For instance, the following (where the minor is the first premise and the major the second one, as in Avicenna's frame) is clearly invalid:

Every stone is a body
Some humans are bodies
Therefore, Some stones are human

Avicenna provides several proofs for each mood, even when it is very simple to prove, as the moods that are proved by conversion. However, since E-conversion, for instance, holds only under some conditions in his frame, he always wants to make sure that this condition is fulfilled. For this reason, the propositions used in the moods of the second and the third figures are the ones that contain the condition "as long as it is S."

[38] In stating these examples, we have started with the minor premise, as Avicenna does.

The first mood *Cesare* is thus expressed as follows:

Cesare

- Every C is B [as long as it is C]
- No A is B [as long as it is A]

Therefore - No C is A [as long as it is C] ([34], 114.15–115.2)

It is proved first by the conversion of the major, which leads to "No B is A (as long as it is B)," which when added to the minor, leads to "No C is A (as long as it is C)" [by *Celarent*] ([34], 114.6–8). So the first proof of this mood reduces it to *Celarent*.

Its proof *by reductio ad absurdum* (*bi al-khalf*) runs as follows: "Suppose the conclusion is false, then Some C's are A's is true; but we have 'No A is B'; so by *Ferio*, we deduce 'Not every C is B; but this contradicts the first premise, i.e.'Every C is B (as long as it is C)', which is absurd." ([34], 114.8–10).

However, the mood and its proofs are valid only when one adds the condition "as long as it is S" (S being the subject), for Avicenna says explicitly that "No A is B" means "No A is B (as long as it is A) (*mā dāma [A]*" ([34], 114.14) and he adds "Likewise (*wa kadhālika*) when we say Every C is B we mean Every C is B (as long as it is C)" ([34], 114.15).

In *al-Najāt*, he even says that if the absolute does not contain this condition "so that [the proposition] converts as itself as in the true doctrine (*fī al-madhhab al-ḥaqq*), no conclusion follows in the second figure from two absolutes, exactly like nothing follows from two possible [premises] as we will show" ([34], 34.11–12). This means that the absolute propositions used in the second figure cannot be of the kind of the general absolutes containing the condition "at some times," which are compared to the possible propositions. They are rather of the kind of the propositions containing the condition "as long as it is S" or the condition "as long as it exists" which are convertible.

Let us present the proofs of the other moods first as they are stated by Avicenna, who does *not* always *explicitly* add the condition "as long as it is S," then by including these conditions in the moods and the proofs by *reductio ad absurdum* and by conversion just to show what kind of propositions is needed and exhibited in these proofs. This will clarify the parallelism with the modal moods and their proofs. Before presenting these proofs, we have to note that the propositions containing the condition "as long as they are S" (S being the subject) imply the propositions (of the same quality and quantity) which contain the condition "at some time while S," just because "as long as S" means "at *all* times while S" and because "all" implies "some".

Camestres	*Camestres* (with the conditions)
- No C is B	No C is B (as long as they are C)
- Every A is B	Every A is B (as long as it is A)
Therefore - No C is A	Therefore No C is A (as long as they are C)

Proof (*by reductio ad absurdum*): "If some C's are A's, and Every A is B; then [by *Darii*], Some C's are B's" ([34], 116.2); but this contradicts "No C is B," which is not acceptable.

The proof by *reductio ad absurdum* runs as follows with the conditions: If the conclusion is false, then we have

Some C is A (at some time while C)

But we had: Every A is B (as long as it is A) (major premise)

It follows: Some C is B (at some time while C) (by *Darii*)

This conclusion contradicts "No C is B (as long as they are C)," i.e., the minor premise, which is absurd. Here the conclusion and the major premise contain the condition "at some time while S," while the propositions in *Darii* and the other moods of the first figure contain the condition "as long as they are S," but as we said above, the propositions containing "at some time while S" are implied by those containing "as long as they are S."

The proof by conversion is the same as that of al-Fārābī, for the conversion is made on the minor premise (i.e., the universal negative) and the conclusion (which is also a universal negative) and reduces it to *Celarent*, except that the condition "as long as it is S" is added in the propositions.

This proof is the following:

Camestres (with the conditions)	*Celarent* (with the conditions)
No C is B (as long as they are C)	No B is C (as long as it is B)
Every A is B (as long as it is A)	Every A is B (as long as it is A)
Therefore No C is A (as long as it is C)	Therefore No A is C (as long as it is A)

[then by the conversion of this conclusion: No C is A (as long as it is C)]

The third mood is *Festino*, which we will state with and without the conditions.

- Some C's are B's	Some C's are B's (as long as they are B)
- No A is B	No A is B (as long as they are A)
Therefore - Not every C is A	Therefore Not every C is A (as long as they are C)

The first proof is by the conversion of the major, which becomes "No B is A" (or "No B is A (as long as it is B)" with the condition) and when added to the minor particular affirmative, leads to the conclusion by *Ferio*.

The proof by *reductio ad absurdum* is the following: "If every C is A, and No A is B, then by *Celarent*, we deduce No C is B; but this contradicts Some C's are B's," which is not acceptable ([34], 116.5–6).

When we add the conditions (which are presupposed by Avicenna but not explicitly stated), the proof runs as follows:

If Every C is A (at some time while C) [the contradictory of the conclusion]
And No A is B (as long as they are A) (minor premise)
Then No C is B (at some time while C) (by *Celarent*)

But this contradicts "Some C's are B (as long as they are C)," the major premise of *Camestres*, which is not acceptable.

The fourth mood, *Baroco*, is proved both by *reductio ad absurdum* and by *ekthesis* (as it is in al-Fārābī's frame and in Alexander of Aphrodisias). It is expressed as follows:

Mood 4: *Baroco*	*Baroco* (with the conditions)
- Not every C is B	Not every C is B (as long as they are C)
- Every A is B	Every A is B (as long as it is A)
Therefore - Not every C is A	Therefore Not every C is A (as long as it is C)

The proof by *reductio ad absurdum* is the following: "If every C is A, and every C is B; then [by *Barbara]*, every C is B; but this contradicts Not every C is B [the minor]," which is not acceptable ([34], 116.9–10).

With the conditions, the proof runs as follows: if the conclusion is false, then

Every C is A (at some time while C) [the contradictory of the conclusion]
And Every A is B (as long as it is A) (major premise)
Therefore Every C is B (at some time while C) (by *Barbara)*

This contradicts the premise "Not every C is B (as long as they are C)," which is not acceptable.

As to the proof by *ekthesis*, it runs as follows: "Let us take some C that are not B and call this some, D; then No D is B, and every A is B, therefore no D is A, and some C is D, which reduces it to the first" ([34], 116.10–12).

Is this proof the same as that of al-Fārābī? Let us write down its steps:

1. Some C's are not B (minor premise)	[Some C are not B (as long as C)]
2. No D is B (from 1, assumption)	No D is B (as long as D)
But 3. Every A is B (major premise)	Every A is B (as long as A)
Therefore 4. No D is A (from 2, 3, by *Camestres*)	No D is A (as long as D)
And 5. Some C is D (assumption)	Some C is D (as long as C)
Therefore 6. Not every C is A (from 4, 5, by *Ferio*)	Not every C is A (as long as C)]

This proof is not exactly the same as al-Fārābī's one, although it is very close to it, because (apart from the conditions added in all propositions) it is shorter and uses directly *Camestres* to get the crucial premise "No D is A" to which *Ferio* is applied, while al-Fārābi evokes also *Celarent* and obtains "No D is A" only by converting the conclusion of *Celarent* (see Sect. 3.3.1). Thus Avicenna avoids the superfluous step that we have found in al-Fārābī's proof, by using directly *Camestres* to obtain the proposition that he needs to apply *Ferio*. But like al-Fārābī's proof, it uses only universal terms and is entirely syllogistic.

Now what about the third figure moods? What are the rules stated by Avicenna which apply to the third figure, and the proofs of these moods?

The rules stated are the following:

- *The third figure produces only particular propositions* ([34], 116.14–15).
- *The minor of a third figure mood must be affirmative* ([34], 116.15).
- *One of the premises of a third figure mood must be universal* ([34], 116.15).

He justifies these rules in the sequel by saying that if both premises are negative, given that they have the same subject (by the definition of the third figure), then "it is not necessary (*lam yajib*) for the two [predicates] (*'amrāni*) that negate the same thing [the same subject] to be compatible (*muttafiqayni*) or different (*mukhtalifayni*)" ([34], 116.16–117.1). So they could be compatible, and when they are compatible, the mood is clearly invalid. To illustrate this, take, for instance, the following terms: man, bird, and flying. With these terms we can construct the following combinations:

(1)	No human is a bird	or	(2) No human is a bird
	No human is flying		No human is flying
Therefore	No bird is flying		Therefore Some birds are not flying

In both cases, whether the conclusion is universal or particular, the syllogism is invalid. This is sufficient to invalidate the whole "mood".

If both premises are particular, apart from the fact that in general two particular premises do not produce any conclusion in whatever figure, in the third figure in particular, since the subject is the same in both premises, the propositions containing one predicate and its contrary or its contradictory negation could be true together ([34], 117.1–3). For instance, the following sentences: "Some men are seeing," "Some men are not seeing," and "Some men are blind," are all true. So, if the terms of the syllogism are man, seeing and blind, the two premises would be true but the conclusion "Some seeing [things] are blind" would be false.

The third non-conclusive case evoked is the one where the minor is negative, for

> "if the minor is negative, it is not necessary, if something [P] is negated from some thing [S], that what [R] holds for that thing [S], holds also for that other thing [P] as well, or that it is negated by that other [P] as well [*wa in kānat al-soghrā sāliba lam yajib idhā suliba shay'un 'an 'amrin 'an yūjada lahu mā yūjadu li dhālika al-'ākhari aw yuslaba 'anhu*]" ([34], 117.4–5)

Let us illustrate this by an example, and represent *shay'un* by P, *'amrun* by S, and *mā* (= what) by R, then we can say that if P is negated of S, it is not necessary that what (that is, R) holds for S, also holds for P. If S: man, R: animal, and P: horse, we have thus the following syllogisms (with variables and with concrete terms):

No S is P (minor)	No human is a horse
Every S is R	Every human is an animal
Therefore No P is R	Therefore No horse is an animal

where the conclusion is obviously false, while the two premises are true.

The first mood is *Darapti*, that is, the following:

Darapti	*Darapti* (with the conditions)
Every B is C (minor)	Every B is C (as long as they are B)
Every B is A	Every B is A (as long as they are B)
Therefore Some C is A	Therefore Some C is A (at some time while C)

([34], 117.6–7)

But before presenting his proof(s), he first explains why this mood does not produce a universal conclusion (given that its two premises are universal). According to him, the reason is that the minor term (= C) may be more general (*'a'amm*), that is, it can have a larger extension than the middle (= B), while the major (= A) may have either an equal extension or a smaller extension than the minor (= C). This is so because, in the third figure, the hierarchy of terms is not necessarily the same as in *Barbara* of the first figure.

For instance, if B: human, C: animal, and A: rational, we have the following syllogism:

Every human is an animal (minor)
Every human is rational
Therefore Every animal is rational

Where the two premises are true while the conclusion is false.

He then provides three proofs for that mood: the first one is by *ekthesis*, the second one by the conversion of the minor, and the last one is by *reductio ad absurdum*.

The second proof is very usual and has already been used by al-Fārābī. It converts "Every B is C" to "Some C is B," and reduces the mood to *Darii* of the first figure. For if every B is A, and some C is B; therefore some C is A. When we add the conditions and apply **A**-conversion to the minor, we get the following:

Some C is B (at some time while C)
Every B is A (as long as it is B)
Therefore Some C is A (at some time while C) (by *Darii*)

This proof uses **A**-conversion, which in modern logic raises problems because it holds only when **A** has an import. The question is then the following: is Avicenna aware of this problem? The answer might be maybe yes, although he does not raise the problem *explicitly* as *we now* raise it. But in any case, his opinion is that all affirmative propositions cannot be true if their subjects do not exist. Therefore, they have an import in his frame. This being so, the use of **A**-conversion is legitimate and validates indeed *Darapti* just because according to Avicenna, the case of an **A** being true when its subject is non-existent is just not conceivable.

The same can be said when **A** is expressed by "Every A is B (as long as it is A)," for in that case too, **A** converts because it has an import, being affirmative, which Avicenna says explicitly when he claims that if the subject is non-existent, an affirmative proposition could never be true (for instance, in ([35], 79.13)). Given that insight, we may say that **A**-conversion holds when the proposition contains the condition "as long as it is S." Although the **A** propositions containing "as long as it is S" and "as long as it exists" are different, they both have an import, because the existential augment is added to both of them.

Let us now go back to the first proof. This one uses *ekthesis* and we don't find it in al-Fārābī. It runs as follows:

> If Every B is C and Every B is A, it does not follow that Every C is A, because C can be more general than B and what holds for every B can be either equal to C or less general than C. But it must follow that "Some C is B"; let us call this some, B. This is *ekthesis* ([34], 117.6–9)

So, by *ekthesis*, B is some C, that is, "Some C is B." Therefore, we have

1. Every B is C (premise 1)
2. Every B is A (premise 2)
3. Some C is B (assumption)

Therefore 4. Some C is A (from 2, 3, by *Darii*)

As to the last proof, which is by *reductio ad absurdum*, it is the following: "If No C is A, and Every B is C, then No B is A; but we had Every B is A; this is absurd." ([34], 117.10–11).

So if the conclusion "Some C is A" is false, then we must hold its contradictory, i.e.,

1. No C is A

But 2. Every B is C (premise 1)

Therefore 3. No B is A (1,2, by *Celarent*)

But 4. Every B is A (premise 2)

This is absurd because "No B is A" and "Every B is A" are *contrary* and cannot be true together. Consequently "No C is A" is false and "Some C is A" is true. Therefore, the mood is valid, given that its conclusion cannot be false, when its two premises are true.

With the conditions the proof runs as follows: If the conclusion is false, then:

No C is A (at some time while C) [the contradictory of the conclusion]
And Every B is C (as long as they are B) (minor premise)
Therefore No B is A (at some time while B) (by *Celarent*)

This is contrary to "Every B is A (as long as they are B)," the major premise; which is not acceptable.

The second mood is *Felapto*n, which is also proved in three ways. The first one is by *ekthesis*, the second one is by conversion, and the third one is by *reductio ad absurdum*.

The two first proofs are grouped and presented in the following:

> "[If] every B is C and no B is A, it does not follow from this that no C is A, for maybe C is more general than the two others. Rather what follows is not every C is A. Let B be that some, or let us convert the minor." ([34], 117.14–118.1)

I will present both proofs in the left and the right in order to show their closeness:

By *ekthesis*	By the conversion of the minor
1. Every B is C (premise 1)	1. Every B is C (premise 1)
2. *No B is A* (premise 2)	2. No B is A (premise 2)
3. B is some C (= *Some C is B*) (*ekthesis*)	3. Some C is B (conversion of 1)
Therefore 4. Some C is not A (2, 3, by *Ferio*);	Therefore 4. Some C is not A (*Ferio* 2, 3)

As we can see, here and previously in the case of *Darapti*, the proofs by *ekthesis* and by conversion amount to the same, given that the propositions and the steps are the same, even if the rules used are not the same. This is so because, in these two cases, the *ekthesis* does not introduce a new term (called D, in general), but uses the very term already subject of both premises (i.e., B). This specific application of *ekthesis* was not found in al-Fārābī or in Aristotle, despite the fact that Aristotle uses *ekthesis* when proving *Darapti*.

Aristotle's proof by *ekthesis* for *Darapti* is made by introducing a new term (N) which is part of S (the middle). We can express Aristotle's proof as follows:

We have the following premises:

1. Every S is P
2. Every S is R

From which we want to deduce

- Some R is P

Suppose N is some S (*ekthesis*), then we have

3. Some S is N (by *ekthesis*)

\+ { 4. Every N is P (assumption)
  5. Every N is R (assumption)

Therefore 6. Some S is R (3, 5, by *Darii*)

7. Some R is S (6, by **I**-conversion)

Therefore 8. Some R is P (1, 7, by *Darii*) ([24], I, 6, 28a18–28a26)

But the particular way of applying *ekthesis* by Avicenna is different from the usual way used by Aristotle, for Avicenna does not introduce a new term. Maybe his aim is to shorten the proof, which is rather long in its Aristotelian form.

As to the proof by *reductio ad absurdum*, it is expressed as follows: "Or let us say: if this were not the case, every C is A and no B is A, then no B is C; but we said every B is C; this is absurd."

The premises are the following:

1. Every B is C (minor premise)
2. No B is A (major premise)

From which we want to deduce

- Some C is not A

If this conclusion is false, then we have its contradictory, that is,

3. Every C is A (assumption)

But then 4. No B is C (2, 3 by *Camestres*)

which is incompatible with the minor premise "Every B is C" since "Every B is C" and "No B is C" are contrary. So the supposition of the falsity of the conclusion leads to an absurdity; therefore, the conclusion itself is true, which makes the mood valid, since the conclusion cannot be false when the two premises are true.

This proof relies on the contrariety, i.e., the incompatibility between the contradictory of the conclusion and one of the premises. This contrariety shows that what is deducible from the assumption is false, and therefore the assumption itself is false, given the truth of the other premise (the major 2); therefore, its contradictory is true and deducible from the two premises.

Every C is A (at some time while C)
And No B is A (as long as they are B)
Therefore No B is C (at some time while B) (by *Camestres*)

This is contrary to “Every B is C” (as long as they are B), which is not acceptable.

As to *Datisi*, the third mood of this figure, it is proved by the conversion of the minor, which reduces it to *Darii*. Its proof by conversion is the following:

Datisi (with the conditions)	*Darii*
Some C is B (as long as they are C)	Some B is C (at some time while B)
Every C is A (as long as they are C)	Every C is A (as long as they are C)
Therefore Some B is A (at some time while B)	Therefore Some B is A (at some time while B)

Since the conversion of a particular proposition with the condition “as long as S” leads to a particular proposition containing the condition “at some time while S.”

But we can provide its proof by *reductio ad absurdum*, when the propositions contain the conditions. The mood is expressed as follows in that case:

Datisi (with the conditions)
Some C’s are B’s (as long as they are C)
Every C is A (as long as they are C)
Therefore Some B are A (at some time while B)

If the conclusion is false, then we have:

No B is A (as long as it is B)
And Some C’s are B’s (as long as they are C)
Therefore Not every C is A (as long as they are C) (by *Ferio*)

This is contrary to the premise “Every C is A” (as long as they are C), which is not acceptable.

As to *Disamis*, it is proved by conversion, *ekthesis* and *per impossibile*.

The proof by *ekthesis*, which is the first presented, proceeds as follows:

Disamis is the following syllogism:

1. Every B is C (minor premise)
2. Some B is A (major premise)

Therefore Some C is A

By *ekthesis*, let us designate that some which is B and also A, and call it D.

Then, we have

3. Every D is A (assumption)
4. Every D is B (assumption)
Therefore 5. Every D is C (1, 4, by *Barbara*)
Then 6. Some C is A (3, 5, by *Darapti*) ([34], 118.6–9)

Its proof by conversion reduces it to *Darii* by converting the major premise and the conclusion:

Every B is C
Some A is B (by the conversion of the major (2) above)
Therefore Some A is C (by *Darii*)
Then Some C is A (by converting that conclusion)

When we add the conditions, the proof runs as follows:

Every B is C (as long as it is B)
Some A is B (at some time while A) (by conversion)
Therefore Some A is C (at some time while A) (by *Darii*)
Then Some C is A (at some time while C) (by the conversion of the conclusion).

Note that the **I** propositions containing the condition “at some times while S” convert simply and lead to **I** propositions containing the same condition. This justifies the conversion of the conclusion in this mood.

As to the proof by *reductio ad absurdum*, it proceeds as follows:

If the conclusion “Some C is A” is false, then its contradictory “No C is A” would be true. But if it is true, we have

7. No C is A (assumption)
And 1. Every B is C (minor premise)
Then 8. No B is A (7, 1, by *Celarent*)
But 2. Some B is A (major premise)

The assumption (7) contradicts (2) the major premise, which is absurd. Therefore, the assumption leads to a contradiction; consequently, it is itself false given the truth of premise 1, and its contradictory is true. So the mood is valid, given that its conclusion cannot be false when its premises are true.

With the conditions *Disamis* is stated as follows:

Every B is C (as long as they are B)
Some B is A (as long as it is B)
Therefore Some C is A (at some times while C)

If the conclusion is false, then we have

No C is A (as long as it is C)
But we had Every B is C (as long as it is B)
Therefore No B is A (as long as it is B) (by *Celarent*)

But we had Every B is C (as long as it is B)
Therefore No B is A (as long as it is B) (by *Celarent*)
This is contrary to "Some B is A (as long as it is B)," which is not acceptable.

As to *Bocardo*, it cannot be proved by conversion, since **O** does not convert and if **A** is converted, it would lead to a particular proposition ([34], 118.14–119.1). But no mood with two particular premises is conclusive. Therefore, one must use *ekthesis* to prove it.

Bocardo is the following mood:	*Bocardo* (with the conditions)
Every B is C	Every B is C (as long as they are B)
Not every B is A	Not every B is A (as long as they are B)
Therefore Not every C is A	Therefore Not every C is A (at some time while C)

Its proof by *ekthesis* runs as follows: "Let us designate the thing that is B and is not A and call it D, then as you know, Every D is C, and No D is A" ([34], 119.1–2). This gives the following steps:

Every D is C (as long as they are D)
No D is A (as long as they are D)
Therefore Not every C is A (at some times while C) (by *Felapton*)

So *Bocardo* is reduced to *Felapton*. But although Avicenna does not say it explicitly, he seems to suggest by saying "as you know," that *Barbara* is also used in this proof, for it is by *Barbara* that he could arrive to "Every D is C" starting from "Every B is C" (minor premise) and "Every D is B" (assumption). This being so, the proof is exactly equivalent to that provided by al-Fārābī, who uses *Barbara* and *Felapton* too, and it is different from Aristotle's proof (see Sect. 3.3.1).

Now Avicenna, unlike al-Fārābī, provides another proof to *Bocardo* which is by *reductio ad absurdum*. This proof is the following: "If Every C is A and Not every B is A, then Every B is C, which is absurd" ([34], 119.3). Thus, if we suppose the conclusion is false, then we would have the following:

Every C is A (assumption)
But we had Not every B is A (major premise)
Therefore Not every B is C (by *Baroco*)

But this conclusion is incompatible with "Every B is C (as long as they are B)," the minor premise of *Bocardo*, which is assumed to be true. Therefore, it cannot be true. Consequently, the assumption itself is not true, and its contradictory, i.e.,"Not every C is A" is true, which makes the mood valid since its conclusion cannot be false when its premises are true.

With the conditions, the proof runs as follows:

Every C is A (as long as it is C) (assumption)
But we had Not every B is A (as long as they are B) (major premise)
Therefore Not every B is C (as long as they are B) (by *Baroco*),

which is incompatible to "every B is C (as long as they are B)".

Finally, he proves *Ferison*, the last mood, both by conversion and by *reductio ad absurdum*.

Ferison is the following mood:	*Ferison* (with the conditions)
Some B is C	Some B is C (as long as it is B)
No B is A	No B is A (as long as it is B)
Therefore Not every C is A	Therefore Not every C is A (at some times while C)

The conversion is applied to the particular premise which leads to "Some C is B (at some times while C)." With this premise, the mood is reduced to *Ferio* of the first figure, but the conclusion is "not every C is A (at some times while C)," given the conversion of the particular premise.

As to the proof by *reductio ad absurdum*, it is the following: If the conclusion is false "then Every C is A, and No B is A, it follows that No B is C, but we had Some B is C, which is absurd." ([34], 119. 7–8). We can write the steps as follows:

Every C is A
No B is A
Therefore No B is C (by *Camestres*).

But this conclusion contradicts "Some B is C," the minor premise, which is assumed to be true; therefore, it is false and "Every C is A" is false too, since "No B is A" is true. Consequently, "Not every C is A" is true and the mood is valid, since its conclusion cannot be false when its premises are true.

With the conditions the proof runs as follows: If the conclusion is false, then

Every C is A (at some time while C)
But we had No B is A (as long as it is B)
Therefore No B is C (at some time while B) (by *Camestres*)

This contradicts "Some B is C (as long as it is B)," which is not acceptable.

Avicenna adds a further paragraph to explain the usefulness of the two last figures, despite the fact that all their moods are reducible to those of the first figure. He says that the particular propositions are more natural when their subject has a larger extension (are more general) that their predicate. For even if it is true that "some men are animals," for instance, it is more natural to say instead "some animals are men" ([34], 120. 4–6), for the extension of animal is larger than that of man, so that the class of men is only one subclass of the class of animals. This is why, when one uses the word "some", it is more natural to start from the larger class.

In the same way, the negative propositions are sometimes more natural than the affirmative ones. For instance, when one says "The sky is not heavy or it is not light" or "The soul is not mortal" ([34], 119.12–14) which are used in the sciences and are more natural than the affirmative ones, which need further proofs.

This justifies, according to him, the relevance and the usefulness of the second and the third figures, whose majors are sometimes particular or negative premises.

We have to add that the proofs by *reductio ad absurdum* already show the validity of some moods which will be used in modal logic as we will see in Sect. 3.4. These moods contain propositions with the conditions "at some time while S." They follow from the moods of the first figure which contain the condition "as long as it is S," because "as long as S" implies "at some time while S." Likewise, we can say that "as long as S exists" implies "at some time while S exists," just because "all" implies "some". In addition, the proofs above sometimes use the contrariety between some kinds of propositions and show that the *universal* propositions containing the condition "as long as S" are contrary to the *universal* propositions containing the condition "at some times while S" when *their quality is not the same*. This contrariety is parallel to the contrariety between Necessary **A** and possible **E** in modal logic, for instance, since "as long as" corresponds to "all" which corresponds to "necessary" while "at some time" corresponds to "some", which corresponds to "possible". So the logical behavior of these conditions is similar to that of the quantifiers and of the modal operators. The same can be said of the conditions "as long as it exists" and "at some time while it exists." We will return to this point in Sect. 3.4.

Let us now turn to Averroes' syllogistic.

3.3.3 Averroes' Syllogistic

Averroes' logic is different from Avicenna's one. In fact, Averroes wants to be as faithful as possible to Aristotle himself, for his aim is to return back to the authentic Aristotelian views. His definition of the syllogism is not, however, very different from his predecessors' ones, who themselves use the Aristotelian definition. In Averroes' text, this definition becomes the following: "As to the syllogism, it is a discourse in which if more than one thing is posited, a conclusion other than these things follows necessarily, not accidentally, from these very things posited" ([26], 139.16–17).

In this analysis of this definition, he makes some points, such as

1. The discourse mentioned here is "meant to be statement-making (*qawlun jāzimun*)" (idem, p. 139.18).
2. Its premises are posited, i.e., admitted as true ([26], 139.20).
3. Their number must be more than one, so that there is no syllogism with only one premise ([26], 139.21).
4. The conclusion (*natīja*) must be different from the premises ([26], 139.22).
5. The necessity between the premises and the conclusion is what distinguishes "the syllogism from the induction (*'istiqrā*), the analogy (*tamthīl*) and the reasonings (*maqāyīs*) that produce negative conclusions sometimes and particular ones other times" ([26], 139.25–140.2). So the syllogism must produce *only one conclusion.*
6. The syllogism deduces the conclusion from the premises only and nothing else; so it is "complete" ([26], 140.2).
7. "Not accidentally" means that the definition rules out "the moods that produce a conclusion in some matters [...] as, for example, the moods containing two affirmative premises in the second figure if their predicates are equal to their subjects" ([26], 140.3–5). So the syllogism does not depend on the matter of the propositions or on an unusual relation between the subject and the predicate to be valid.
8. The syllogism is said to be *jazmī* (statement-making), which means, according to Averroes (as in Aristotle), that this discourse is either "true or false" ([26], 140.7). So the propositions involved in the syllogism must have a truth value. This being so, he says that the syllogism (*qiyās*) differs from the demonstration (*burhān*) in that the syllogism may be conclusive even with false premises, while the demonstration must involve true premises ([26], 151).
9. Finally, he clarifies what is meant by "what is said of every" (*al-maqūlu 'alā al-kull*) and "what is said of none" (*al-maqūlu wa lā 'alā wāḥid*) ([26], 140.22–141.6). This refers to what the medieval logicians afterward called the *dictum de omni et de nullo* that qualifies, according to some people, the following text of Aristotle: "That one term should be in another as in a whole is the same as for the other to be predicated of all of the first. And we say that one term is predicated of all of another, whenever nothing can be found of which the other term cannot be asserted; 'to be predicated of none' must be understood in the same way" ([24], I, 1, 24b25–30). Averroes says what follows:

> "As to "what is said of every" (*al-maqūlu 'alā al-kull*) and to "what is said of none" (*al-maqūl wa lā 'alā wāḥid*), he means by it: if the predicate is attributed to everything that is present in all the subject (*idhā lam yūjad shay'un fī kull al-mawḍū' illā wa yuḥmalu 'alayhi al-maḥmūl*), so that the predicate is satisfied by all the subject and all what is characterized by the subject and is present in it (*wa dhālika bi-an yakūna al-maḥmūlu mawjūdan li-kull al-mawḍū' wa li-kull mā yattaṣifu bi-al-mawḍū' wa yūjadu fīhi*), so that our sentence: "every animal is a body", when we give it the meaning of "what is said of every" does not signify that "every single animal is a body" (*ḥatta yakūna qawlunā: 'kull mā huwa ḥayawānun fa-huwa jismun', idhā aradnā bihi ma'nā "al-maqūl 'alā al-kull" laysa ma'nāhu kull wāḥid min al-hayawānāti fa-huwa jismun*), what it signifies is that every

single animal and all what is characterized by any of them [= any of the animals] is a body (*bal kull wāḥid min al-hayawānāti wa kull mā yattaṣifu bi kull wāḥid minhā fa-huwa jismun*)" ([26], 140.25–141.3).

So when the *dictum de omni* means is that the predicate is true of the whole of the subject, but also of all of what the subject is true of, that is, all the subclasses of the subject. So if the subject is "animal" and the predicate is "body", being a body is true of all animals and all what can be described as an animal, e.g., cats or dogs or horses, and so on, since the term "animal" is true of cats, and it is true of dogs, etc.

Likewise, the "said of none" means that "the predicate is negated from the whole of the subject and all the things of which the subject is true (*kull al 'ashyā al-mawjūdu fīhā al-mawḍū'*)" ([26], 141.5). According to him, this is a principle, which is different from the universal affirmative proposition. As it is expressed, the "said of every" principle says that the predicate can be affirmed of the entire class corresponding to the subject and also all its parts (or subclasses in modern terms). Thus expressed, it justifies *Barbara*, for instance, which says that "if every A is B (that is, if B is said of every single A), and every C is A, then every C is B," in other words, "if B is said of every A, then it is said of every subclass of A; since C is a subclass of A (A being said of every C), it follows that B is said of every C too, that is, of that subclass too". The same could be said of the principle involving "none", instead of "every".

Now, as it is expressed in *Prior Analytics*, it is clear that this principle involves extensions, i.e., classes (species and genus) and is founded on the relation of inclusion between these extensions. This shows according to many commentators[39] that the syllogistic privileges the extensions of the terms. Averroes will use this principle to justify the syllogisms of the first figure, which, as far as we can tell, neither al-Fārābī nor Avicenna explicitly did, at least in the syllogistics of *categorical* propositions.

However, it has a correspondent *dictum* involving attributes, which is claimed by Aristotle in the *Categories* where he says:

"Whenever one thing is predicated of another as of a subject, all things said of what is predicated will be said of the subject also. For example, man is predicated of the individual man, and animal of man; so animal will be predicated of the individual man also-for the individual man is both a man and an animal" ([22], I, 3, 1b10–1b15).

So if the attribute man is predicated of the individual man (e.g., Socrates) and animal is predicated of man, then animal is predicated of the individual man who will then possess both attributes.

Now how does he define the terms of the syllogism? His definition relies on the question: is it true that "Every C is A"? ([26], 151.10). This question is called the

---

[39]The commentary that we can find in the French Edition of *Prior Analytics* mentions many sources, where this extensionalist view is criticized. The editor says: "The modern logicians of the school of Lachelier, Rodier and Hamelin (*Le système d'Aristote*, pp. 178ff) blame Aristotle for having abandoned the comprehensivist point of view in favor of the extensionalist view." ([21], note 1, p. 2, my translation)

maṭlūb (= what is asked for, or the "objective"). The answer will have that proposition as a conclusion and it is expressed in a form close to a *Barbara* syllogism as follows: "If C is B and B is A, it follows that A is in C (= C is A)" ([26], 151.15). In this example, the conclusion (*maṭlūb*) is "C is A", and the subject of that conclusion is called the *minor*, while its predicate is called the *major*, while the term which is shared between both premises (*al-ḥadd al-mushtarak baynahumā*) is called the *middle*. This definition seems to be close to that of al-Fārābī and Avicenna, since it also defines the minor and the major by evoking their places in the conclusion. But it is less general than al-Fārābī's and Avicenna's definitions and closer to Aristotle's one because it relies on the place of the minor and the major in a *Barbara* syllogism, not in any syllogism. For this reason, the minor appears to have the smallest extension and the major the largest one, while the middle has an intermediary one.

Thus, it appears that Averroes is closer to Aristotle's text than his predecessors. As a matter of fact, he follows Aristotle very faithfully in all his definitions and proofs for he explicitly aims at being as faithful as possible to Aristotle's authentic views.

For instance, when talking about the assertoric propositions, he criticizes the conditions which, he says, have been added by Alexander of Aphrodisias and Theophrastus (explicitly evoked in his text). Thus, the following two conditions "as long as the subject exists" (*mā dāma al-mawḍū' mawjūdan*) and "as long as the predicate exists" (*mā dāma al-maḥmūl mawjūdan*) ([26], 143.15) are attributed to Alexander. These conditions, as noted by one reviewer, are ambiguous in Theophrastus and Themistius' writings, because these authors did not specify clearly whether what is meant by the condition (involving the subject for instance) is "the subject term has non empty extensions" (i.e., there are individuals satisfying the subject term), or "the subject individual exists" or "the subject individual satisfies the subject term." Avicenna claimed that he has clarified the issue and removed the ambiguity by separating between several kinds of conditions as we can read in what follows:

> "It is not the case that if every C is a B all the time it is a C, then necessarily it is a B for as long as its essence is satisfied. You have already learned this. But it wasn't possible for our predecessors to take the same view as us, given their examples and usages" ([36], i.5.2)

Here Avicenna separates between the two conditions "as long as S exists" ("as long as its essence is satisfied") and "as long as it is S" ("all the time it is S"), which we have already analyzed in Sect. 3.3.2. It seems then that he was the first author to make a difference between these two conditions, and that his predecessors were not as clear as him. Averroes, however, does not evoke Avicenna in this context and does not distinguish between his views and those of Theophrastus and Alexander.

Now according to Averroes, in Aristotle's theory the assertoric proposition is a factual one, which is actually true, and differs from both the necessary and the possible propositions. While the necessary proposition holds "at all times" ([26], 143), the assertoric one is "true in fact" (*mawjūda bi-l-fi'l*) and holds only "most of the time" ([26], 143).

Let us turn to the conversion rules. **E**-conversion is proved by *ekthesis*, as in Aristotle, together with a *reductio ad absurdum*, as follows:

E-conversion is the following rule:

- If No A is B then No B is A.

If it is false, this means that "No A is B" is true while "No B is A" is false; therefore, its contradictory, i.e.

- Some B is A

will be true. Now "let this 'some B' be sensitive (*shay'an maḥsūsan*) and call it C, for instance. Then C, which is 'some B', will be perceptively (*bi-l-ḥissi*) in A, so it is some A" ([26], 145); so B is some A, that is,

- Some A is B

But we said:

- No A is B

"which contradicts 'Some A is B'. This is unacceptable, therefore 'Some B is A' is false, consequently 'No B is A' is true" ([26], 145).

This proof is almost the same as Aristotle's one, except that it contains the word "perceptively", maybe to avoid using **I**-conversion explicitly, for if "Some B is A" is true because it is perceived, "Some A is B" will be true for the same reason. Therefore one does not need to deduce "Some A is B" from "Some B is A" by conversion, since both are held true by perception. So **I**-conversion is not really used.

This notion of perception is evoked by Aristotle very briefly in *Prior Analytics*, where he says: "We use the process of setting out terms like perception by sense, in the interests of the student—not as though it were impossible to demonstrate without them, as it is to demonstrate without the premises of the deduction" ([24], I, 41, 50b1–10). From this passage, one can understand that the use of perception in proofs has a pedagogical interest ("in the interest of students") and can be dispensed with ("not as though it were impossible to demonstrate with them"), if there is another way of proving.

But one may be unsatisfied with this appeal to perception, which is presumably made to avoid using **I**-conversion and thus falling into a circle [since **E**-conversion will be used afterward in the proof of **I**-conversion]. For the propositions are not all empirical or perceivable by the senses. They may talk about theoretic entities such as numbers, which are not perceivable, and say, for instance, "Some numbers are even." In that case, the proposition is not held true because we see or perceive something, but for purely theoretical reasons, such as the mathematical definitions. Its converse "Some even [entities] are numbers" is not perceivable too, but held true by definition. The same may be said about all kinds of non-perceptual sentences, such as "Some people are just," for instance.

Consequently, the proof provided by Averroes is restrictive in some way and does not account for all kinds of absolute propositions.

The proofs of **A**-conversion and **I**-conversion are almost the same as Aristotle's proofs. They both use **E**-conversion and are made by *reductio ad absurdum*.

Now, what about the moods of the different figures? Averroes, like his two predecessors, rejects the fourth figure because of its unnaturalness, for the fact that "the middle is predicated to the major and is the subject of the minor" is not a construction that "the mind makes naturally" ([26], 151).

Like Avicenna, in stating the moods of the three figures, he starts by the minor premise (containing the minor term), which is thus the first one while the major is the second one. He also starts in all his formulations by the subject term and the quantifier and does not use two formulations as al-Fārābī did. In this respect, he is not entirely faithful to Aristotle himself.

Let us consider the moods of the first figure.

Averroes says that *Barbara* is justified by the principle of "what is said of every" [*dictum de omni* in Medieval Latin words] already mentioned, which makes this specific syllogism evident and natural. For in *Barbar*a [which says: Every A is B, Every B is C; therefore Every A is C], the conclusion affirms that if C is true of the middle (B), it is also true of the minor (A) which is the sub-kind of B, since B is true of it.

Likewise, *Celarent* is justified by the principle of "what is said of none" [*dictum de nullo* in Medieval Latin words]. For *Celarent* [that is, Every A is B, No B is C; therefore No A is C] says that what is not true of the whole middle (B) is not true of the minor (A), which is its sub-kind.

Darii is also justified by the same principle (*dictum de omni*) in the following way:

> If we say Some C is B and Every B is A, then it follows necessarily that Some C is A. And this is clear from "what is said of every" because our discourse "Every B is A", as we said several times, [means] that all what is described as B affirmatively, is A, and some C is described as B, therefore it is necessary that this some is described as every A. ([26], 155).

However, as we saw above the minor term in *Darii* need not be a sub-kind of the middle, nor even of the major, for its extension can be larger than both of them. In that case, what is true (A) of the whole middle (B) would be true of that part of the minor (C) which satisfies the middle and that part is not necessarily a sub-kind of the middle. For instance, we can say:

"Every human is biped, Some animals are humans; therefore Some animals are biped."

Here biped (the major) is true of the whole middle (human), but it is true of only the part of the minor (animal) which satisfies the middle. But this part is not a subclass of the middle, given that it extends it; it is rather a subclass of animal, which is here the minor. So here we cannot say that what is true of the class "human" is true of its subclass, since "animal" is not its subclass; rather it extends both the middle and the major.

As to *Ferio*, it is also justified by the principle of "what is said of none," for "if Some C is B, and No B is A, then it follows that Some C is not A," since what is not true of the whole middle will not be true of the part of the minor that satisfies the middle. But here too, this part is not perforce a subclass of the middle or even of the major, since it can be independent of both of them.

In addition, he rules out several possible moods in this figure by providing counterexamples as those that we find in Aristotle. Some moods in that figure may have sometimes a negative conclusion and other times an affirmative one, which makes them non-conclusive. For instance, when the major is affirmative and the minor is negative, we may have the following: "No horse is a man, every man is living, it follows every horse is living" (since "no horse is living" is false), but if the terms are man, stone, and animal, the mood would be expressed as follows: "No animal is a stone, every man is an animal, therefore no stone is a man" ([26], 154). Consequently, this mood is not valid, because it does not produce a single conclusion following necessarily from the premises. Note that the example is Aristotelian too.

Other moods are ruled out by the same method or by some obvious rules such as those stating that "from two negative premises nothing follows" ([26], p. 154) or "from two particulars or two indefinites or from one particular and one indefinite, nothing follows" ([26], 158), or because they do not respect the principle "of what is said of every" or of "what is said of none." This is the case, for instance, when "the universal premise is the minor, whether affirmative or negative, and the major premise is not universal, whether it is indefinite or particular, negative or affirmative" ([26], p. 154.13–14). Note that these "rules" and the *dictum de omni and de nullo* are related, according to him, for he says what follows:

> "the "said of every" requires two conditions: the first one is that the major premise must be universal, whatever its quality is, i.e. whether affirmative or negative, and the second is that the minor premise must be affirmative, whatever quantify it may have, i.e. whether it is universal or particular" ([26], 158.5–7)

Now what about the moods of the second and the third figures?

The second figure is also natural, according to him, for people can use the moods of this figure spontaneously as when someone deduces that "this new-born is not alive" from "All alive (babies) cry" and "this new-born did not cry" ([26], 159.8–10). Here, the verb "to cry" is the predicate in both premises, which is the feature of the second figure and shows that this figure can be used in natural contexts.

Averroes proves *Cesare* by the conversion of the universal negative, which reduces it to *Celarent*. As to *Camestres*, it is proved by the conversion of the universal negative to obtain *Celarent* of the first figure, then by converting the conclusion in *Celarent* to obtain the conclusion of *Camestres* ([26], 160.10–15).

The third mood, *Festino*, is proved by the conversion of the universal negative which reduces it to *Ferio* of the first figure ([26], 161.13–17).

As to *Baroco*, the last mood of this figure, it is proved by *reductio ad absurdum* as follows: If *Baroco* [i.e., the following: Some C is not B, Every A is B; therefore Some C is not A] is not valid, then the two premises are true and the conclusion is false. "If it is false, then the following is true:

Every C is A

But we had also:

Every A is B

So by *Barbara*, we deduce:

Every C is B

But this contradicts

Some C is not B

the minor premise of *Baroco*, which was admitted as true. This cannot be admitted. Therefore "Every C is B" is false, and since it is false, what leads to it, that is, "Every C is A" must be false too. Consequently its contradictory "Some C is not A" is true" ([26], 161.21–162.3).

This proof follows Aristotle and is different from the one provided by al-Fārābī who uses *ekthesis* instead, but is the same as the proof by *reductio ad absurdum* provided by Avicenna, who adds a proof by *ekthesis* too.

In this figure too, he provides counterexamples of the non-conclusive moods, by showing that in some cases depending on the matter and on the terms chosen, from the same set of premises, we may arrive at a different conclusion. This shows that the "conclusions" are in fact "pseudo-conclusions" which do not really follow from the premises and consequently that the combinations considered are not valid. So "when the major premise is particular and the minor premise is universal and both do not have the same quality, there is no [conclusive] syllogism" ([26], 162.4–6), for instance, if the major premise is a particular negative and the minor is a universal affirmative, and the terms are the following: "living being, raven and substance" or the following: "raven, living being and white." In the first case, we have the following combination: "Every raven is a living being, and some substances are not living, therefore every raven is a substance" ([26], 162.8–10), where the "pseudo-conclusion" is an affirmative proposition. In the second case, the combination would be the following: "Every raven is a living being, and some white (things) are not living; therefore no raven is white" ([26], 162.10–12) where the "pseudo-conclusion" is a negative universal proposition. These pseudo-conclusions used in the counterexamples are given precisely to show that the combination being considered is non-productive.

Note that these counterexamples too are Aristotelian (See [24], I, 5, 27b 4–6). What they show is that when a set of premises gives rise to different "pseudo-conclusions" as Wilfrid Hodges calls them, this set is not productive and is not a syllogism, since the pseudo-conclusions cannot be said to really follow from their premises, because when a premise set is productive, it leads to one unique conclusion and cannot lead to two different (and even contrary) propositions.

Averroes shows also that in this figure, no conclusive mood can have two affirmative premises ([26], 163.17–22 + 163.3–8), or two negative premises ([26], 162.23–26) or two particulars or two indefinites or one particular and one indefinite ([26], 163.11–18) by using the same method of counterexamples.

As to the moods of the third figure, he first proves *Darapti* by the conversion of the minor which reduces it to *Darii* of the first figure. But here, he also adds a proof by *reductio ad absurdum*, and a proof by *ekthesis*.

The first proof runs as follows: if *Darapti* [that is, every B is C and every B is A; therefore some C is A] is not valid then its conclusion is false, therefore the contradictory of the conclusion, that is,

No C is A

will be true.

But we had:

Every B is C (or else Every B is A)

From which it follows:

No B is A (or else No B is C) (by *Celarent*).

which is not acceptable, since "this *contradicts* (*yalzamu 'anhumā naqīḍ al-muqaddima al-thānya*) the second premise, and what produces a falsity is itself false" ([26], 165.18–166.1, emphasis added). This proof is not in Aristotle, who does not provide the proof by *reductio ad absurdum* although he evokes it (see [24], I, 6, 28a18–28a26). It is correct except for one thing, which is that the proposition deduced by *Celarent*, which is a universal negative, is *not* the *contradictory* of the universal affirmative premise, as is said by Averroes; rather it is its *contrary*. The result is the same, since the contraries are incompatible and cannot be true together, which makes the proof correct, but what Averroes *literally says* is not true, since there is no contradiction between two universal propositions.

As to the proof by *ekthesis*, which is long but very shortly developed (or summarized) in Aristotle, it is expressed as follows by Averroes:

> Or [we can prove it] by *ekthesis*. Let us call Z, that some B. Since C is in every B and Z is part of B, then Z is by necessity part of C, and since A is in every B and Z is part of B, then Z is by necessity part of A. But it was part of C, then Some C is A. ([26], 166.1–3)

We can compare this proof to Aristotle's one (below, in the left), for there are correspondences between the terms, given that S (in Aristotle's text) corresponds to B (in Averroes' text), N corresponds to Z, R corresponds to C, and P corresponds to A. This being so, the two proofs are presented as follows:

Aristotle's proof	**Averroes' proof**
1. Every S is P	1. Every B is A
2. Every S is R	2. Every B is C
From which we want to deduce	from which we must deduce
- Some R is P	- Some C is A
Suppose N is some S (*ekthesis*), then we have	Z is some B (*ekthesis*)
3. Some S is N (by *ekthesis*)	3. Some B is Z
+ { 4. Every N is P (assumption)	4. Every Z is A ("Z is part of A")
5. Every N is R (assumption) }	5. Every Z is C ("Z is part of C")
Therefore 6. Some S is R (3, 5, by *Darii*)	So 6. Some B is C (3,5 by *Darii*)
7. Some R is S (6, by **I**-conversion)	7. Some C is B (6, by **I**-conversion)
Therefore 8. Some R is P (1,7, by *Darii*)	8. Some C is A (1, 7, by *Darii*)
([24], I, 6, 28a18–28a26)	

As we have seen above (Sect. 3.3.2), Avicenna's proof is much shorter than this one. But Avicenna uses *ekthesis* in an original way, while there is no original insight in Averroes, who just reproduces what we can already find in Aristotle.

As to *Felapton* [that is, "Every B is C and No B is A; therefore Some C is not A"], it is proved by the conversion of the minor premise **A**, which reduces it to *Ferio* from the first figure, which is thus the following: "Some C is B and No B is A; therefore Some C is not A." This proof is not in Aristotle, but it is in al-Fārābī as we saw above (Sect. 3.3.1). It is correct provided that **A** has an import.

The proof of *Disamis* is made by the conversion of the major (particular) premise to reduce it to *Darii* and then convert the conclusion.

As to *Datisi* [that is, Some B is C and Every B is A; therefore Some C is A], it is proved by the conversion of the minor particular premise first to reduce it to *Darii*, then by *ekthesis* and by *reductio ad absurdum*. The proof by *ekthesis* is expressed as follows: "Take some B to be Z, then every Z is C and every Z is A, it follows that Some C is A" ([26], 167.13–15). We thus have the following:

1. Some B is Z (assumption, *ekthesis*)
2. Every Z is C (assumption)
3. Every Z is A (assumption)
4. Some B is C (minor premise)
5. Every B is A (major premise)
6. Some C is B (conversion of 4)
7. Some C is A (5, 6 *Darii*)

Bocardo [that is, Every B is C and some B is not A; therefore Some C is not A] is proved by *reductio ad absurdum*, which supposes the conclusion to be false. Therefore, the following will be true:

1. Every C is A
But we have 2. Every B is C (minor premise)
It follows 3. Every B is A (1, 2, by *Barbara*)

But this contradicts the major premise "Some B is not A." Therefore, since the premise is true, its contradictory will be false. Consequently, the assumption itself is false, which means that "Some C is not A" is true and follows from the premises of that mood.

He also provides a proof by *ekthesis* for that mood, which is the following:

> "Let B be some 'sensible' (*maḥsūsun*) thing and let us call it Z so that we have No Z is A and Every Z is C, for Z is part of B. So it is reduced to the conclusive mood of this figure, I mean the one which contains two universal premises, the major being negative and the minor affirmative [*Felapton*], which produces Some C is not A" ([26], 168.9–11)

This proof is almost the same as that of al-Fārābī and Avicenna, which uses both *Barbara* and *Felapton*. *Barbara* is used here to get "Every Z is C" starting from "Every B is C" (minor premise) and "Every Z is B" (assumption "Z is part of B"), while *Felapton* is used to get "Some C is not A" starting from "Every Z is C" and "No Z is A."

The last mood, *Ferison*, is proved simply by the conversion of its particular affirmative minor premise, which reduces it to *Ferio* of the first figure ([26], 168.14–16).

He adds some counterexamples which rule out other possible moods in that figure and some general rules, which state the conditions under which the moods are conclusive in each figure:

- In the first figure, the moods must respect the principle of "what is said of every" and of "what is said of none" (*dictum de omni et de nullo*).
- In this figure, the moods are obvious and need no further proof. Their conclusions are unique and should not lead to another conclusion by conversion, for instance.
- In the second figure, "the major premise [must] be universal and the minor [must] differ from it in quality" ([26], 164).
- "In this figure the conclusion is never affirmative, and could be either universal or particular."
- In the third figure, "if the minor premise is affirmative and the major or the minor are universal, the syllogism is conclusive" ([26], 170).
- In this figure, no conclusion is universal, whether it is affirmative or negative ([26], 170).

3.4 Further Developments

3.4.1 The Fourth Figure in the Arabic Tradition

As reported by Nicholas Rescher, who devoted a study to Galen and the fourth figure and analyzed the link between Galen and the Arabic writings in this respect, the authors of the Arabic tradition tended either to reject the fourth figure because of its unnaturalness or to give it less importance than the other figures or even to just ignore it as was the case with al-Fārābī, for instance. But he also says that Avicenna for instance, who rejects the fourth figure because of its unnaturalness, and despite the fact that he "downgrades it" ([126], 10) is also the one author who "Distinguishing the figures in terms of the placement in the premises (as subject or predicate) of the subject and predicate of the conclusion" "arrives at the four theoretically possible figures in the usual way. He explicitly attributes this procedure to Galen, but dismisses the fourth figure as unnatural ..." ([126], 10). So according to Rescher, Avicenna makes it possible to admit the fourth figure as all other ones, because of his general definition of the figures and the placement of the terms in the premises *and* the conclusion. In other words, he is the one who prepares the general background for the admission of the fourth figure, even if he himself does not admit it in his own frame. He is also the first author in the Arabic tradition to relate this figure explicitly to Galen, although the most significant attribution of the fourth figure to Galen has been made by Averroes, according to N. Rescher, who says what follows:

> "We shall endeavor to show that datum (1), the long-mistrusted report of Averroes, is in fact the crucial consideration and must be given a probative weight far beyond that which any writer on the problem has to date accorded it. It is our thesis that Averroes' report – seemingly late, isolated and dubious – becomes in fact the most significant and telling item of evidence we possess, when once it is viewed against the background of its context in the Arabic logical tradition" ([126], 3–4)

Averroes's opinion is said to have been based on a long tradition which started from the very beginning of the Arabic tradition, since Rescher, who refers to the report of Ḥunayn Ibn Isḥāq, says "Galen's logical writings were available to the earliest generation of Arabic logicians" ([126], 6). And he adds that, given the huge influence of al-Fārābī on the Western authors of the Arabic tradition (i.e., those of "Muslim Spain", namely, essentially "Ibn Bājjah, Maimonides and Averroes" ([126], 8)

> "It is as certain as circumstantial evidence can make it that Averroes' information about Galen and the fourth figure derives directly from al-Fārābī's (lost) great commentary on *Prior Analytics* – and, as was indicated above, it is virtually incredible that al-Fārābī could have been mistaken about this matter. Thus wholly apart from the question of further Arabic data on the issue, Averroes' report is by itself, I submit, an item of evidence to which great probative weight must be assigned" ([126], 9)

Now the fourth figure has indeed been admitted afterward by the logicians in the Arabic tradition. Some of them admit it without commentary and treat it just as one of the figures, while other ones "downgrade it" as Nicholas Rescher says. Nicholas

Rescher provides a complete listing of these authors in his book but gives a special attention to one of them, namely, Ibn al-Ṣalāh (ca.1090–ca.1153), a Persian "Mathematician-physician" ([126], 12) who wrote a treatise entitled *Maqāla fī al-shakl al-rābi' min ashkāl al-qiyās* (*Treatise on the fourth figure of the figures of the syllogism*), which Rescher presents, provides entirely, and translates in his book *Galen and the Syllogism* ([126], 49–87). Among the other authors we find, for instance, Abū 'l-Barakāt Ibn Mālka (ca. 1075–ca.1170), Al-Abharī (ca. 1190–1265), Al-Qazwīnī al-Kātibī (ca.1220–1292), who wrote a very famous and influential book entitled *al-Risāla al-Shamsīyya*, and much later Al-Akhdarī (ca.1514–1546), who was influenced by al-Abharī and wrote a "versification of al-Abhari's *Kitāb al-Isaghūjī*" ([126], 11) entitled *Al-Sullam al-murawniq fī al-mantiq*. In what follows, we will discuss only Ibn al-Ṣalāh's treatment of the fourth figure and the moods he admits, since his analysis is the most elaborate and the most convincing.

In examining Ibn al-Ṣalāh's treatment of the fourth figure, we will rely on the Arabic text and the translation provided by Nicholas Rescher in *Galen and the Syllogism*.

3.4.2 Ibn al-Ṣalāh on the Fourth Figure

In his treatise *Maqāla fī al-shakl al-rābi' min ashkāl al-qiyās*, Ibn al-Ṣalāh presents his views about the fourth figure, which express a very positive opinion about this figure, unlike Avicenna's and other authors' opinions. He complains about the rejection of that figure and provides several arguments to prove that the fourth figure should not be rejected, for it is closer to the first figure than the second and third figures.

At the beginning of his treatise, he attributes explicitly the fourth figure to Galen and gives interesting historical details about the knowledge of the fourth figure and its attribution to Galen in the Arabic area. Among these details, N. Rescher reports the following:

> "… (2) An unnamed Syrian scholar contemporary with al-Kindī (d.ca.870) told him that he had, in Syriac translation, a treatise of Galen's in which the fourth figure was discussed. (3) Al-Fārābī (d. 950) discussed and rejected the fourth figure (presumably in his *Great Commentary on "Prior Analytics"*)…. (4) Ibn al-Ṣalāh himself had in hand an Arabic version of a treatise by an obscure scholar, named Dinḥā the Priest (*Dinḥā al-qass*), a Syrian Christian who flourished around 800. His work dealt *knowledgeably* and *sympathetically* with the topic of "The Fourth Figure of Galen"…" ([126], 13).

So Ibn al-Ṣalāh seems to have had serious historical evidence about the fourth figure and its attribution to Galen. But in his treatise he complains about the negative opinion endorsed by most scholars about the fourth figure, and he shows, against these previous scholars, that this figure should not be under-evaluated. In doing so, he analyzes the general conditions of productivity and unproductivity of all figures and distinguishes between those that are common to the four figures and those which are specific to each one. Given these conditions, he says, the fourth figure deserves to be considered as the second one rather than the fourth one. He classifies the figures by the place of the middle term in both premises and says that the fourth figure is precisely the one where "the middle term is subject in the minor and predicate in the major" ([98], in [126], 77). As such, it is close to the first figure which is the one "where the middle term is predicate in the minor and subject in the major" ([98], in [126], 77). This is why according to him, the fourth figure "should be the second one in the classification" ([98], in [126], 77) because it shares with the first figure this feature related to the middle term which is both subject and predicate in the premises, unlike the other figures where the middle term is either subject in both premises or predicate in both premises. Besides that, he says, the fourth figure has some other features which make it closer to the first figure.

The general conditions of unproductivity of all figures are the following:

1. "No syllogistic mood can have two negative premises" ([98], in [126], 78.123b–13).
2. "No syllogistic mood can have two particular premises" ([98], in [126], 78.123b–13).
3. "No syllogistic mood can have a negative minor and a particular major" ([98], in [126], 78.13).

As to the specific conditions related to each figure they differ from one figure to another. For the first figure, the following condition holds:

4. In the first figure, "the *minor* premise must be affirmative and the *major* premise must be universal" ([98], in [126], 78.123b–14, emphasis added).

Likewise, the fourth (or "additional" (= *mazīd*)) figure, the following (parallel, though not identical) condition holds:

5. In the additional (= fourth) figure, "*one of the* premises must be affirmative while *the other one* must be universal" ([98], in [126], 78.123b–15, emphasis added). For this reason, in the fourth figure, "no premise is particular negative" ([98], in [126], 78.123b–15).

So the only difference between this condition and the condition related to the first figure is that the premise that should be universal is not specified and likewise for the premise which should be affirmative, while for the first figure, both are specified.

The second condition related to the fourth figure is the following:

6. In the fourth figure, "no mood can have a minor affirmative particular and a major universal affirmative" ([98], in [126], 78.123b–15–16).

Furthermore, he also evokes the following specific conditions related to the second and the third figures:

7. In the second figure, "the premises should have a different quality" ([98], in [126], 81.125a–23).
8. In the third figure, "the minor must be affirmative" ([98], in [126], 81.125b–2).

According to our author, these are the "two conditions which make the fourth figure different from the three other figures" ([98], in [126], 78.123b–16), for its second condition (condition 6) makes it different from both Figs. 3.1 and 3.3, since in Fig. 3.1, *Darii* is productive, while in Fig. 3.3, *Datisi* is productive. By this condition, he says, the fourth figure shares something with the second figure, since no mood of the second figure has two affirmative premises ([98], in [126], 78.124a.1). As to the first condition (condition 5), it makes it different from Figs. 3.2 and 3.3, since in Fig. 3.2, *Baroco* is productive, while in Fig. 3.3, *Bocardo* is productive. With this condition, the fourth figure shares a common feature with the first figure, which does not have any mood with a particular negative premise either.

Another difference between the fourth figure and the other ones has to do with the kinds of conclusions this figure admits. As is well known, the first figure admits all kinds (i.e., the four kinds) of conclusions, since *Barbara* has a universal affirmative conclusion, *Celarent* has a universal negative conclusion, *Darii* has a particular affirmative conclusion and *Ferio* has a particular negative conclusion. As to the second figure, it admits only negative conclusions, whether universal or particular, as we can see from its moods (*Cesare*, *Camestres*, *Festino,* and *Baroco*), while the third figure admits only particular conclusions, whether affirmative or negative, as shown by its productive moods which are the following: *Darapti*, *Felapton*, *Datisi*, *Disamis*, *Ferison,* and *Bocardo*. So as noted by Ibn al-Ṣalāh, they both admit "only *two* (kinds of) conclusions" ([98], in [126], 79.124a–7): the particular affirmative and the particular negative for the third figure, and the universal negative and the particular negative for the second figure. As to the fourth figure, it "produces three (kinds of) conclusions: the negative universal, the negative particular and the affirmative particular" ([98], in [126], 79.124a–10). In this respect, it does better—so to speak—than the second and the third figures, since it produces three kinds of conclusions rather than two, and also because "the second figure does not produce any affirmative conclusion, and the third figure does not produce any universal conclusion" ([98], in [126], 79.124a–12–13), while the fourth figure can produce the two kinds of negative propositions and one kind of affirmatives.

According to him, in the fourth figure, there are five productive moods, unlike the first and the second figures where the number of productive moods is 4, and the third figure where this number is 6. This makes a difference with the other figures, but we cannot say that it is decisive, since the undoubtedly most perfect figure, namely, the first one, contains only four productive moods. So the number of moods cannot be considered as a criterion determining the degree of perfection of each figure.

Nevertheless, Ibn al-Ṣalāh endorses the opinion that the fourth figure should be considered as the second one rather than the fourth one, for it is closer to the first figure than the two other figures, for two reasons:

1. The first one is that the middle term in its moods can be either subject or predicate in both premises, which makes the fourth figure and the first figure be something like "two [varieties] of the same species (*naw'ayni qasīmayni*)" ([98], in [126], p. 79.124a–20), while the two other figures are more like "two other species of the genus of the first and fourth figures" ([98], in [126], 79124a–21). Since the fourth figure is of the same species as the first figure, it is closer to it than the other figures which pertain to the same genus but not to the same species as the first figure, just as, for instance, "humans and horses are closer to each other than humans and plants" ([98], in [126], 79.124a–22–23). This is why he considers the fourth figure as deserving the second place in the classification of figures.
2. Another reason why the fourth figure should be placed just after the first figure, according to him, is that in the fourth figure, "three kinds of conclusions are produced," while in the second and the third figures, only "two kinds of conclusions are produced" ([98], in [126], 79.124b–1–2). Since according to him, this criterion was the one that made "the Philosopher (i.e. Aristotle) priviledge the first figure over the two other ones because it produces four kinds of conclusions" ([98], in [126], 79.124b–4–5), likewise one should consider this same criterion as a sufficient reason to privilege the fourth figure over the second and the third figures, by analogy with what Aristotle did when he privileged the first figure.

Now it could be objected that the fourth figure is "less natural" than the other ones and that "it requires the conversion of the two premises" ([98], in [126], 80.124b–1). But Ibn al-Ṣalāh answers this objection by saying that some moods of the third figure, for instance, require also two conversions, namely, the conversion of one premise and of the conclusion. So there is no harm in converting two premises or one premise and the conclusion, since it is a method already used in syllogistic. Another objection according to which three conversions are needed for the moods of the fourth figure is considered next and answered by arguing that the three alleged conversions are not needed at all, since all what is needed is just the transposition of the premises and afterward the conversion ([98], in [126], 80.125a–1).

Besides that, *no mood of the fourth figure* is proved by *reductio ad absurdum*, unlike what happens in the second and the third figures, where *Baroco* ("the fourth mood of the second figure") and *Bocardo* ("the sixth mood of the third figure") are proved by *reductio ad absurdum*. This is an advantage of the fourth figure over the second and the third figures, according to him, since the proof by *reductio ad absurdum* is "less natural (*ab'adu 'an al-ṭab'i*)" ([98], in [126], 80.124b–19) than the proofs which use conversion, given its indirect character. However, this answer seems to show that Ibn al-Ṣalāh either does not have sufficient knowledge of al-Fārābī's and Avicenna's proofs by *ecthesis* of *Baroco* and *Bocardo*, which we have analyzed above, or that he knows them but does not take them into account. This seems strange because he evokes nominatively both al-Fārābī and Avicenna (among other authors) and he also refers to Avicenna's "*al-Shifā, al-Qiyās*, section iv.1" ([98], in [126], 76) at the beginning of his treatise. But it seems that despite these explicit references to his Arabic predecessors, in his analysis of the figures and their characteristics, he considers much more Aristotle himself, to whom he refers several times (for instance, at p. 79.124b–4–5, where he evokes "the Philosopher" (= Aristotle), and p. 80.125a.5, where he evokes "*Analytica al-awwal*" (*Prior Analytics*)).

He then enumerates all the 16 possible combinations of premises between the quantified propositions (the unquantified ones being ruled out, because they are treated "as particulars" ([98], in [126], 81.125b–10)). These 16 combinations of premises are the following: "(a) two universal affirmatives [**AA**], (b) two universal negatives [**EE**], (c) one universal affirmative and one universal negative [**AE**], (d) one universal negative and one universal affirmative [**EA**], (e) two particular affirmatives [**II**], (f) two particular negatives [**OO**], (g) one particular affirmative and one particular negative [**IO**], (h) one particular negative and one particular affirmative [**OI**], (i) one universal affirmative and one particular affirmative [**AI**], (j) one universal negative and one particular negative [**EO**], (k) one universal affirmative and one particular negative [**AO**], (l) one universal negative and one particular affirmative [**EI**], (m) one minor particular affirmative and one universal affirmative [**IA**], (n) one minore particular negative and one universal negative [**OE**], (o) one particular affirmative and one universal negative [**IE**], (p) one particular negative and one universal affirmative [**OA**]" ([98], in [126], 82.125b–18–126a–4).

Among these combinations, some are very easily ruled out by the general conditions related to all the figures, that is, (**EE**), (**II**), (**OO**), (**IO**), (**OI**), (**EO**), (**OE**), and (**IE**). Then by the specific conditions of the fourth figure, three other combinations are ruled out, namely, (**AO**), (**OA**), and (**AI**) ([98], in [126], 82.–83).

Once all these combinations are ruled out, what remains are the following combinations: (**AA**), (**AE**), (**EA**), (**IA**), (**EI**), that is, five combinations which could lead to five productive moods. These moods are presented and proved as follows in the rest of the treatise:

I/ "From two universal affirmative premises, it follows one particular affirmative conclusion" ([98], in [126], 84.126a–15), that is, one **AAI** (*Bamalip*) mood, the analogue of *Darapti* from the third figure. This mood is stated as follows:

"Every A is B (minor premise)
Every C is A (major premise)
Therefore Some B is C" ([98], in [126], 84.126a–15–16)

It is proved first by the transposition of the premises, so that the minor becomes the major and the major becomes the minor, which leads to

1. Every C is A (transposition of the major → minor)
2. Every A is B (transposition of the minor → major)

Therefore 3. Every C is B" (from 1, 2, by *Barbara*)
Then 4. Some B is C (from 3, by conversion) ([98], in [126], 84)

This proof shows that Ibn al-Ṣalāh introduces a new method to deduce the conclusion apart from the usual methods (conversion and *ecthesis*). This new method that we don't find in any of al-Fārābī's, Avicenna's, or Averroes' writings is the *transposition* of premises. Absolutely speaking, it is legitimate from a modern viewpoint, since the two premises are related by a conjunction, and the conjunction is commutative, so that starting with the first premise or with the second one does not affect the deducibility of the conclusion. However, when one changes the order of the premises, one changes the place of the middle term in the premises, which leads to a change of the figure itself. But it is precisely this reason that made Ibn al-Ṣalāh use this method, since he wanted to reduce the fourth figure mood to one mood of the first figure, namely, here, to *Barbara* and then to get the desired conclusion by conversion, which permutes the subject and the predicate of the conclusion and turns back the mood to the fourth figure. He also wanted to use conversion as less as possible and to avoid using the *reductio ad absurdum* method, which is seen as "less natural" than the direct methods of proof.

II/ The second productive mood is **AEE** (*Camenes*). This mood is stated as follows:

"No A is B (minor premise)
Every C is A (major premise)
Therefore No B is C" ([98], in [126], 84)

This is also proved by the transposition of the premises first, then by conversion as follows:

1. Every C is A (transposition of the major → minor)
2. No A is B (transposition of the minor → major)

Therefore 3. No C is B (from 1, 2 by Camestres)
Then 4. No B is C (from 3, by conversion) ([98], in [126], 84)

III/ The third productive mood is the following **EAO** (*Fesapo*), the analogue of *Felapton* from the third figure. It is stated as follows:

"Every A is B (minor premise)
No C is A (major premise)
Therefore Not every B is C" ([98], in [126], 84)

This is proved by the conversion of the two premises, so that we get

Some B is A (conversion of the minor)
No A is C (conversion of the major)
Therefore Not every B is C (by *Ferio*) ([98], in [126], 84)

IV/ The fourth productive mood is the following **IAI** (*Dimaris*). It is stated as follows:

"Every A is B (minor)
Some C is A (major)
Therefore Some B is C" ([98], in [126], 85)

This is also proved by transposition of both premises, so that the minor becomes the major and the major becomes the minor as follows:

1. Some C is A (transposition of the major → minor)
2. Every A is B (transposition of the minor → major)

Therefore 3. Some C is B (from 1, 2, by *Darii*)
Then 4. Some B is C (from 3, by conversion) ([98], in [126], 85)

V/ The fifth and last productive mood is an **EIO** mood, that is, *Fresison*. It is stated as follows:

"Some A is B (minor premise)
No C is A (major premise)
Therefore Not every B is C" ([98], in [126], 85)

This is proved by the conversion of the two premises, which leads to the following:

1. Some B is A (conversion of the minor)
2. No A is C (conversion of the major)

Therefore 3. Not every B is C (1, 2 by *Ferio*) ([98], in [126], 87)

Now, *Camenes*, *Dimaris* and *Fresison* do not raise any problem and they are admitted even by the modern logicians too, for their validity is not doubtful and can be shown very easily by a simple truth table. But *Bamalip* and *Fesapo*, which are the analogues, respectively, of *Darapti* and *Felapton* raise the same problems as these third figure moods, which are valid only when we add the augments to the affirmative universals. So they are not likely to be simply valid without any addition. Would they be valid if we add the augments, as was the case with *Darapti* and *Felapton*? To answer this question, let us formalize these two moods and check their validity by truth tables. *Bamalip* can be formalized as follows, when we add the augments:

$$\{[(\exists x)Ax \wedge (\forall x)(Ax \supset Bx)] \wedge [(\exists x)Cx \wedge (\forall x)(Cx \supset Ax)]\} \rightarrow (\exists x)(Bx \wedge Cx)$$

This formula is valid. So the mood is productive provided we add the augments. We can also add that unlike *Darapti*, which is valid when we add the augment to *only one* (universal affirmative) premise, this one is *not* valid if we add the augment only to *one* of the premises: we have to add the augments to *both* (universal affirmative) premises. So the validity of *Bamalip* is conditioned by the addition of these augments. But Ibn al-Ṣalāh does not say anything about the existential augments of the premises, and nothing is said also about the existential import of the universal affirmative propositions in general. If his opinion on the import is the same as al-Fārābī's and Avicenna's one, to the effect that all affirmative propositions have an import, then he is right in admitting *Bamalip*; but if the import of the **A** propositions is not clear in his theory, then this mood is not valid. However, since almost all traditional logicians, whether in the Arabic tradition or in the Latin West, tend to give an import to all affirmative propositions, *let us* credit him with that opinion and admit *Bramantip* as a valid mood!

The same problem is raised with *Fesapo* which would be formalized as follows:

$$\{[(\exists x)Ax \wedge (\forall x)(Ax \supset Bx)] \wedge (\forall x)(Cx \supset \sim Ax)]\} \rightarrow (\exists x)(Bx \wedge \sim Cx) \quad .$$

This formula is valid. But it would have been invalid if the affirmative premise did not have an import. So here too, its validity is conditioned by the existential import of the affirmative proposition.

3.5 Conclusion

Al-Fārābī and Averroes are very close while Avicenna presents a different conception of the absolute propositions. However, Averroes is the closest one to Aristotle, unlike al-Fārābī who introduces some original proofs, such as the proof by *ekthesis* of *Baroco*. This departs from Aristotle, who proves *Baroco* by *reductio ad absurdum*, even if elsewhere, as one reviewer stresses, Aristotle says that a proof by *ekhtesis* could be provided for *Baroco* (as for *Bocardo*) with necessary premises, for instance in the following passage:

"But in the middle figure when the universal is affirmative, and the particular negative, and again in the third figure when the universal is affirmative and the particular negative, the demonstration will not take the same form, but it is necessary by the exposition of a part of the subject, to which in each case the predicate does not belong, to make the deduction in reference to this: with terms so chosen the conclusion will be necessary. But if the relation is necessary in respect of the part exposed, it must hold of some of that term in which this part is included; for the part exposed is just some of that. And each of the resulting deductions is in the appropriate figure" ([24], I, 8, 30a 6–14).

Averroes follows very faithfully Aristotle and applies the same methods in his syllogistic. As to Avicenna, he is the most prolific one in this respect for he provides several proofs for each mood.

All three authors evoke Alexander of Aphrodisias and are more or less influenced by his commentaries. But this influence is more important with regard to al-Fārābī and Avicenna than it is for Averroes whose aim is explicitly to return back to Aristotle's text itself, which makes him criticize some of the developments and precisions introduced by Alexander and others.

But Avicenna's syllogistics contains some precisions that we don't find in the other two systems, for he introduces temporal precisions and other conditions in the categorical absolute propositions, which modify his analysis of the conversions and his proofs of the moods. These conditions lead to a different account of the oppositions between the propositions in Avicenna's system, since the perpetual sentences (or their disjunction) are the real contradictories of the absolutes (general or special).

As to the non-conclusive moods, al-Fārābī rules them out by relying on calculations and some obvious rules, either general or related to each figure which may be verified by examples as we have shown, while Avicenna starts by stating the rules related to each figure, which show the reasons why the non-conclusive moods are not valid by relying on the definitions and the very structures of each figure, before proving any of the valid moods. Averroes gives like Aristotle counterexamples to illustrate the non-conclusive moods. Unlike his predecessors, he evokes explicitly the principles of "what is said of every" and "what is said of none," called afterward by the medieval logicians *dictum de omni et de nullo*, to justify the moods of the first figure.

Al-Fārābī and Avicenna define the extremes (major and minor terms) by means of their places in the conclusion and not by means of their respective extensions or the presumed inclusion of the minor inside the middle which in turn is included inside the major. Their definitions are thus both different and more general than that of Aristotle, for they apply to all moods of whatever figure. Avicenna even adds that the minor's extension might be in some cases larger than that of the middle and the major, which means that the above *dictum*, which is based on these inclusions, is not so fundamental in his theory.

The three systems, however, agree in rejecting the fourth figure because of its "unnaturalness". But as we saw above, some later authors like Ibn al-Ṣalāh and others after him admit the fourth figure and prove their moods by using some new methods such as the transposition of premises. In Ibn al-Salah's text, these proofs are presented with some detail. Let us now turn to modal logic.

References

1. *Alexander of Aphrodisias on Aristotle's Prior Analytics 1.1–7*, translation J. Barnes et al., 1991.
2. Al-Fārābī, Abū Naṣr. 1960. *Şharh al-Fārābī li kitāb Aristūṭālīs fī al-'Ibāra*, 2nd edn, ed. Wilhelm Kutch and Stanley Marrow. Beirut: Dar el-Mashriq.
3. Al-Fārābī, Abū Naṣr. 1968. *Iḥṣā al-'Ulūm*, edited and introduced by Uthman Amin, Maktabat al-anjelu al-misriyya, Cairo.
4. Al-Fārābī, Abū Naṣr. 1968. *al-Alfāḍ al musta'mala fi l-manṭiq*, second edition, ed. Mohsen Mahdi. Beirut: Dar el Machriq.
5. Al-Fārābī, Abū Naṣr. 1985. *Kitāb al-Tanbīh 'Ala Sabīl as-Sa'āda*, edited and introduced by Jafar Al Yasin, Dar al-Manahil, Beiruth.
6. Al-Fārābī, Abu Nasr. al-Risālah allatī ṣadara bihā al-Manṭiq (or "al-Tawṭi'a"). In *al-Manṭiq 'inda al-Fārābī*, vol. 1, ed. Rafik Al Ajam, 55–62. Beirut: Dar el Machriq.
7. Al-Fārābī, Abū Naṣr. 1986. *Al-Fuṣūl al-Khamsa*. In *al-Manṭiq 'inda al-Fārābī*, vol. 1, ed. Rafik Al Ajam, 65–93. Beirut: Dar el Machriq.
8. Al-Fārābī, Abū Naṣr. 1986. *Kitāb al-'Ibāra*. In *al-Manṭiq 'inda al-Fārābī*, vol. 1, ed. Rafik Al Ajam, 133–164. Beirut: Dar el Machriq.
9. Al-Fārābī, Abū Naṣr. 1986. *Kitāb al-Maqūlāt*. In *al-Manṭiq 'inda al-Fārābī*, vol. 1, ed. Rafik Al Ajam, 89–132. Beirut: Dar el Machriq.
10. Al-Fārābī, Abū Naṣr. 1986. *Kitāb al-Qiyās*. In *al-Manṭiq 'inda al-Fārābī*, vol. 2, ed. Rafik Al Ajam, 11–64. Beirut: Dar el Machriq.
11. Al-Fārābī, Abū Naṣr. 1986. Kitāb al-Qiyās al-Ṣaghīr 'alā ṭarīqati al-mutakallimīn. In *al Mantiq 'inda al-Fārābī*, vol. 2, ed. Rafik Al Ajam, 65–93. Beirut: Dar el Machriq.
12. Al-Fārābī, Abū Naṣr. 1988. *Sharh al-'Ibāra*". In *al-Manṭiqiyāt li-al-Farābi*, vol. 2, texts published by Mohamed Teki Danesh Pazuh, Edition Qom, 1409 of Hegira.
14. Al-Fārābī, Abū Naṣr. 1988. *al-Qawl fī al-'Ibāra*, in *al-Manṭiqiyāt li-al-Fārābi*, volume 1, texts published by Mohamed Teki Danesh Pazuh, Edition Qom, 83–114.
15. Al-Fārābī, Abū Naṣr. 1988. Kitāb al Qiyās. In *al-Manṭiqiyāt li-al-Fārābi*, vol. 1, texts published by Mohamed Teki Danesh Pazuh, Edition Qom, 115–151.
16. Al-Fārābī, Abū Naṣr. 1988. al-Qiyās al-Ṣaghīr. In *al-Manṭiqiyāt li-al-Fārābi*, vol. 1, texts published by Mohamed Teki Danesh Pazuh, Edition Qom, 152–194.
17. Al-Fārābī, Abū Naṣr. 1988. *Kitāb al-Burhān*, in *al-Manṭiqiyāt li-al-Fārābī*, vol. 1, texts published by Mohamed Teki Danesh Pazuh, Edition Qom, 267–349.
18. Al-Fārābī, Abū Naṣr. 1988. Mā yanbaghī an yuqaddama qabla ta allum al-falsafa. In *al-Mantiqiyyāt li-al-Fārābī*, texts published by Mohamed Teki Danesh Pazuh, Edition Qom, 1–10.
21. Aristote. 1971. *Premiers Analytiques*, Translated by J. Tricot, Librairie philosophie J. Vrin, Paris.
22. Aristotle. 1991. *Categories*. In *The complete works of aristotle*, vol. 1, ed. Jonathan Barnes. The Revised Oxford Edition.
23. Aristotle. 1991. *De Interpretatione*, in *The Complete Works of Aristotle*, the Revised Oxford Edition, vol. 1, ed. Jonathan Barnes.
24. Aristotle. 1991. Prior analytics. In *The complete works of Aristotle*, the Revised Oxford Edition, ed. Jonathan Barnes, vol. 1.
25. Arnault, Antoine, and Pierre Nicole. 1970. *La logique ou l'art de penser*, Editions Flammarion.
26. Averroes. 1982. *Talkhīṣ Manṭiq Arisṭu* (*Paraphrase de la logique d'Aristote*), volume 1: *Kitāb Al-Maqūlāt* (pp. 3–77), *Kitāb al-'Ibāra* (pp. 81–141), *Kitāb al-Qiyās* (pp. 143–366), ed. by Gérard Jehamy. Manshūrāt al-Jāmi a al-lubnānīya, al-Maktaba al-sharqiyya, Beirut.
31. Avicenna. 1910. *Manṭiq al-Mashriqiyīn*, Muḥyī al-Dīn al Khatīb and 'Abdelfattāh al Qatlane, Cairo.

32. Avicenna. 1938. *al-Najāt*, Muḥyi al-Dīn Sabrī al-Kurdī, second edition, Library Mustapha al Bab al Hilbi, Cairo.Avicenna.
33. Avicenna. 1959. *al- Shifā', al-Manṭiq* 2: *al-Maqūlāt*, ed. G. Anawati, M. El Khodeiri, A.F. El-Ehwani, S. Zayed, rev. and intr. by I. Madkour, Cairo.
34. Avicenna. 1964. *al-Shifā', al-Manṭiq* 4: *al-Qiyās*, ed. S. Zayed, rev. and intr. by I. Madkour. Cairo.
35. Avicenna. 1970. *al-Shifā', al-Manṭiq 3*: *al- 'Ibāra*, ed M. El Khodeiri, rev and intr. by I. Madkour, Cairo.
36. Avicenna. 1971. *Al-Ishārāt wa l–tanbīhāt, with the commentary of N. Ṭūsi*, intr by Dr. Seliman Donya, Part 1, third edition, Cairo: Dar al Ma'arif.
37. Avicenna. 1982. *Manṭiq al-Mashriqiyīn*, ed. Shokri Najjar, Dār al Ḥadātha, Beirut.
38. Avicenna. 2017. *Al-mukhtaṣar al-awsaṭ fī al-manṭiq*, ed. Seyyed Mahmoud Yousofsani, Muassasah-i Pizh ūhishī-i Ḥikmat va Falsafan-i Īrān, Tehran (hij. 1396).
40. Badawi, Abderrahman. 1980. *Manṭiq Arisṭu*, vols. 1 & 2, Beirut: Dar al Kalam.
44. Blanché, Robert. 1970. *La logique et son histoire, d'Aristote à Russell*. Paris: Armand Colin.
50. Chatti, Saloua. 2012. Logical oppositions in Arabic logic, Avicenna and Averroes. In *Around and beyond the square of opposition*, ed. J. Y. Béziau, and D. Jacquette. Basel: Springer.
51. Chatti, Saloua, and Fabien Schang. 2013. The cube, the square and the problem of existential import. *History and Philosophy of Logic* 34 (2): 101–132.
52. Chatti, Saloua. 2014. Syncategoremata in Arabic logic, al-Fārābi and Avicenna. *History and Philosophy of Logic* 35 (2): 167–197.
53. Chatti, Saloua. 2014. Avicenna on possibility and necessity. *History and Philosophy of Logic* 35 (4): 332–353.
54. Chatti, Saloua. 2015. Les carrés d'Avicenne. In *Le carré et ses extensions. Approches théoriques, pratiques et historiques*, ed. Hmaid Ben Aziza and Saloua Chatti. Publications de la Faculté des Sciences Humaines et Sociales de Tunis.
55. Chatti, Saloua. 2016. Existential import in Avicenna's modal logic. *Arabic Sciences and Philosophy* 26 (1): 45–71 (Cambridge University Press).
57. Chatti, Saloua. 2016. Avicenna (Ibn Sīnā): Logic. In *Encyclopedia of logic, internet encyclopedia of philosophy*. College Publications, www.iep.utm.edu/av-logic/.
58. Chatti, Saloua. 2017. On the asymmetry between the four corners of the square. Available online in the site 'Cercle Ferdinad de Saussure', proceedings of the workshop 'The Arbitrariness of the Sign', organized by Prof. J-Y. Beziau in the congress *Le Cours de Linguistique Générale*, 1916–2016. L'émergence, Geneva, 9–13 janvier 2017. https://www.clg2016.org/contribution/281.html.
60. Chatti, Saloua. 2019. Logical consequence in Avicenna's theory. *Logica Universalis* 13: 101–133.
61. Chatti, Saloua. 2019. The logic of Avicenna, between *al-Qiyās* and *Manṭiq al-Mashriqiyyīn*. *Arabic Sciences and Philosohy* 29 (1): 109–131.
62. Couturat, Louis. 1901. *La logique de Leibniz*, Georg Olms Verlagsbuchhandlung Hildesheim, New Edition (1969).
72. El-Rouayheb, Khaled. 2012. Post-Avicennan logicians on the subject matter of logic: some thirteenth—and fourteenth—century discussions. *Arabic Sciences and Philosophy* 22 (1).
77. Gutas, Dimitri. 1988. *Avicenna and the Aristotelian tradition: Introduction to reading Avicenna's philosophical works*, 1st ed. Leiden: E.J. Brill.
78. Gutas, Dimitri. 2014. *Avicenna and the Aristotelian tradition: Introduction to reading Avicenna's philosophical works*, 2nd ed. Leiden: Brill.
84. Hodges, Wilfrid. 2012. Affirmative and negative in Ibn Sīnā. In *Insolubles and consequences; essays in honour of Stephen Read*, ed. Catarina Dutilh Novaes and Ole Hjortland Thomassen. UK: College Publications, Lightning Source, Milton Keynes.
87. Hodges, Wilfrid. 2018. Proofs as cognitive or computational: Ibn Sīnā's innovations. *Philosophy and Technology* 31 (1): 131–153.

89. Hodges, Wilfrid. 2018. Nonproductivity proofs from Alexander to Abū al-Barakāt:1. Aristotelian and logical background, http://wilfridhodges.co.uk/history26.pdf.
90. Hodges, Wilfrid, and Druart Thérèse-Anne. 2018. Al-Fārābī's philosophy of logic and language. In *Stanford encyclopedia of philosophy*.
92. Hodges, Wilfrid. forthcoming. *Mathematical background to the logic of Avicenna*. http://wilfridhodges.co.uk/arabic44.pdf.
95. Horn, Laurence. 2001. A natural history of negation. In *The David Hume series*. Stanford: CSLI, University of Chicago Press.
97. Ibn al-Muqaffa. 1978. *'Al-manṭiq*, ed. M. T. Dāneshpazhūh, Iranian Institute of Philosophy, Tehran.
98. Ibn al-Ṣalāh. 1966. *Maqāla fī al-shakl al-rābi' min ashkāl al-qiyās*. In *Galen and the syllogism*, ed. N. Rescher, 76–87.
99. Ibn al-Sikkit. 1956. *Iṣlāḥ al-manṭiq*, ed. Aḥmad M. Shākir, and 'Abd-al-Salām M. Mārūn, Dār al-Ma'ārif, Cairo.
102. Klima, Gyula. 2006. Syncategoremata. In *Encyclopedia of language & linguistics*, 2nd edn, vol. 12, ed. Keith Brown, 353–356. Oxford: Elsevier.
108. Lameer, Joep. 1994. *al-Fārābī and Aristotelian syllogistics; Greak theory and Islamic practice*, Brill Edition.
110. Lee, Tae-Soo. 1984. *Die Griechische Tradition der Aristotelischen Syllogistik in der Spätantike*. Göttingen: Vandenhoeck & Ruprecht.
112. Łukasiewicz, Jan. 1972. *La syllogistique d'Aristote dans la perspective de la logique formelle moderne*, French translation by Françoise Zaslawsky, Librairie Armand Colin, Paris (1951).
120. Movahed, Zia. 2010. De re and de dicto modality in the Islamic traditional logic. *Sophia Perennis* 2 (2): 5–19.
121. Parsons, Terence. 2006. The traditional Square of Opposition. In *Stanford encyclopedia of philosophy*, ed. Edward N. Zalta. Stanford: Metaphysics Research lab, CSLI, http://plato.stanford.edu/entries/square/index.html.
122. Philoponus, John. 1905. *In Aristotelis Analytica Priora Commentaria*, ed. M. Wallies, Reimer, Berlin.
127. Rescher, Nicholas. *The development of Arabic logic*, University of Pittsburgh Press, Arabic translation by Mohamed Mahrān. Cairo: Dar el Ma'ārif (1964), Arabic translation (1985).
126. Rescher, Nicholas. 1966. *Galen and the syllogism*. Pittsburgh: University of Pittsburgh Press.
130. Ryle, Gilbert. 1966. *The concept of mind*. Harmondsworth: Penguin.
131. Sabra, A. I. 1980. Avicenna on the subject matter of logic. *Journal of Philosophy* 77: 746–764.
133. Spruyt, Joke. 2007. Peter of Spain. In *The Stanford encyclopedia of philosophy*, ed. Edward. N. Zalta, http://plato.stanford.edu/entries/peter-spain/.
135. Street, Tony. 2001. 'The eminent later scholar' in Avicenna's Book of the Syllogism. *Arabic Sciences and Philosophy* 11: 205–218.
136. Street, Tony. 2002. An outline of Avicenna's syllogistic. *Archiv für Geschichte der Philosophie* 84 (2): 129–160.
137. Street, Tony. 2004. Arabic logic. In *Handbook of the history of logic*, vol. 1, ed. Dov Gabbay and John Woods. Elsevier, BV.
139. Street, Tony, 2008. Arabic and Islamic philosophy of language and logic. In *Stanford encyclopedia of philosophy*, ed. Edward N. Zalta. Stanford University, http://plato.stanford.edu/entries/arabic-islamic-language/, New Edition (2013).
138. Street, Tony. 2010. Appendix: Readings of the subject term. *Arabic Sciences and Philosophy* 29: 119–124.
141. Street, Tony. 2016. Kātibī (d. 1277), Taḥtānī (d. 1365) and the *Shamsiyya*. *The Oxford Handbook of Islamic Philosophy* 348.
142. Strobino, Riccardo. 2018. Ibn Sīnā's logic. *The Stanford Encyclopedia of Philosophy*, https://plato.stanford.edu/archives/fall2018/entries/ibn-sina-logic.

143. Thom, Paul. 2008. al-Fārābī on indefinite and privative names. *Arabic Sciences and Philosophy* 18 (2): 193–209.
145. Thom, Paul. 2010. al-Fārābī on the number of categories. http://paulthom.net/Papers-on-Arabiclogic.html.
146. Thom, Paul. 2010. Abharī on the logic of conjunctive terms. *Arabic Sciences and Philosophy* 20: 105–117.
147. Türker, Sadik. 2007. The Arabico-Islamic background of al-Fārābī's logic. *History and Philosophy of Logic* 28 (3): 183–255.
148. Versteegh, C.H.M. 1977. *Greek elements in Arabic linguistic thinking*. Leiden: Brill.
149. Versteegh, Kees. 1997. The debate between logic and grammar. In *Landmarks in linguistic thought III, The Arabic linguistic tradition*, ed. Kees Versteegh, 52–63. London: Routledge.
150. Zimmermann, F.W. 1972. Some observations on Al-Fārābī and logical tradition. In *Islamic philosophy and the classical tradition, Essays presented by his friends and pupils to Richard Walzer on his seventieth birthday*, ed. S.M. Stern, Albert Hourani, and Vivian Brown, 517–546. Oxford: Cassirer.

Chapter 4
Modal Logic

4.1 The Modal Propositions

4.1.1 The Modal Propositions in al-Fārābī's Frame

Let us start by al-Fārābī's classification of the modal propositions. Al-Fārābī talks about these propositions in two writings: *Sharh al-ʿIbāra* ([12]) and *al-Qawl fī al-ʿIbāra* ([14]) (the latter is called *Kitāb al-Ibāra* in Rafik al Ajam ([8])). Both correspond to the Aristotelian treatise *De Interpretatione,* but the latter is much shorter than the former, which is a long commentary and has been also published under the title *Sharḥ al-Fārābī li-kitāb arisṭūṭālīs li-al-ʿIbāra* ([2]).

In *Sharh al-ʿIbāra* ([12]), which is the long commentary on Aristotle's *De Interpretatione*, he comments Aristotle's text and follows it faithfully.

Aristotle distinguishes between:

- Necessary to be,
- Necessary not to be (= impossible to be),
- Possible to be (+ admissible to be), and
- Not necessary to be (= possible not to be).

We find the same notions in al-Fārābī's text but al-Fārābī gives up the word "admissible" (used by Aristotle as a synonymous of "possible") and starts, in his table, by necessity instead of possibility.

the following:

"Necessary to be / Not necessary to be
Not possible not to be / Possible not to be
Impossible not to be / Not impossible not to be

Necessary not to be / Not necessary not to be
Not possible to be / Possible to be
Impossible to be / Not impossible to be" ([12], 217)

S. Chatti, *Arabic Logic from al-Fārābī to Averroes*, Studies in Universal Logic,
https://doi.org/10.1007/978-3-030-27466-5_4

We can note here that al-Fārābī, unlike Aristotle, starts with the necessary, but the equivalences he admits are the same as the Aristotelian ones, for they are the following:

- Necessary = Not possible not = Impossible not (in the top left column),
- Impossible = Not possible = Necessary not (in the bottom left column),
- Possible = Not necessary not = Not impossible (in the top right column), and
- Not necessary = Possible not = Not impossible not (in the bottom right column).

This means that al-Fārābī, as well as Aristotle, admits the duality laws, i.e., $\Box\sim\ = \sim\Diamond$, hence $\sim\Box\sim\ = \sim\sim\Diamond = \Diamond$ and $\Diamond\sim\ = \sim\Box$, hence $\sim\Diamond\sim\ = \sim\sim\Box = \Box$[1].

Note also that each *line* contains contradictory modal operators. The possibility operator expressed in this table is one-sided. But in al-Fārābī's account, the possibility may also be two-sided.

The bilateral or two-sided meaning, which is expressed by al-Fārābī as follows: "Possible to be and Possible not to be" is contrary to both necessity and impossibility as witnessed by the following quotation:

> "As to « necessary to be » and « necessary not to be » , none of them can be true when the conjunction of the two others, which is « possible to be and [possible] not to be » is true. Because when « necessary to be » is true, « possible not to be » is not true and when « necessary not to be » is true « possible to be » is not true "([12], 204, my translation).

These modalities are *de dicto*, which means that the modal operator is external and puts on the whole proposition. But al-Fārābī uses also *de re* modalities, i.e., internal modalities when he talks about the quantified propositions. However, the words *de dicto* and *de re*, which are Latin, are not used by al-Fārābī, and the distinction made is not as complex as that of the medieval logicians for it is only a distinction of scope. The *de dicto* modalities are external because they put on the whole proposition, while the *de re* ones are internal, because they put on the predicate.

In *Al-Qawl fi al 'Ibāra* ([14]) (that is, ([8]), which is the Lebanese edition of it), al-Fārābī studies all kinds of modal propositions, that is, the singulars, the indefinites, and the quantified propositions. The modalities considered are internal, for they are put inside the proposition, in front of the predicate as witnessed by all the examples given, e.g., "Humans are possibly just."

As to the negation, it is put either in front of the predicate or in front of the modality itself, which leads to different negative propositions in each case.

In the indefinite and singular propositions, when the negation puts on the modality itself, we obtain

---

[1]See ([76]) for the symbols and the equivalences. I have introduced these modern symbols for convenience. These modern symbols can be found in [49] and [96], for instance.

- "Humans are not necessarily just" or "Humans are not possibly just."
- "Zayd does not possibly walk" or "Zayd does not necessarily walk" ([12], 106–107, ([8]), 155.10–11).

When the negation puts on the predicate (or the copula), we obtain

- "Humans are necessarily not savant."
- "Zayd is possibly not savant" ([12], 106–107, ([8]), 155.17).

The indefinites and the singulars function in the same way for their real negations are obtained when the negation puts on the modality itself. When the negation puts on the predicate, the propositions obtained are called by al-Fārābī the "possible negative" or the "necessary negative" propositions, depending on the modality contained in the proposition.

The "possible negatives" and the "necessary negatives" are different from the "negation of possibility" and the "negation of necessity" ([12], 110), for instance,

- "Zayd is possibly not a writer" is a possible negative and
- "Humans are necessarily not ducks" is a necessary negative.

While

- "Zayd is not possibly a duck" expresses the negation of possibility

and

- "Humans are not necessarily just" expresses the negation of necessity.

Note that all these examples involve the unilateral or one-sided kind of possibility.

As to the quantified propositions, their real (i.e., contradictory) negations may be obtained by negating the quantifier itself. For instance, "Not every human possibly walks" is the contradictory negation of "Every human possibly walks," while "No human possibly walks" is the contrary of the above universal affirmative proposition ([12], 105–106, ([8]), 155. 18–20). In general, the contradictories and the contraries of the modal affirmative propositions differ from their assertoric correspondents only by the presence of the modalities.

The negative propositions are expressed in the following way:

1. Not every human possibly walks (= Not every A is $\Diamond$B = Some A's are $\Box\sim$B's).
2. Every human possibly does not walk (= Every A is $\Diamond\sim$B).
3. No human possibly walks (= No A is $\Diamond$B = Every A is $\Box\sim$B).
4. Not every human is necessarily just (= Not every A is $\Box$B = Some A's are $\Diamond\sim$B's).
5. No human is necessarily just (= No A is $\Box$B = Every A is $\Diamond\sim$B).
6. Every human is necessarily not just (= Every A is $\Box\sim$B).

As we can see, (3) is equivalent to (6) and (2) is equivalent to (5), by the duality laws which are already held by al-Fārābī as we saw above. If we rule out these redundancies, we have only four propositions, which are the following:

(6) Every human is necessarily not just: **E**$\Box$
(2) Every human possibly does not walk: **E**$\Diamond$
(1) Not every human possibly walks: **O**$\Box$

And (4) Not every human is necessarily just: **O**$\Diamond$.

If we add the four modal affirmative quantified propositions to this set, we get the eight usual vertices of the modal octagon, which are the following:

(1) Every A is necessarily B (= $(\forall x)(Ax \supset \Box Bx)$ = **A**$\Box$).
(2) Every A is possibly B (= $(\forall x)(Ax \supset \Diamond Bx)$ = **A**$\Diamond$).
(3) Some A are necessarily B (= $(\exists x)(Ax \supset \Box Bx)$ = **I**$\Box$).
(4) Some A are possibly B (= $(\exists x)(Ax \wedge \Diamond Bx)$ = **I**$\Diamond$).
(5) Every A is necessarily not B (= $(\forall x)(Ax \supset \Box {\sim} Bx)$ = **E**$\Box$).
(6) Every A is possibly not B (= $(\forall x)(Ax \supset \Diamond {\sim} Bx)$ = **E**$\Diamond$).
(7) Some A are necessarily not B (= $(\exists x)(Ax \wedge \Box {\sim} Bx)$ = **O**$\Box$).
(8) Some A are possibly not B (= $(\exists x)(Ax \wedge \Diamond {\sim} Bx)$ = **O**$\Diamond$).

In that treatise, he considers the universal affirmative propositions with an affirmative predicate or with a negative predicate and provides systematically their contradictories and their contraries. This gives him the propositions (2), (1), (3), (5), (6), and (7). And he says that one can do the same with the necessary propositions just by replacing "possibly" by "necessarily" in all cases. This leads to the propositions (4) and (8) ([8], 155.18–157.12) and completes the whole set of modal quantified propositions. The relations between these propositions give rise to a modal Octagon which will be drawn by Buridan[2] among others. We will see later that these relations are partly stated by al-Fārābī.

4.1.2 The Modal Propositions in Avicenna's Frame

As to Avicenna, he expresses the modal propositions in various ways, for he considers three kinds of possibilities which are the following:

1. The usual possible = what is not impossible (this is the unilateral meaning).
2. The "narrow-possible" = what is neither necessary, nor impossible (= bilateral).
3. The "narrowest-possible" = what is neither actual nor necessary, nor impossible.

[2]On the modal Octagon of Buridan see [124]. On Buridan's logic in general see [47], [101], and [103]. On Medieval logic, see [106] and on Medieval modal logic see [104].

Avicenna defines the third meaning as follows: "the third meaning is that which is not actual and is not necessary in the future" ([35], 118). Thus, it qualifies "what is not existent (*ma'dūm*), but is not necessarily existent or non-existent in the future" ([35], 117).

Consequently, his account of the modal propositions is more complex than al-Fārābī's one, who focuses mainly on the one-sided possible, although he evokes in some cases the bilateral one. But he says explicitly that the genuine meaning of possibility is the second one, i.e., the bilateral meaning which is different from the first and the third meanings. His account of the bilateral meaning is more complete than that of Aristotle and al-Fārābī, for unlike his predecessors, Avicenna provides explicitly the negation of that bilateral meaning as appeared in the following quotation:

> "If we intend by it [i.e. the possible] the narrow meaning, then everything is either possible, or impossible or necessary [*wājib*], and what is not possible will not be the impossible, rather what is *not possible* is necessary [*ḍarūr*i] either *in existence or in non existence"* ([35], 117, emphasis added).

As appears clearly in this quotation, the negation of the narrow-possible is: "necessary to be *or* necessary not to be" (= necessary in existence *or* in non-existence). Note, here, that "*wājib*" applies only to the affirmatives, while "*ḍarūri*" (which usually means the same as *wājib* in Arabic) applies to both affirmatives and negatives.

More importantly, we can also note that he holds the following equivalence:

Not (possible α *and* possible not α) $\equiv$ (*not* possible α *or not* possible $\sim\alpha$).

That is, $\sim(\Diamond\alpha \wedge \Diamond\sim\alpha) \equiv (\Box\sim\alpha \vee \Box\alpha) \equiv (\sim\Diamond\alpha \vee \sim\Diamond\sim\alpha)$ [by the definitions of the modalities].

This is another application of the first De Morgan's law, i.e., "$\sim(P \wedge Q) \equiv (\sim P \vee \sim Q)$, which, as we noticed above (Sect. 3.2.2), Avicenna had already used when stating the contradictory of **A** special absolute. He will apply this equivalence to the quantified modal propositions too, as we will see below. So we can say that his use of this law is quite explicit, even if the law is not stated as it is stated by De Morgan himself and the modern logicians afterward.

The modal propositions may contain the same conditions as the absolute ones, that is, "as long as it is S," "as long as it exists," "as long as it is P," "at a determined time") (e.g., "The moon eclipses at a determined time"), "at regular times" (e.g., "Men are breathing (at regular times)"), and so on. When there is no condition at all, necessity is absolute, as in "God is necessarily living," in which case necessity is close to eternity. But according to Avicenna, necessity is not equivalent to permanence, for some things may be permanent (may happen permanently) without being really necessary, for permanence may be "accidental (*ittifāqan*)" ([34], 48.16) too. For instance, we may say: "some things are permanently white," but they are not necessarily white, since their color is not an essential property. On the contrary, some necessary events are not permanent (for instance, the eclipse of the moon is necessary at some determined times, due to the laws of astronomy, but it does not happen permanently, i.e., at all times).

These modalities are either *internal* or *external*. The internal modalities are put in front of the predicate, as in "Zayd is possibly a writer" or "Humans are necessarily animals," while the external modalities are put in the beginning of the propositions as in "Necessarily every human is an animal" or "Possibly Zayd is just." What I call internal modalities are comparable to the medieval *de re* modalities, while the external modalities would be comparable to *de dicto* medieval modalities. In medieval writings, *de re* means that the modality puts on the objects (*res* = object), while *de dicto* means that the modality puts on the sentence (*dicto* = what is said). According to Zia Movahed ([120], 6), Avicenna was the first author to draw this distinction clearly, long before the medieval logicians. He did that in *al-'Ibāra* where he distinguished between two kinds of sentences containing the possibility operator. This distinction is expressed as follows:

> 'If we say that 'every human is possibly a writer', this is natural, and means that 'every single person is possibly a writer', ..., but if we say that 'possibly every human is a writer', and if the possibility is put in front of the universal quantifier, then this would be doubtful (*fa-qad yushakku fīhi*)' ([35], 115.2–9).

According to what is said in this quotation the internal (*de re*) reading is more natural and more likely to be true than the external (*de dicto*) reading, since as Avicenna adds in the sequel, many people think that "it is impossible that all humans are writers... so that every single person would be a writer" ([35], 115.9–11). This being so, the two readings are very different semantically and with regard to their truth conditions. However, as noted by Zia Movahed, the *de dicto* reading implies the *de re* one in Avicenna's frame, for he says:

> 'Here by reading the first example appropriately as 'every human being x, possibly x is a writer', we can symbolize it as: $\forall x \Diamond W(x)$, and his second example can be symbolized as: $\Diamond \forall x W(x)$. Furthermore from what he says, though he does not explicitly says so, one can infer that according to him [**]: $\Diamond \forall x W(x)$ implies [*]:$\forall x \Diamond W(x)$ but not vice versa. So Ibn Sīnā accepts the so-called Buridan formula and rejects the converse Buridan formula' ([118], 253)

Now Buridan's formulas are the following:

BUF: $\Diamond \forall x Fx \rightarrow \forall x \Diamond Fx$ (Buridan formula)
CBUF: $\forall x \Diamond Fx \rightarrow \Diamond \forall x Fx$ (Converse Buridan formula)

According to Zia Movahed, the first formula is endorsed by Avicenna while the second one is not admitted by him. He adds that this second formula is not admitted in modern quantified modal logic either since "it is not a theorem of any standard QML" ([118], 251, see also [65]).

In addition, two more theorems of quantified modal logic seem also to be endorsed by Avicenna, according to Z. Movahed, namely, the following two formulas generally attributed to the contemporary logician Ruth Barcan Marcus:

BF: $\forall x \Box Fx \rightarrow \Box \forall x Fx$ (Barcan formula)

CBF: $\Box \forall x Fx \rightarrow \forall x \Box Fx$ (Converse Barcan formula) ([118], 248)

For he quotes Avicenna who says what follows:

> 'But to say that: some people possibly are not writers is modally the same as saying that: possibly some people are not writers and although one implies the other, the meaning of the one may be opposite to the other [*wa amma qawlunā ba'ḍ annasi yumkinu an lā yakūna kātiban, fa-innahu qad yusāwī min jihatin qawlanā yumkinu an lā yakāna ba'ḍ annāsi kātiban, wa qad yukhālifuhu wa in lāzamahu*]' ([35], 116.9–10, quoted and translated by Z. Movaheb in ([118]), Arabic text added).

This passage of Avicenna is interpreted as follows by Z. Movahed:

> "This is accordingly translated as:
>
> $\exists x \Diamond \neg Wx \leftrightarrow \Diamond \exists x \neg Wx$
>
> Which is logically equivalent to:
>
> $\forall x \Box W(x) \leftrightarrow \Box \forall x\, W(x)$
>
> This is the conjunction of BF and CBF in one biconditional. So Ibn Sīnā discovers and endorses both BF and CBF while admitting that there are differences of meaning between the antecedent and the consequent of each conditional. This is an insight of genius into the subject… ' ([118], 253).

Zia Movahed is right when he stresses the closeness between Avicenna's text and the Barcan formulas, although saying that Avicenna "discovers and endorses" these formulas could be seen as exaggerated, since Avicenna does not distinguish between different kinds of quantification, and could not be said to endorse Ruth Barcan's interpretation of the quantification, which, in her system, is not referential but rather substitutional. However, the equivalence that Movahed mentions can be found in Avicenna's text and can be proved by considering the duality laws and the principle of contraposition according to which "$(p \supset q) \equiv (\sim q \supset \sim p)$". Thus starting from Avicenna's formulas, that is,

$\exists x \Diamond \neg Wx \leftrightarrow \Diamond \exists x \neg Wx$.

We can by contraposition state the following equivalence (= double implication):

$\neg \Diamond \exists x \neg Wx \leftrightarrow \neg(\exists x) \Diamond \neg Wx$,

which by the laws of duality of the modal operators and of the quantifiers amounts to

$\Box\neg(\exists x) \neg Wx \leftrightarrow (\forall x) \neg \Diamond \neg Wx$

itself equivalent to the following:

$\Box \forall x\, Wx \leftrightarrow (\forall x)\Box Wx$,

since "$\neg\Diamond = \Box\neg$", $\neg(\exists x) = (\forall x)\neg$", "$\neg\Diamond\neg \equiv \Box$", and "$\neg\exists\neg \equiv \forall$".

This formula is equivalent to "$\forall x\ \Box W(x) \leftrightarrow \Box\forall x\ W(x)$" cited by Z. Movahed, since the biconditional is symmetric.

So it is true as Zia Movahed emphasizes that Avicenna uses and endorses both *de re* and *de dicto* modalities unlike al-Fārābī who uses only internal, i.e., *de re* modalities. However, when stating the relations between modal propositions, Avicenna uses preferably external (*de dicto*) formulations ([34], 49), while in his syllogistic, the *de re* formulations seem to be privileged. But as we will see in the sequel, Avicenna's calculations of the oppositions (contradictions especially) of the *de re* bilateral possible propositions are not always adequate, although his calculations of the oppositions between the *unilateral* possible propositions, whether *de re* or *de dicto*, are all correct.

The formulations provided by Avicenna are the following for the singular propositions, where the modalities are external (*de dicto*):

1. Possibly Zayd is a writer.
2. It is not possible that Zayd is a writer.
3. Possibly Zayd is not a writer.

(1) and (3) are true, while (2) is false.

4. Necessarily Zayd is a writer: false.
5. Necessarily Zayd is not a writer: false.
6. Not necessarily Zayd is a writer: true.

The same may be said about the indefinite propositions which function in the same way as the singulars. The singulars as well as the indefinites may be formalized by $\Diamond\alpha$ or $\Box\ \alpha$ when the modality is external. The negations of these propositions are, respectively, $\sim\Diamond\alpha$ (= $\Box\sim\alpha$) and $\sim\Box\alpha$ (= $\Diamond\sim\alpha$).

As to the quantified modal propositions, they may be expressed in two ways, whether the modality is possibility or necessity. When the modalities are external, the modal quantified propositions are expressed as follows:

- Necessarily every A is B = $\Box$**A**.
- Possibly every A is B = $\Diamond$**A**.
- Necessarily no A is B = $\Box$**E**.
- Possibly no A is B = $\Diamond$**E**.
- Necessarily some A's are B's = $\Box$**I**.
- Possibly some A's are B's = $\Diamond$**I**.
- Necessarily not every A is B = $\Box$**O**.
- Possibly not every A is B = $\Diamond$**O**.

Those containing internal modalities may be formalized as follows:

- Every A is necessarily B = **A**$\Box$.
- Every A is possibly B = **A**$\Diamond$.
- Every A is necessarily not B = **E**$\Box$.
- Every A is possibly not B = **E**$\Diamond$.

- Some A's are necessarily B's = **I□**.
- Some A's are possibly B's = **I◇**.
- Some A's are necessarily not B's = **O□**.
- Some A's are possibly not B's = **O◇**.

However, Avicenna seems to privilege the external formulations when he states the oppositions between the modal propositions in *al-Qiyās*, for unlike al-Fārābī, he frequently starts by the modal words in these formulations, but he does not neglect the internal ones, which he explicitly uses in his syllogistic, for instance.

Sometimes, he puts the modality at the end of the propositions by stating the proposition as follows: "Every A is B necessarily." This formulation seems external, since the word "necessarily" is at the end of the sentence; hence, it is not attached directly to the predicate as in al-Fārābī's formulations which are clearly internal. But one should treat them as internal rather than external formulations, because in the real external ones, the modality puts on the whole sentence, which therefore follows it. In this external formulation, the modality affects the sentence and says of it that it is necessary.

When the modalities are external, the bilateral (two-sided) possible propositions are expressed as follows:

- **A** (bilateral possibility): ◇ every B is A ∧ ◇~(every B is A)=**◇A** ∧ **◇O**.
- **E** (bilateral possibility): ◇ No B is A ∧ ◇ ~ (no B is A) = **◇E** ∧ **◇I**.
- **I** (bilateral possibility): ◇ some A's are B's ∧ ◇ ~ (some A's are B's) = **◇I** ∧ **◇E**.
- **O** (bilateral possibility): ◇ ~ (every A is B) ∧ ◇ ~ (not every A is B) = **◇O** ∧ **◇A**.

As we can see, **A** in its bilateral possibility meaning is equivalent to **O** in its bilateral possibility meaning. Likewise **E** and **I** in the bilateral possibility meaning are also equivalent, which Avicenna recognizes explicitly when he says: "What possibly is all so and so, is possibly not all so and so too" ([34], 175.4).

However, things are not so clear when it comes to internal modalities, for the negative propositions (**E** and **O**) with internal bilateral modalities are different from the affirmative ones, unlike what Avicenna seems to believe. Let us see how Avicenna interprets these internal bilateral possible propositions.

Avicenna does not say explicitly in *al-Qiyās*, at least in the chapter where he analyzes the modal propositions and their oppositional relations (pp. 49–50) how he interprets **E** narrow-possible and **O** narrow-possible when the modality is internal, for he provides the external formulations rather than the internal ones. But he does interpret the negative special absolutes (**E** and **O** special absolutes) which are internal and contain the expression "sometimes but not permanently" in a way that separates the two sides of the temporal modalities ([34], 38–50, [36], 310–311). From this, we can assume that he would do the same with the alethic modalities, which are parallel to the temporal ones.

When the modality is internal, the affirmative narrow-possible propositions are expressed as follows:

- **A** (bilateral possible): Every A is $\Diamond$B $\wedge$ $\Diamond\sim$B = Every A is $\Diamond$B $\wedge$ Every A is $\Diamond\sim$B = **A**$\Diamond$ $\wedge$ **E**$\Diamond$.
- **I** (bilateral possible): Some A's are $\Diamond$B's $\wedge$ $\Diamond\sim$B's.

However, the negative ones, i.e., **E** (bilateral possible) and **O** (bilateral possible), should be different because of the bilateral nature of possibility, which groups the two elementary possibilities (*possibly* and *possibly not*) under the scope of the negation.

For how could one express **E** (bilateral possible), for instance? Should one say that **E** (bilateral possible) means that "Every A is possibly not B and possibly not not B"? In that case, the formalization would be the following:

- **E** (bilateral possible): Every A is $\Diamond\sim$B $\wedge$ $\Diamond\sim\sim$B = Every S is $\Diamond\sim$B $\wedge$ Every A is $\Diamond$B = **E**$\Diamond$ $\wedge$ **A**$\Diamond$.

And the proposition **E** narrow-possible would be equivalent to **A** narrow-possible, which is the opinion that Avicenna seems to endorse when he says: "It seems that for both the narrow-possible and the narrowest-possible, the affirmative is the same as the negative" ([34], 174.16). For "what possibly occurs for every one (*li kulli wāḥidin*) possibly does not occur for any one, and what possibly occurs for some, possibly does not occur for that some" ([34], 175.1–3).

If we take into account this quotation, according to him,

A (bilateral possible) = **E** (bilateral possible)
and **I** (bilateral possible) = **O** (bilateral possible).

Consequently, **E** (bilateral possible) should be rendered as **E**$\Diamond$ $\wedge$ **A**$\Diamond$ (= every B is possibly not A and every B is possibly A); likewise, **O** (bilateral possible) would be equivalent to **I** (bilateral possible) and consequently would be expressed as "Some A's are $\Diamond\sim$B $\wedge$ $\Diamond$B."

So according to him, we would have the following equivalences:

A (bilateral possible) = **E** (bilateral possible) = **E**$\Diamond$ $\wedge$ **A**$\Diamond$ (= Every B is possibly not A and every B is possibly A).

I (bilateral possible) = **O** (bilateral possible) = Some A's are $\Diamond\sim$B $\wedge$ $\Diamond$B.

However, this opinion raises problems, for one could ask: is this the way one should express **E** (bilateral possible) and **O** (bilateral possible)? If we consider the text, we note that Avicenna expresses these propositions in a way that makes the negation put on *each part* of the modal predicate *separately*, i.e., on $\Diamond$B on the one hand and $\Diamond\sim$B on the other hand. But this cannot be so, for the negation should put on the whole conjunction "$\Diamond$B $\wedge$ $\Diamond\sim$B", which is, *as a whole*, the predicate of such a proposition. So the negation in **E** should put on the entire conjunction, because what it says is the following: "Every A is not (possibly B and possibly not B)," given that what the narrow-possible means is that both sides of possibility (the affirmative and the negative sides) should be grouped, not separated, while the formalization above separates them.

The same could be said about **O** (bilateral possible), which ***should not*** be expressed as follows:

- **O** (bilateral possible): Some A's are $\Diamond \sim$B's $\wedge$ $\sim\Box\sim$B's (Some A's are possibly not B and not necessarily not B).

Rather, it shoud be expressed as follows:

- **O** (bilateral possible): Some A's are not B (possibly but not necessarily) = Some A's are not (possibly B and possibly not B).

We could also arrive at the same result just by negating **A** (bilateral possible) and **I** (bilateral possible), which are expressible in a clear way. So since **A** (bilateral possible) is expressed by "Every S is $\Diamond$P and $\Diamond\sim$P," i.e., when formalized "$(\forall x)[Sx \supset (\Diamond Px \wedge \Diamond\sim Px)]$", its negation (= **O**) is "$\sim(\forall x)[Sx \supset (\Diamond P \wedge \Diamond\sim P)] = (\exists x)[Sx \wedge \sim(\Diamond Px \wedge \Diamond\sim Px)] = (\exists x)[Sx \wedge (\sim\Diamond Px \vee \sim\Diamond\sim Px)] = (\exists x)[Sx \wedge (\Box\sim Px \vee \Box Px)]$" = *Some S are either necessarily not P or necessarily P*. This **O**, which is the contradictory of **A** (bilateral possible), does not really contain the possibility operator; rather it contains a disjunction between two necessary propositions.

The same can be said about **I** (bilateral possible) and its negation (= **E**), which can be expressed as follows: "Some S are $\Diamond P \wedge \Diamond\sim P$," i.e., $(\exists x)[Sx \wedge (\Diamond Px \wedge \Diamond\sim Px)]$. Its negation is then: "$\sim(\exists x)[Sx \wedge (\Diamond Px \wedge \Diamond\sim Px)] = (\forall x)[Sx \supset \sim(\Diamond Px \wedge \Diamond\sim Px)] = (\forall x)[Sx \supset (\sim\Diamond Px \vee \sim\Diamond\sim Px)] = (\forall x)[Sx \supset (\Box\sim Px \vee \Box Px)]$" = *Every S is either necessarily not P or necessarily P*. Here too, the negation of **I** (bilateral possible), which is an **E** proposition, contains a disjunction of two necessary propositions.

As a result, all four propositions are distinct from each other, for they are and should be expressed, respectively, as follows:

A (bilateral possible) = Every S is $\Diamond P \wedge \Diamond\sim P$= every S is $\Diamond\sim$P and every S is $\Diamond$P = **E**$\Diamond$ $\wedge$ **A**$\Diamond$.

E (bilateral possible) = Every S is $\Box\sim P \vee \Box P$.

I (bilateral possible) = Some S are $\Diamond P \wedge \Diamond\sim P$.

O (bilateral possible) = Some S are $\Box\sim P \vee \Box P$ = Some S are $\Box\sim$P $\vee$ Some S are $\Box$Px = **O**$\Box$ $\vee$ **I**$\Box$.

As we can see, thus expressed, all four propositions are different from each other. So we cannot say, as Avicenna claims, that **A** (bilateral possible) and **E** (bilateral possible) are equivalent, nor that **I** (bilateral possible) and **O** (bilateral possible) are equivalent. The calculus shows that they are not equivalent.[3]

[3]See S. Chatti ([55], 45–71), for a full analysis of these modal propositions.

This being so, what Avicenna says about the moods containing negative bilateral propositions, if there are any such moods, should be revised accordingly by taking into account the right formulations of these negative propositions.

These equivalences and their contradictories are what *Avicenna claims*, but as we just showed, the *couples of equivalences for internal modalities are wrong as well as their contradictories*. However, we have to take into account what *he himself thinks* in determining whether the modalities in his syllogistic are internal or external. For it is only by considering how *he himself interprets* the contradictories of the bilateral modalities that we can know how he interprets these modalities, since *only in that case*, the internal modalities and the external modalities are differently expressed as well as their contradictories.

As to the tables he provides in *al- 'Ibāra,* which contain the bilateral possible too, they externalare or *de dicto*, as they are in Aristotle's *De Interpretatione*.

Let us now turn to Averroes' analysis of the modal propositions.

4.1.3 The Modal Propositions in Averroes' Frame

As to Averroes, he expresses the modal propositions as follows:

- Every A is B by necessity (*bi-ḍ-ḍarūra*).
- No A is B by necessity.
- Some As are Bs by necessity.
- Some As are not Bs by necessity.
- Every A is B by possibility (*bi-l- 'imkān*).
- No A is B by possibility.
- Some As are Bs by possibility.
- Some As are not Bs by possibility ([26], 147–148).

He seems to use internal formulations, although the modal words ("by possibility" and "by necessity") are always in the end of the propositions, i.e., neither in the beginning, nor in front of the predicate, for he says:

> "The modality is comparable to the copula in the assertoric propositions… It is related to the predicate as the negation is related to the predicate in the assertorics, for instance, "this is possible not to be in any of that" and "this is not possible to be in some of that"" ([26], 149)

This confirms what we said above to the effect that when the modality is at the end of the proposition, it is internal rather than external, for it does not put on the whole sentence, but is related to the predicate.

In *Talkhīs al 'Ibāra* (his commentory of *De Interpretatione*), he presents the usual equivalences in tables which are almost identical to Aristotle's ones ([26], 120) except that, like al-Fārābī and Avicenna, he rules out the word "admissible" used by Aristotle.

The table he first provides is the following:

"Possible to be / Not possible to be
Not necessary to be / Necessary not to be
Not Impossible to be / Impossible to be

Possible not to be / Not possible not to be
Not necessary not to be / Necessary to be
Not Impossible not to be / Impossible not to be" ([26], 120)

But this table is incorrect like Aristotle's first table for "not necessary to be" does not follow from "possible to be," when possibility has its unilateral affirmative meaning. These errors are corrected one page later where he says:

"What is implied (*lāzim*) by the phrase (*qawl*) 'possible to be' is the phrase 'not necessary not to be', which is the contradictory of 'necessary not to be', itself implied by 'not possible to be', and not the phrase 'not necessary to be'; as to what is implied by 'possible not to be', it is the phrase 'not necessary to be', and not the phrase 'not necessary not to be' as we have supposed at first sight" ([26], 121).

These corrections apply to the usual unilateral meaning of possibility, which does not admit degrees. For only in that case, "possible to be" is equivalent to "not necessary not to be," for instance.

However, when distinguishing, in his *Talkhīs al-Qiyās*, between the different kinds of possibility, he says that the possible may have the three following meanings:

- The possible as likely, that is, as frequent (*'alā al-akthar*) or as Aristotle ([24], I, 13, 32b4–32b13) says as what "happens for the most part." For instance, the possibility for a man to "turn grey" when old ([26], 189).
- The possible as unlikely, that is, as seldom (*'alā al-aqall*) which is opposed to the above.
- The bilateral possible (= equally possible, for it can equally happen or not happen), which is usually formalized as $\Diamond\alpha \wedge \Diamond\sim\alpha$.

For this last kind, the affirmative bilateral possible is equivalent to its negation for: $\Diamond\alpha \wedge \Diamond\sim\alpha \equiv \Diamond\sim\alpha \wedge \Diamond\alpha$.

But this equivalence holds only when the possibility is *de dicto* or external, as we saw above when analyzing the modal quantified propositions in Avicenna's frame.

The two other kinds are to be found in Aristotle's text which Averroes develops by providing more details. For instance, he holds the following equivalences:

- Likely α implies Unlikely $\sim\alpha$, given that if "it is likely for men to turn grey when old," "it is unlikely for them not to turn grey when old."
- Likely $\sim\alpha$ implies Unlikely α.
- Unlikely α implies Likely $\sim\alpha$.
- Unlikely $\sim\alpha$ implies Likely α ([26], 189).

From which we can deduce that

- Unlikely $\alpha \equiv$ Likely $\sim\alpha$.

And

- Likely $\alpha \equiv$ Unlikely $\sim\alpha$.

Consequently,

- $\sim$Likely $\alpha \equiv \sim$Unlikely $\sim\alpha$.
- $\sim$Unlikely $\alpha \equiv \sim$Likely $\sim\alpha$.

But this makes his account different from the usual account of possibility and its negation. For usually, $\sim\Diamond\alpha = \Box\sim\alpha$ and $\Diamond\sim\alpha = \sim\Box\alpha$. But "not unlikely not" is not equivalent to "necessarily not." So the negation does not produce the same propositions in both cases, which shows that its effect depends also on the meaning of the modality.

This replacing of "possible" by "likely" or by "unlikely" changes the meaning of possibility which becomes closer to probability or improbability (in their imprecise meanings). The main difference between possibility and probability (even when it has its imprecise meaning) is that probability admits degrees, while possibility does not admit degrees. This major difference does not seem to have been noticed by Averroes.

4.2 The Modal Oppositions

4.2.1 The Modal Oppositions in al-Fārābī's Frame

What are the oppositional relations between the modal propositions? Let us start by Al-Fārābī's frame. Those between the singular (and indefinite) modal propositions are the following:

The two couples of contradictories are

- possibly α / not necessarily α (= possibly not α).
- possibly α / not possibly α (= necessarily not α).

The two contraries are

- necessarily α / necessarily not α.

The two subcontraries are

- possibly α / possibly not α.

The two couples of subalterns are

- necessarily α / possibly α.
- necessarily not α / possibly not α.

In addition

- "possibly $\alpha \wedge$ possibly not α" is contrary to both "necessarily α" and "necessarily not α."

All these relations are expressed in the tables presented by al-Fārābī in *Kitāb al-'Ibāra* and in *Sharḥ al-'Ibāra*, where he talks about the contrarieties between the bilateral possible, the necessary and the impossible.

As a consequence, al-Fārābī defends both the modal square of oppositions and the triangle of contrarieties, which makes his theory close to that of Aristotle except that the ambiguity related to the meaning of possibility is already present in his *Kitāb al-'Ibāra* (the correspondent of *De Interpretatione*), while Aristotle defends the square in *De Interpretatione* and the triangle of contrarieties in *Prior Analytics*.

In *al-Qawl fi al 'Ibāra* ([14]) (= *Kitāb al-'Ibāra* ([8])), al-Fārābī provides the contradictory and the contrary of each of the eight quantified propositions.

The propositions and their contradictories are the following:

- Every A is possibly B / Not every A is possibly B (= Some A are necessarily not B).
- Every A is possibly not B / Not every A is possibly not B (= Some A are necessarily B).
- Every A is necessarily B / Not every A is necessarily B (= Some A are possibly not B).
- Every A is necessarily not B / Not every A is necessarily not B (= Some A are possibly B).

While the contrary propositions are the following:

- Every A is possibly B / No A is possibly B (= Every A is necessarily not B).
- Every A is possibly not B / No A is possibly not B (= Every A is necessarily B).
- Every A is necessarily B / No A is necessarily B (= Every A is possibly not B).
- Every A is necessarily not B / No A is necessarily not B (= Every A is possibly B) ([14], 105–106, [8], 155–157)[4]

In addition, we may include the subcontrarieties between the two possible particulars because al-Fārābī compares the assertoric particulars with the indefinites which are subcontrary ([15], 122). From that, we can deduce that the particular possible propositions, whether negative or affirmative, are subcontraries too.

As to the possible singulars, they are also subcontrary, for they may be true together as is the case with the two propositions "Zayd is possibly learned" and "Zayd is possibly not learned." The subcontrariety between these two kinds of propositions is already claimed by al-Fārābī in his long commentary of *De Interpretatione* (*Sharh al-'Ibāra*) as witnessed by the following quotations: "*possible to be* and *possible not to be* are true of the same subject," because "what sees

[4]See Chatti [56] in [117] for a full analysis of these propositions and their relations.

now may also not see" ([12], 191). It is still defended in the short treatise *Kitāb al-'Ibāra*.

The subalternations between the necessary propositions and the possible ones (whether affirmative or negative) are also explicitly held by al-Fārābī since he defends the modal square as appears in the tables he holds in the long commentary on *De Interpretatione*.

We might thus say that the vertices of the modal octagon were already present in al-Fārābī's text, but not all the relations between these vertices are provided by al-Fārābī, since he provides only the contradictions and some contrarieties. Besides that, al-Fārābī does not draw the octagon or any other figure.

4.2.2 The Modal Oppositions in Avicenna's System

As to Avicenna's theory, it is more elaborate than al-Fārābī's one, for he admits more propositions and more oppositional relations between them, although, unlike al-Fārābī, he does not include the assertoric inside the realm of modalities.

He presents several tables where he expresses these oppositions and the equivalences between the modal propositions.

In the first table, he presents the order of the necessary and its contradictory, the non-necessary. This table presented in ([35], 122) contains two sides, where the left side shows that "necessary α" is equivalent to "impossible not α" and to "not possible not α," and the right side shows that "not necessary α" is equivalent to "not impossible not α," itself equivalent to "possible not α." This table expresses then one of the duality laws, since necessary = not possible not, and not necessary = possible not.

The second table ([35], 122) expresses the orders of the general (i.e., the unilateral) possible and its contradictory the impossible. The equivalences are the following: necessary not α = impossible α = not possible α in the left, and not necessary not α = not impossible α = possible α in the right. This table thus expresses the other duality law, since possible = not necessary not, and not possible = necessary not.

Then he presents the order of the narrow-possible (i.e., the bilateral possible) about which he says that "it does not follow anything else but what is from its same nature" ([35], 122). Consequently, we have the following equivalences:

- narrow-possible to be = narrow-possible not to be = $\Diamond\alpha \wedge \Diamond{\sim}\alpha$,

while

- not narrow-possible to be = not narrow-possible not to be = ${\sim}(\Diamond\alpha \wedge \Diamond{\sim}\alpha) = {\sim}\Diamond\alpha \vee {\sim}\Diamond{\sim}\alpha = \Box{\sim}\alpha \vee \Box\alpha$.

The third table is the most interesting one since it is not in Aristotle's *De Interpretatione*, nor in al-Fārābī's *Kitāb al-'Ibāra*. This table is the following:

"(A) The order of the Necessary and what goes with it (*wa mā ma'ahā*)

(B) As to the order of the Non necessary, nothing follows from it except its converse.

(C) As to the order of Necessary not to be, what follows from it is:

Not Impossible to be	Not impossible not to be
General possible to be	General possible not to be
Not narrow-possible to be	Not narrow-possible to be
Not narrow-possible not to be	Not narrow-possible not to be" ([35], 122–123).

It presents the order of the impossible, i.e., "Necessary not to be" and what follows from it, but *without* being equivalent to it, and presumably also from "Necessary to be."

This table is somewhat ambiguous, for one could think, given the sentence that immediately precedes it (= C) that the whole table follows from "Necessary not to be." But this opinion is untenable, for the left side does not follow from the impossible; rather, it has to follow from the necessary. So the only solution is to relate the left side of the table to (A) and the right side to the impossible expressed by (C).

In the right side, we have the following correct implications:

1. "necessary not α" implies "not not possible not α," for $\sim\sim\Diamond\sim\alpha = \sim\Box\sim\sim\alpha = \sim\Box\alpha$.
2. "necessary not α" implies "possible not α."
3. "necessary not α" implies also "$\sim(\Diamond\alpha \wedge \Diamond\sim\alpha)$" = "$\sim\Diamond\alpha \vee \sim\Diamond\sim\alpha$" = "$\Box\sim\alpha \vee \Box\alpha$."

Since "not narrow-possible not to be" is equivalent to "not narrow-possible to be," they are both implied by $\Box\sim\alpha$.

In the left side, although the text is not very clear, we have to say that the propositions are implied by $\Box\alpha$, because it is the only acceptable solution.

This left side shows that $\Box\alpha$ implies $\Diamond\alpha$, and also $\Box\sim\alpha \vee \Box\alpha$, i.e., "not narrow-possible to be" = "not narrow-possible not to be."

Finally, he considers the fourth table, which contains the propositions *implied* by "narrow-possible to be" (but not equivalent by it). This table is the following:

"Not necessary to be	Not necessary not to be
Not impossible to be	Not impossible not to be
General possible to be	General possible not to be" ([35], 123)

All the implications are valid, for in the left side, $\sim\Box\alpha$ (= $\Diamond\sim\alpha$) follows from $\Diamond\alpha \wedge \Diamond\sim\alpha$. The same may be said about the second line, since "not impossible α" (= $\sim\sim\Diamond\alpha = \Diamond\alpha$) is also implied by $\Diamond\alpha \wedge \Diamond\sim\alpha$, and about the third line which just expresses $\Diamond\alpha$.

As to the right side, the first line, i.e., "not necessary not α" (= $\sim\Box\sim\alpha = \Diamond\alpha$), is also implied by $\Diamond\alpha \wedge \Diamond\sim\alpha$. The second line, i.e., $\sim\ \sim\Diamond\ \sim\alpha = \Diamond\sim\alpha$, is implied by $\Diamond\alpha \wedge \Diamond\ \sim\alpha$, while in the third line, $\Diamond\ \sim\ \alpha$ is implied by $\Diamond\alpha \wedge \Diamond\ \sim\ \alpha$.[5]

Now what about the contradictions between these modal propositions?

The contradictions between the general (unilateral) possible and necessary propositions are expressed as follows: **□A / ◇O; □E / ◇I; □O / ◇A; □I / ◇E.**

The bilateral possibility propositions and their contradictories are the following: Bilateral possibility **A** (= bilateral possibility **O**) and its contradictory:

◇A ∧ **◇O / □A** ∨ **□O.**

Bilateral possibility **E** (= bilateral possibility **I**) and its contradictory:

◇E ∧ **◇I / □E** ∨ **□I.**

Now what about the other oppositions?

Let us start by the contrarieties. When the propositions are singular or indefinite, the contrary propositions are the following: $\Box\alpha\ /\ \Box\ \sim\ \alpha\ /\ \Diamond\alpha \wedge \Diamond\ \sim\ \alpha$.

This gives rise to the triangle of contrarieties.

When it comes to the quantified propositions, the contrarieties explicitly provided by Avicenna are the following: **□A / □O / ◇A** ∧ **◇O; □E / □I / ◇E** ∧ **◇I.**

But one can deduce many other contrarieties by using the very relations held by Avicenna.[6]

The subcontrariety between the singular and indefinite propositions held by Avicenna is the following: $\Diamond\alpha\ /\ \Diamond\ \sim\ \alpha$.

Those between the quantified propositions are the following: **◇A / ◇O** and **◇E / ◇I**. But here too, one may add many other subcontrarieties by deduction from the very oppositional relations held in Avicenna's system (see the article cited in the preceding note).

As to the subalternations, they are the following when the propositions are singular or indefinite: $\Box\alpha \rightarrow \Box\alpha \vee \Box \sim \alpha$; $\Box \sim \alpha \rightarrow \Box\alpha \vee \Box \sim \alpha$; $\Box\alpha \rightarrow \Diamond\alpha$; $\Box \sim \alpha \rightarrow \Diamond \sim \alpha$; $\Diamond\alpha \wedge \Diamond \sim \alpha \rightarrow \Diamond\alpha$; $\Diamond\alpha \wedge \Diamond \sim \alpha \rightarrow \Diamond \sim \alpha$.

When the propositions are quantified, the subalternations explicitly held are the following: **□A** → **◇A; □O** → **◇O; □A** → **□A** ∨ **□O; □O** → **□A** ∨ **□O; ◇A** ∧ **◇O** → **◇A; ◇A** ∧ **◇O** → **◇O; □E** → **◇E; □E** → **□E** ∨ **□I; □I** → **◇I; □I** → **□E** ∨ **□I; ◇E** ∧ **◇I** → **◇E; ◇E** ∧ **◇I** → **◇I** (See [53], Sect. 4).

But he also holds explicitly the subalternations between the universal and the particular propositions as we saw in Sect. 3.2.2. This gives us the following further set of subalternations: **□E** → **□O; □A** → **□I; ◇A** → **◇I** and **◇E** → **◇O** and also these other ones: **□A** → **◇I** and **□E** → **◇O**, given that necessary implies general possible and universals imply particulars. The remaining subalternations may also be deduced as shown in ([53], Sect. 4).

---

[5]See ([53], 339–340), for a full analysis of these tables and these relations.

[6]See ([53], 332–353).

This gives rise to a dodecagon when one groups all the quantified propositions and states their mutual relations.

4.2.3 The Modal Oppositions in Averroes' Frame

As we saw above, Averroes defends the modal square of oppositions since he presents the corrected tables expressing the orders of the necessary, the possible and the impossible. He thus admits the duality laws expressed by these tables. He also recognizes the bilateral possible which is defined as what is equally possible, thus opposed both to necessity and impossibility. So he admits also the triangle of contrarieties in *Talkhīs al-'Ibāra*, for he says that the impossible is the contrary of the necessary ("*inna al-mumtani' huwa ḍidd al-wājib al-wujūd*") ([26], 121.9).

The bilateral possible is applied to the human activities like walking, while the unilateral possible is applied to the potential behavior of the "irrational" things, like the fire, of which one can say that it possibly heats ([26], 124. 1–2). This last kind of possibility is implied by the necessary, unlike the bilateral possible which is contrary to it.

But when he analyzes the possibility in *Talkhīs al-Qiyās*, he says that there are three kinds of possibility: the bilateral possible, the possible as likely (*'alā al-akthar* = most of the time), and the possible as unlikely (*'alā al-aqall* = at rare times).

Unfortunately, when the possibility is interpreted as likeliness, the duality laws do not hold any more, given that the possible as "likely" is not equivalent to "not necessary not" because "not likely" is not equivalent to "necessary not." When interpreted as likeliness or as unlikeliness, the possibility is closer to the imprecise probabilities than it is to the real possibility, because it admits degrees, while the real possibility does not admit degrees. This makes the oppositions involving the possible as likely and unlikely unclear.

Despite these confused additions, however, we may say that he admits the above contradictions between the modal propositions, for one can generalize the relations of the modal square to the quantified modal propositions and obtain the same relations, that is, the contradictions between **A** necessary and **O** possible, **E** necessary and **I** possible, **I** necessary and **E** possible, **O** necessary and **A** possible, provided that the possibility is interpreted as unilateral.

But Averroes does not provide a detailed analysis of the contrarieties and the subcontrarieties between the modal quantified propositions. He just says that the possible and the necessary are contraries, which means that he privileges the bilateral meaning of possibility at least in his *Talkhīs al-Qiyās*. We will enter into more details when analyzing the modal syllogistic moods and their proofs, which rely on the contradictions and the conversions between the modal propositions.

4.3 The Modal Syllogistic

In this section, we will not analyze al-Fārābī's theory, because it is not available. Unfortunately, the sections of his *Kitāb al-Qiyās* where the modal syllogistic would have been presented and developed are now lost as told by Lameer, for instance.

However, we could find some ideas and comments in what Averroes says about him in his different treatises. Averroes talks about al-Fārābī several times both in his *Talkhīs al-Qiyās* (the middle commentary of *Prior Analytics*) and in his other essays published under the title *al-masā'il fī al-mantiq wa al-'ilm aṭ-ṭabī'ī wa aṭ-ṭib* ([30]), which was first published by Jameleddine al-Alawi under the title *Maqālāt fī al-mantiq wa al-'ilm aṭ-ṭabī'ī* ([29]). He is mainly interested by al-Fārābī's opinion about the principle that the Latin logicians will call afterward the *dictum de omni et de nullo*, and also by the kind of modal moods admitted by al-Fārābī in all figures, especially the first figure, since this principle is the basis of the first figure moods. He analyzes the theories presented by several commentators, Greek and Arabic, and distinguishes between two distinct trends: those who reject Aristotle's opinion and those who endorse it but try to interpret it in a way that validates the moods which Aristotle considers as productive. According to him, Al-Fārābī belongs to the second trend or group together with Alexander of Aphrodisias, while in the first group we find Eudemus, Theophrastus and Themistius ([29], 123, in [69], 59).

According to Averroes, the mixed moods of the first figure in Aristotle's theory are governed by the following rule which determines the kind of conclusion they should lead to: "the modality of the conclusion of the moods containing one necessary premise and one assertoric premise is the modality of the major premise" ([69], 58). Thus in his Opuscule n° 5 entitled "Discourse on the modalities of the conclusions in the composed (*murakkaba*) syllogistic moods and on the meaning of the *dictum de omni*" ([29], 123–138, [30], 93–120), where he considers that Aristotle's opinion is "the true doctrine" and that "no other doctrine is true" ([29], 123), Averroes says that in the first figure, when the premises are necessary and assertoric, the conclusion follows the major premise, i.e., "if the major premise is necessary, the conclusion will be necessary and if the major premise is assertoric, the conclusion will be assertoric (*wujūdī*, in Averroes' wording, which means as he says "actual", ("*mawjūda bi-l-fi'l*")) ([29], 124). So Aristotle admits the following combinations in the first figure: **LXL** and **XLX** (where **L**: necessary, and **X**: assertoric (*wujūdī* = non-modalized, or "actual")).

When the premises are either "necessary and possible" or "assertoric and possible," then we have the following cases: "when the major is possible, then the conclusion, according to him, must be possible, whether the minor is necessary or assertoric, and whether the possible is negative or affirmative. And he says that this kind of moods is perfect, that is, it is clearly productive by means of the dictum de omni that we find in these [moods] (*al-mawjūdu fīhā*)" ([29], 124–125). So according to Averroes, Aristotle admits the following (productive) moods in the first figure: **MXM** (where **M**: unilateral possible) [or maybe **QXQ**? where **Q**: bilateral possible), (Averroes does not specify what kind of possibility is meant, at this stage)], and also **MLM** (or maybe **QLQ**?). According to Averroes, Aristotle considers these moods as perfect. Abdelali Elamrani-Jamal summarizes these combinations as follows:

"(a1) maj. Poss.
min.assert.
concl. Poss.
(a2) maj. Poss.
min. necess.
concl. Poss." ([69], 58).

If "the possible premise is the minor, and the major is affirmative and assertoric (*mūjiba wujūdīyya*)," then "the conclusion is possible" ([29], 125), but "this combination is not perfect (*ghair tāmm*), and he proved it by *reductio ad absurdum*" ([29], 125). But when "the assertoric major is negative, [and the minor possible], then the conclusion is sometimes a negative possible and other times a negative necessary" ([29], 125). This gives the following combinations: with affirmative **X**: **XMM** (or **XQQ**?), and with negative **X**: **XMM** or **XML** (**XQQ** or **XQL**?). This also is "not perfect" ([29], 125). This raises a problem, since in the second combination, the conclusion is said to be necessary, while none of the premises is necessary, so that the conclusion should be assertoric if it follows the major, or possible if it follows the least premise. This objection has been made by some "ancients (*qudamā*)" ([29], 126.1) according to Averroes, who thus reject this particular combination. If "the minor is possible, and the major is necessary, then if the major is affirmative, then the conclusion… is possible. And if [the major] is negative, then the conclusion….is either negative possible or negative necessary (*sāliba ḍarūrīyya*)" ([29], 126). So here we have the following combinations: with affirmative **L**, **LMM** (**LQQ**?) and with negative **L**, **LMM** or **LML** (**LQQ** or **LQL**?).

These are summarized as follows by Abdelali Elamrani-Jamal:

"(b1) maj. Affirm. Assert
Min. poss.
Concl. Affirm. Poss
(b2) maj. neg. assert.
min. poss.
concl. neg. poss., or neg. necess.
(b3) maj. affirm. necess.
min. poss.
concl. affirm. poss.
(b4) maj. neg. necess.
min. poss.
concl. neg. poss. or neg. assert[7]." ([69], 59)

[7]Here what should be written is "neg. necess" instead of "neg. assert" (see [29], 126, and [30], 100).

Now what does al-Fārābī say about these combinations? According to Averroes, while Eudemus, Themistius, and Theophrastus on the basis of (b1) and (b3) (in Jamel's classification), say that the conclusion should follow "the least premise" ([69], 59) and thus reject Aristotle's opinion to the effect that it follows the major, al-Fārābī and Alexander do not reject Aristotle's opinion and search for a plausible interpretation that could validate all these combinations in a "coherent" ([69], 59) way. But their opinions are not exactly the same, for Alexander considers that the *dictum de omni et de nullo*, governing the moods of the first figure should assert that "A is attributed affirmatively or negatively, according to some modality, to all what is qualified as B and to which B is attributed affirmatively and actually (en acte)" ([69], 60, my translation), where "B is the middle term," leading to the following opinion considered as "coherent" with Aristotle's claims:

> "when the assertoric premise is mixed with a necessary premise, then the conclusion follows the major; when the major possible is mixed with a minor assertoric or necessary, then the modality of the conclusion follows that of the major.... In all these cases, the *dictum de omni et de nullo* as stated above is applied. If the major is assertoric or necessary and the minor possible, then the *dictum* is not applied and the syllogism is not perfect" ([69], 60–61, my translation).

But al-Fārābī "objects" to this interpretation of the principle that if it were to be understood that way, then "the syllogisms whose two premises are possible would not be perfect and their conclusions would not be obvious by themselves, unlike what is said in Aristotle's text (Cf. *APr.*, I, 14, 32b38–40)" ([69], 61). This is why he replaces the Alexandrian formulation of the principle by another one, which is considered as more in accordance with Aristotle's claims, and which Averroes reports as follows: "A will be attributed affirmatively or negatively, whether [the premise] is assertoric or possible or necessary, to all what is qualified as B in an assertoric, or a possible or a necessary (way)" ([29], 146).

As to Averroes, he criticizes this interpretation of the principle, which according to him does not account for the moods where the minor is assertoric or necessary. In these moods, the minor cannot be possible, since "A is predicated necessarily or actually to all what is actually B" ([29], 146). We find also this criticism in *Talkhīṣ Kitāb al-Qiyās*, where Averroes says that according to al-Fārābī, what is generally claimed in the *dictum* in "all kinds of premises" is "'Everything that is B is A by possibility' is true of what is *potentially* or *actually* B" ([26], 182.8–9, my emphasis). But this does not apply to the moods where the minor is assertoric or necessary, since in an assertoric or necessary premise, the predicate A does not apply to what is *potentially* B; it only applies to what is *actually* B. And what applies to what is actually so-and-so does not necessarily apply to what is potentially so-and-so. If so, the *dictum* applies only to some matters, not to all of them, namely, those where the middle applies to what is "potentially or actually so-and-so." Given that in the moods **LXL** and **XLX**, the minor is assertoric or necessary, the middle in these moods would apply only to what is actually so-and-so, it would not apply to what is "potentially so-and-so." Take, for instance, the following *Barbara* **LXL**:

Every B is A by necessity (L)
Every C is B by actuality (X)
Therefore Every C is A by necessity (L)

By the *dictum de omni* which justifies it, the mood says that "A is (necessarily) true of what B is (actually) true of, namely of C (which is said to be actually B)." Now can we say that the *dictum* justifying this mood should be expressed as follows: "A is (necessarily) true of what B is (actually or potentially) true of, namely of C"? In other words, could the minor say "Every C is B (potentially or actually)"? According to Averroes, this cannot be admitted because if something is true of what is actually B, this something is not necessarily true of what is potentially B.

At page 196 of his *Talkhīṣ Kitāb al-Qiyās,* he gives examples showing that the *dictum* in this interpretation could not be applied to necessary and assertoric premises. Consider, for instance, the two sentences: "Every human is walking" and "every human is speaking," "these premises are true of what is actually human, they are not true of what is potentially human" ([26], 196), while if we consider the following sentences "Every moving is a body," this can be "true of what is actually moving (*mutaḥarrik bi al-fiʿl*) and of what is potentially moving (*mutaḥarrik bi al-quwwa*)" ([26], 196). So generally speaking, what is common to both cases is the fact of "being actually" so-and-so, since "potentially" cannot be present in all cases. This is why according to him, al-Fārābī is wrong when he says that the *dictum* should contain "actually or potentially" and can apply to the combinations where "the major is assertoric or necessary and the minor is possible" ([26], 196.14), because stating the necessary or assertoric major premise as "Every (potentially or actually) B is necessarily A" is bizarre and does not correspond to what Aristotle claimed. , this combination (major assertoric, or necessary and minor possible) [i.e., (b1) and (b3) in Elamrani-Jamal's classification above] is "imperfect" in Aristotle's frame. This is why he says "there is no *dictum de omni* in this combination, because the *dictum de omni* is something that always exists in every matter of a single combination, so al-Fārābī's opinion according to which the *dictum* applies to this combination does not make sense" ([26], 196.11). This means that since the *dictum* is a general principle, it should not depend on the different matters, which is what happens when one states it in the way al-Fārābī does.

It seems thus that, if what Averroes is saying is right, and if my own report is correct, al-Fārābī defends a theory which is not entirely faithful to Aristotle's claims, since some of the moods considered as imperfect by Aristotle, are considered by al-Fārābī as justified by the *dictum de omni,* while the obvious moods that are justified by the *dictum* in Aristotle's frame could hardly be justified by al-Fārābī's interpretation of that same *dictum.* This is criticized by Averroes, who finds this position inadequate and not faithful to Aristotle's authentic position and claims.

However, these fragments can hardly reveal the main features of al-Fārābī's modal syllogistic although they tell us that he adds some clarifications to the text and tries to make Aristotle's claims coherent. We cannot, on the basis of these remarks, say what are the characteristics of al-Fārābī's modal syllogistic, what moods he admits, how he defines the conversions, how he proves the moods, etc. We will thus consider that the present report is insufficient to account for al-Fārābī's modal syllogistic.

As to the remaining fragments of al-Fārābī's long commentary on *Prior Analytics*, which have been published by Daneshpazuh (*al-Mantiqiyāt lil-Fārābī*, vol 2, Iran, [12]), it does not really contain anything related to the modal syllogistic as such. Al-Fārābī is commenting on other sections of Aristotle's treatise and seems much more interested by the contrary premises than by the modal ones.

Let us now turn to Avicenna and Averroes who both do present a modal syllogistic. Their theories are very different though, for Averroes wants to be as faithful as possible to Aristotle while Avicenna is not so much faithful to the Aristotelian opinions. These differences appear from the very beginning when both authors present their rules of conversion, which are stated differently in the two frames. We will thus focus on these differences and their motivations while presenting the two modal syllogistics. Let us start by Avicenna's theory, then we will present Averroes' one.

4.3.1 Avicenna's Modal Syllogistic

In modal logic, the rules used in the reduction of the syllogisms are not so different from those used in categorical syllogistic. These rules are the conversion rules which concern the universal negative propositions, the universal affirmatives, and the particular affirmative ones, whether modal or absolute, since some modal syllogistic moods contain absolute propositions together with necessary or possible ones. The absolute propositions used in categorical syllogistic are, as we have seen above, the propositions containing the condition "as long as they are S," S being the subject of the proposition. But sometimes, as we will see in the sequel, Avicenna uses other kinds of absolutes, for instance, those containing the condition "as long as it exists." Both kinds of absolute propositions are convertible.

Now **E**-conversion may concern the necessary propositions and the possible ones, together with the absolute propositions that we already analyzed above. According to Avicenna, as is usually admitted, the conversion holds for necessary **E** and leads to necessary **E** too. For if "Necessarily no C is B" is true, "Necessarily no B is C" is true too ([34], 95). This is usually proved as follows: "otherwise, its contradictory 'Possibly some B is C' will hold, hence 'Possibly some C is B' will also be true" ([34], 95.3), which contradicts the first proposition. However, this proof presupposes and uses the conversion of **I** possible which, according to Avicenna, must also be proved ([34], 95.6–7).

Given this difficulty in the preceding proof, Avicenna provides another one which he considers to be more acceptable. This second proof is the following:

> "But what other people say is better, that is, if "it is possible that some B are C", then its supposition is not impossible (*kāna faraḍuhu ghaira muḥālin*). It may probably be false. If it is false and not impossible, it does not imply the impossible, for what follows from the possible is possible, given that the impossible is never the case. So what is the case only when something impossible follows from it is not the case at all. For how could it be the case if it is conjoined with what could never be the case? A falsity which is not impossible does not imply the impossible. So if we suppose that "Some B is C actually", then "Some C is B actually", in which case "Some C is B" will be false but not impossible. But you said: "Necessarily no C is B", so how could "Some C is B" be not impossible, when it is in fact impossible? But it follows from "Some B is C". So "Some B is C" is false and impossible." ([34], 95.11–96.5, part of the translation, Wilfrid Hodges, [94])

This proof is made by *reductio ad absurdum* and relies on some principles such as "what follows from the possible is possible" and "what is not impossible does not imply the impossible." From the former, we may deduce that the impossible follows the impossible, for if "$\Diamond\alpha \rightarrow \Diamond\beta$" then "$\sim\Diamond\beta \rightarrow \sim\Diamond\alpha$" by contraposition. What is wanted is to prove that "Necessarily no B is C" follows from "Necessarily no C is B." If this entailment does not hold, this means that the contradictory of the conclusion, that is, "Possibly some B is C" is true. This contradictory is a possibility proposition. Being possible, it cannot be impossible and for this reason, it cannot imply anything impossible, according to the above principle. Since what is implied is not impossible, it can be the case. If we suppose that it is the case, then "Some B is C" is the case; therefore "Some C is B" is the case. Hence, it is not impossible. But we said "Necessarily no C is B," which was assumed to be true. Given that "Necessarily no C is B" is equivalent to "it is not possible that some C is B," if the former is true, the latter must be true too. So "Some C is B" *is* impossible. Consequently, what follows from it, that is, "Some B is C" is impossible too by the principles above. But "it is impossible that some B is C" is equivalent to "Necessarily no B is C," which was the proposition to be deduced. Therefore, the conversion for **E** necessary holds.

This proof uses **I**-conversion for absolute propositions which was already proved, together with some principles held true by Avicenna. These principles are plausible and already held by Aristotle in *Prior Analytics* (I, 14, 34a25–34a33), but they are validfor external modalities. So the whole proof is valid for external modalities. Would it be valid for internal ones, that is, for modalities putting on the predicate? Maybe not, for the principles used do not involve internal modalities. So, if the syllogistic uses internal modalities rather than external ones, could one use **E**-conversion, which is valid for external modalities?

As a matter of fact, **E**-conversion for external modalities is formalized thus:

1. $\Box(x)(Sx \supset \sim Px) \rightarrow \Box(x)(Px \supset \sim Sx)$.

While if one uses internal modalities, one has the following:

2. $(x)(Sx \supset \Box \sim Px) \rightarrow (x)(Px \supset \Box \sim Sx)$.

(1) is valid given that its categorical counterpart is valid, which makes it possible to say that $\Box[(x)(Sx \supset \sim Px) \rightarrow (x)(Px \supset \sim Sx)]$ is true and to apply Axiom 1 (or K) of modal logic according to which "$\Box(\phi \rightarrow \psi) \rightarrow (\Box\phi \rightarrow \Box\psi)$" [$\phi$: (x)

$(Sx \supset \sim Px)$, ψ: $(x)(Px \supset \sim Sx)]$. This application of axiom K makes **E**-conversion valid when the modalities are external.

But what about (2)? It would be valid if no case of falsity is possible.

Let us suppose the case of falsity by assuming two entities x_1 and x_2 in the universe and calculating the result with the following lines of the truth table:

$$[(Sx_1 \supset \Box{\sim}Px_1) \rightarrow (Px_1 \supset \Box{\sim}Sx_1)] \wedge [(Sx_2 \supset \Box{\sim}Px_2) \rightarrow (Px_2 \supset \Box{\sim}Sx_2)]$$

(1) 1 0 0 0 1 1 1 0 0 0 1 **0** $\underline{\mathbf{0}}$ 1 0 0 1 **0** 1 0 0 0 $\underline{\mathbf{1}}$

(2) 0 1 0 0 1 ? 1 ? ? 1 0 ? 0 1 0 0 1 ? 1 ? ? 1 0

In line (1), since the antecedent "$(x)(Sx \supset \Box{\sim}Px)$" is false, the whole left side is true. But in the right side the case of falsity leads to a contradiction, given that Sx_2 is both true and false. So the whole formula cannot be false. Still, it is not always true for it can be *undetermined*, since the necessity operator may be undetermined when the proposition on which it puts is true, in which case the whole formula is undetermined too, as shown in line (2). As a result, we can say that unlike the formula with external modalities, this one is not always true, even if it can never be false.

However, if the antecedent is true, the consequent will be true too. Consequently, even if the formula corresponding to **E**-conversion with internal modalities is not a tautology, we may hold it and use it in modal syllogistic, given that in the proofs provided, the antecedents (the premises converted) are held true; consequently what can be deduced by conversion from these premises will be true too. So this rule can also be applied, given that it never leads to a false conclusion.

However, as we already noticed, Avicenna's formulations of **E**-conversion seem to be *de dicto* as he puts the modal operator in the beginning of the propositions (see for instance, [34], 95.1–2). This point has been stressed by Zia Movahed who says what follows:

"Ibn Sīnā and following him, Fakhr al-Rāzī, Naṣīr al-Dīn al-Ṭūsī and Kātibī (in *Ḥikamat al-'ayn*) maintain that the converse of: 'Necessarily no A is B' is 'Necessarily no B is A'. here again the modality of the proposition to be converted can be taken as *rede* or *de dicto*:

$$\Box(x)(Ax \rightarrow \neg Bx)$$
$$(x)(Ax \rightarrow \Box\neg Bx)$$

But none of them has as the converse the proposition:

$$(x)(Bx \rightarrow \Box\neg Ax)$$

And it is only the *de dicto* one which can be converted to:

$$\Box(x)(Bx \rightarrow \neg Ax)$$

So in this case we have no ambiguity. Ibn Sīnā in this case puts modality before the proposition (2) [= ([34], p. 95)]. But here there is a subtil point involved. In this *de dicto* proposition the converse is not really converse. It is contrapositive of the proposition." ([120], 10).

This comment calls for some remarks: first, if the conversion is, as assumed by Z. Movahed and as Avicenna's first formulation seems to show, valid only when the propositions are *de dicto*, then the whole modal syllogistic should use *de dicto* readings, since the main rule is valid only in that reading. Second, the fact that **E**-conversion is a kind of contraposition [since "$(p \supset \sim q) \equiv (q \supset \sim p)$" derives from "$(p \supset q) \equiv (\sim q \supset \sim p)$"], this is true not only in modal logic, but also in categorical logic, since **E**-conversion is also expressed as an equivalence between two conditionals, though without the modal operator. So there is nothing new in this fact, which would be specific to modal logic. Now if in *de re* readings, the predicate is permuted with the modal operator which would be part of it, then **E**-conversion would lead from a necessary proposition to an absolute one, since we would have the following "implication": "$(x)(Ax \rightarrow \Box\neg Bx) \rightarrow (x)(\Box Bx \rightarrow \sim Ax)$". But this is not what Avicenna says about **E**-conversion, which in his view leads from a necessary proposition to a necessary one.[8]

Now what about the other conversions? **A**-conversion for **A** necessary leads to **I** absolute and it (also) leads to **I** possible, according to Avicenna, for **A** necessary does not lead to **I** necessary. Avicenna expresses **A** necessary as follows: "Necessarily every C is B" (*bi-iḍṭirārin 'an yakūna kullu [C] [B]* ([34], 96.8). His opinion about this conversion is different from the usual Aristotelian one, according to which **A** necessary leads when converted to **I** necessary, when the subject and the predicate are permuted. On the contrary, Avicenna says that the particular proposition that follows **A** necessary must be a possibility one; it needs not be a necessity one. He gives the following example to illustrate this opinion: "Every writer is a man by necessity," but "Every man is a writer" and consequently "Some men are writers" are not necessary propositions ([34], 97.1–2), they are absolute and they may also be possible. This opinion is held in *al-Ishārāt* too, where Avicenna says what follows:

> "The universal necessary affirmative converts as a particular affirmative of the general absolute kind (*bi-mā min ḥukmi al-muṭlaq al 'āmm*), but it does not need to convert as a necessary proposition, for it is possible for the converse of the necessary to be possible" ([36], 335.4–8).

Avicenna illustrates this by the following example: one may say "Every laughing [thing] is a human, by necessity" but its converse would be rather "Some humans are laughing [at some times]" or else "Some humans are laughing, by possibility" ([36], 336), given that humans are not necessarily laughing. In this example, the first converse is a general absolute proposition, since the predicate "laughing" used as an example is the one used to illustrate the general absolute propositions in *al-Qiyās* as we saw in Sect. 3.2.2. This predicate is not essential to

[8]For an account of the conversions in some of Avicenna's followers, see [140].

humans for in the list of predicables; it is rather a proper but non-essential predicate, i.e., it is specific to humans but not necessary to them. Avicenna's opinion about the conversion of necessary affirmative propositions is thus the result of his analysis of the general absolute propositions. The second converse is a possibility proposition, which shows that the general absolute propositions are parallel to the possibility ones. So maybe as one reviewer suggests, **A**-conversion leads to a general absolute proposition containing "at some times" when the initial **A** proposition is a *dā'ima* (permanent) proposition (i.e., contains the condition "as long as it exists") and it leads to a possible proposition when the initial **A** proposition contains the necessity operator, given the parallelism between "*dā'im*" and "necessary" on the one hand and "general absolute" and "possible" on the other hand.

In *al-Najāt*, he criticizes Aristotle's opinion about the conversion of **A** necessary by giving the following example: from "necessarily (*bi-ḍ-ḍarūrati*) every writer is a human," one can deduce by conversion the following sentence "some humans are writers" not by necessity, i.e.,

> "not by the necessity that you want, but if [necessity] is still wanted (*bal in kāna wa lā budda*), it is by another necessity which holds for every possible, such as [saying that] 'some humans are writers as long as they are writers', but this is not what we mean by necessity (*lasna naqṣidu bi-ḍ-ḍarūrati mithla hādhā*)" ([32], 30.14–15)

Here too, he denies the fact that **A** necessary leads to **I** necessary by conversion, but he adds that if one really insists that **I** must be a necessary proposition, then this necessity will be related to the fact that it will have the structure "some B's are A's as long as they are A's," i.e., its necessity will be due to the use of the condition "as long as it is P" (P being the *predicate* of the sentence). This condition by itself introduces necessity in all kinds of propositions, because the proposition will be expressed as follows: "some B are [A if (or 'if and only if') they are A]," since in that case, the predicate whether expressed by "A if A" or by "A if and only if A" will be tautological in modern terms. But this, he says, is not what is meant by a necessary proposition in Aristotle's account, nor in his own one. This confirms the fact that the conversion of **A** necessary does not lead to a real necessary proposition, rather it leads to an absolute one, or to a possibility one.

As noted by one referee, "Avicenna says at *Mukhtaṣar* 118.2 that the converse of 'Every B is an A necessarily' can be 'Some A is a B absolutely' and it can be 'Some A is a B possibly'," and he adds "I think the natural interpretation is that if *ḍarūrī* is read as 'not possibly not' then the converse is 'possibly', whereas if it is read as 'perpetual' then the converse is 'absolutely'." This is a plausible interpretation of Avicenna's text, which is not clear in this respect. Thus the conversion of **A** necessary can lead to a general absolute **I** proposition or to a possible **I** proposition. If we use our names above, we would say that the universal affirmative **L** proposition converts to a particular affirmative **M** proposition, while the universal affirmative **Xp** proposition converts as a particular affirmative **Xga** proposition. Still, we will have to check Avicenna's own use of these different kinds of propositions in the different parts of his modal logic.

Now Avicenna says that absoluteness should be taken in its broadest sense in what follows: "and the truth is that this one converts [as an] absolute in the broadest sense, which is Some B is C without adding [any] condition. And it is proved by the two examples provided" ([34], 97.11–12). These examples are the following: "Some bodies are moving by necessity, while some others are moving not by necessity" and "Some bodies are black by necessity, that is, permanently," while "Some bodies are black, not by necessity [that is, not permanently]" ([34], 97.9). So from "Some black things are bodies by necessity" one may deduce "Some bodies are black" not by necessity, that is, by actuality.

The same can be said about the conversion of **I** necessary ([34], 97.5–10, [36], 336), for from "Some B are C by necessity," what follows is not "Some C is B by necessity," but rather "Some C is B" not necessarily (*lā bi-ḍ-ḍarūrati*) ([34], 97.7). As to necessary **O**, it does not convert as is the case with **O** general absolute.

The propositions of the kind "Every S is P (as long as it is S)" [which is absolute, since being S does not always durate the whole time of the existence of an individual, and for this reason "as long as S" and "at some times" are close as Avicenna claims, because "at some times" can be "precisely at the time while the individual is S," Ex: "Every runner is moving as long as they are running," here "the time when they are running" is just "some time in their lifes"] converts to "Some P are S (at some times while P), which is also absolute, but in a more restricted sense.

Now what about the conversion rules for possible propositions? According to Avicenna, possible **E** does not convert, for if "Possibly no human is a writer" is true, "Possibly no writer is a human" is false and "Possibly some writers are not human" is false too ([36], 338.2–5). We can also express the proposition by using internal modalities as follows: "Every human is possibly not a writer" does not lead to "Every writer is possibly not a human" or to "Some writers are possibly not human."

As to possible **A** and possible **I**, they do convert but if the possible is narrow (two-sided), the conversion leads to a general possible (one-sided) proposition. In both cases, the conversion leads to possible **I** ([36], 339.1–2). For instance, if "Every animal is possibly moving spontaneously (*bi al-irāda*) [and possibly not moving spontaneously]," the converse cannot be "Some spontaneously moving things are possibly animals and possibly not animals," given that "animal" is necessarily predicated of the subject "moving spontaneously" ([36], 339.7–8). So according to him, "the unilateral as well as the bilateral kinds of possibility convert as unilateral possibles, so that if 'Every C is B by possibility' or if 'Some C is B by possibility', then 'Some B is C, by unilateral possibility'; otherwise 'Not possibly Every B is C'" ([36], 339.11–13). So if one says "Every human is possibly a writer (and possibly not a writer as well)," then we can deduce "Some writers are possibly humans," which does not deny that they are actually or necessarily humans. In *al-Najāt*, he criticizes what he calls the usual (Aristotelian) opinion (*al-mashhūr*) according to which **A** possible converts to **I** possible where "the possibility is bilateral (*mumkina ḥaqīqīya*)" ([32], 31.8), so that "if 'Every B is A by possibility' then 'Some A are B, by the real (= bilateral) possibility'. Otherwise, 'No A is B by necessity', hence 'No B is A by necessity'; but this is absurd" ([32], 31.7–10). The absurdity is shown by the fact that "No B is A by necessity" is contrary to "Every B is A by possibility," which was the proposition to be converted. But Avicenna says that this proof is not correct because if the possibility of **I** is bilateral, then its contradictory is not "No A is B by necessity," as is assumed in the proof. Rather it "can be (*rubbamā kāna*) 'Every or Some A is B by necessity' as we said" ([32], 31.11). So "No A is B by necessity" is the contradictory of "Some A is B by (the unilateral) possibility," which means that the converse of "Every B is A by possibility" would rather be "Some A is B by the unilateral (= general) possibility." Therefore, according to him, the converse of **A** possible is **I** possible where the possibility is unilateral ([32], 31.13–14).

Note here that Avicenna interestingly says that the contradictory of "Some A is B by the bilateral kind of possibility" is "Necessarily *Every or Some* A is B (*bi-ḍ-ḍarūrati kull aw ba'ḍ [A] [B]*)," which means that this contradictory would be formalized by $\Box\mathbf{A} \vee \Box\mathbf{O}$. But we saw above (Sect. 4.1.2) that this contradictory is precisely the one corresponding to the *external* kind of possibility, the internal one being expressed differently in Avicenna's frame.

Now, to summarize what he says about the conversions, we can say that

A bilateral possible and **A** unilateral possible both lead to **I** unilateral possible.

I bilateral possible and **I** unilateral possible both lead to **I** unilateral possible.

Let us now turn to the modal syllogistic moods as they are presented and proved by Avicenna. Before considering the moods mixing between absolute propositions and necessary or possible ones or necessary propositions and possible ones, let us first consider the moods containing only necessary propositions, which Avicenna does not analyze in full but which have some specificities that are stressed by Avicenna in the end of Sect. 3, book 4.

According to Avicenna, these moods are comparable to the ones containing absolute propositions which we just presented. However, there are two differences between the two kinds of moods. The first one is that the moods with necessary propositions contain a *modal* operator *explicitly stated*, which is "necessarily" (while the absolute propositions do not contain explicit modal operators); the second difference is related to the proofs of the different moods, in particular to the proofs of *Baroco* (= "the fourth mood of the second figure" and *Bocardo* (= "the fifth mood of the third figure") ([34], 121.6; [32], 36.20) with necessary propositions, which, unlike the other moods, cannot be proved by conversion. According to him these two moods cannot either be proved by *reductio ad absurdum*, just because the contradictories of their **O** necessary conclusions are **A** general possible propositions, which when added to one of the necessary premises of *Baroco* or *Bocardo*, lead to moods mixing between necessary and general possible propositions. But these propositions have not been studied yet, so the conclusions of these moods cannot be known yet. This is why he claims what follows about an eventual proof by *reductio ad absurdum*:

"This is so because if the conclusion 'necessarily (*bi-ḍ-ḍarūrati*) not every C is A', either in the second figure or in the third one. Thus if we say: if this is not true, then its contradictory is true, and this contradictory can only be the following 'Not necessarily not every C is A', but this premise is not found [among the available premises] so that one could add one of the premises of the mood to it (*fa-lā tajidu hādhihi al-muqaddama bi-ḥaythu yumkinu an yuḍāfa ilayhā shay'un mimmā fī al-qiyās*); as to the proposition implied by it (*lāzimu dhālika*), namely 'It is possible that every C is A' (*yumkinu an yakūna kull [C][A]*'), this implied proposition is an affirmative general possible proposition. But you don't know how a syllogism with a general possible premise and a necessary one is composed (*yata'allafu*). Therefore there is no way to prove it by *reductio ad absurdum* (*lā sabīla ilā tabyīnihi bi-l-khalf*) before examining the moods mixing between possibility and necessity." ([34], 121.5–11).

This is why Avicenna says that these two moods containing necessary propositions should be proved by *ekthesis* (*bi al-iftirāḍ*). These proofs by *ekthesis* are provided in *al-Qiyās*.

In *al-Qiyās*, Avicenna presents the proof of *Baroco* and of *Bocardo*. *Baroco* with necessary premises is the following[9]:

"Necessarily not every C is B (*bi-ḍ-ḍarūrati laysa kull [C] [B]*)
Necessarily every A is B (*bi-ḍ-ḍarūrati kull [A] [B]*)

It follows: Necessarily not every C is A (*bi-ḍ-ḍarūrati laysa kull [C][A]*)" ([34], 121.13–14).

The proof by *ekthesis* runs as follows: "Let us determine the part which is C by necessity and is not B and call it D. Then if 'necessarily no D is B' and 'necessarily every A is B', then 'necessarily no D (which is some C) is A, therefore 'Some C is not A'" ([34], 121.14–17).

If we write it down, we have the following: Some C = every D and both are not B. So if "necessarily not every C is B," then

1. Necessarily no D is B (assumption)
and 2. Necessarily every A is B (major premise)
then 3. Necessarily no D is A (1, 2 by *Camestres*)
4. Necessarily Some C is D (assumption)
Therefore 5. Necessarily Some C is not A (3, 4, by *Ferio*)

This proof is parallel to the one provided for *Baroco* with absolute propositions. It uses the same moods and the same assumption and could be criticized for the same reasons (see Sect. 3.2.2, plus Sect. 3.2.1, and note 47 for the objection).

As to *Bocardo*, it is the following:

[9]Here, there is an error in the text, for what is written is "necessarily every B is A." Thus written, the letters B and A are not in their right places, given that *Baroco* is a mood of the *second* figure, i.e., the figure PP; so B is the middle term, and the premise should be rather "necessarily every A is B," as we have written.

"Every B is C by necessity
And necessarily not every B is A
It follows: necessarily not every C is A" ([34], 122.1–2)

Its proof is the following: "Let D be that some B which is not A, and is also some C," so that "every D = some B" and also "some C", then

1. Necessarily no D is A (assumption)
2. Necessarily Every D is C (from the minor premise and the assumption)

Therefore 3. Necessarily Some C is not A (1, 2 by *Felapton*) ([34], 122.1–4)

Note also that in all these propositions, the modal operator is always in the beginning of the propositions. We could interpret this phrasing as showing that the modalities are external in Avicenna's frame, but this interpretation is not necessarily the right one for at least two reasons: the first one is that the formulation of the different propositions, in particular the negative ones, is easier (in all ordinary languages) when the modalities are external, given that necessary **E**, i.e., "Necessarily no D is A," for instance, is expressed in a clear way when the modality is external, while if the modality is internal, i.e., when it puts on the *predicate*, one would have to express **E** necessary either by "Every D is necessarily not B" or by "No D is possibly A" (i.e., by using the word "possibly" instead of "necessarily"). In the first case, the proposition does not seem to be a negative one, given the way it is formulated and despite the presence of "not" inside it, while in the second case, its formulation is not obvious and not very intuitive. This is probably why Avicenna puts the modality in the end of the proposition when he intends it to be internal. But even in that case, a real internal modality should be understood as putting on the predicate. The second reason is that to really assume that Avicenna is using external modalities rather than internal ones in his syllogistic, we would have to check the way he expresses the contradictories of the different modal propositions. But the contradictories of the necessary propositions are unilateral possible propositions and when the possibility is unilateral, the contradictories are the same, whether the modalities are external or internal. For instance, necessary **E** is contradicted by possible **I**; likewise **E** necessary is contradicted by **I** possible. This is why we won't assume from now that Avicenna is using external modalities until we find further evidence for that, i.e., until we see how he expresses the contradictories of the *bilateral* possible propositions, which, unlike those of the unilateral possible propositions, are expressed differently in case the modalities are external or in case the modalities are internal.

However, on the other hand, we can note that if the necessity puts on the predicate, then if that same predicate is the middle term of the mood, it can be the subject in the other premise of that mood, and as such, it must contain the expression "necessarily" too in that other premise, in order to be the *same* middle in the whole mood. But it does not seem that Avicenna includes the word "necessarily" inside the *subject* of any premise.

4.3.1.1 The Moods with Absolute and Necessary Propositions

The first figure moods to be presented are those containing one necessary premise and one absolute premise. Before presenting and analyzing these moods, let us first choose the letters by which we will express each kind of modal or absolute propositions. For the modal propositions, we will follow Marko Malink and Paul Thom, for instance, by choosing the following symbols:

L stands for necessary, when "necessary" means "not possibly not," i.e., when the necessity is expressed explicitly by using the words "by necessity" or "necessarily" (*bi-ḍ-ḍarūrati*).

M stands for the unilateral possible propositions (when the word "possibly" is used explicitly).

Q stands for the bilateral possible propositions (when the expression "possibly and possibly not" is used explicitly).

But in his analysis, in particular of the first *Barbara*, where the major is necessary and the minor is absolute, Avicenna considers also another kind of necessary propositions containing the condition "as long as it is S" (S being the subject). We could name this necessary proposition **Ls** ("s" standing for the condition "as long as it is S") and state it as follows:

Ls: A is B necessarily (as long as it is A).

As to the non-modal propositions, we cannot simply use the letter **X** as other authors (for instance, M. Malink and P. Thom) do when expressing Aristotle's assertoric propositions, because Avicenna distinguishes between several kinds of absolutes. Recall that in Aristotle's logic, the assertoric propositions used in the syllogistic can be defined as the *non-modal* quantified predicative propositions *a*, *e*, *i*, and *o*, which can be *either true or false*. They are distinguished from the problematic propositions which say that something is possibly true (or false) and from the apodeictic propositions which say that something is necessarily true (or false). In all cases, assertoric propositions can be either affirmative or negative.

As to Avicenna, he does not merely state the Aristotlean assertoric (called *jāzim* in Arabic) propositions, because he adds various conditions to these assertoric propositions and says that their truth values depend on these conditions. We have thus to determine which condition is added to these non-modal propositions in each case ("non-modal" being understood as "not containing any *alethic* modality"). This is why we will have to use different letters to characterize these non-modal propositions depending on the conditions they contain. In order to do so, we will keep the letter **X**, which names the assertorics and add another symbol to this letter to account for the different conditions added by Avicenna, which are mainly: "as long as it exists," "as long as it is S," "at some times, while it exists" and "at some times while S." Thus we will use "**Xp**" (where "p" stands for permanent) to name the propositions containing the condition "as long as it exists," called *dā'ima* by Avicenna. Then "**Xga**" names the propositions containing the condition "at some time while it exists," which are called "general absolutes" by Avicenna; then we will add "s" ("s" referring to the subject) to **X** to name the propositions containing the condition "as long as it is S," and following W. Hodges, we will use the letter "m" (= *muwāfiqa*) and add it to **X** to name the propositions containing the condition "at some time while S." We will thus use the following names for Avicenna's non-modal propositions, i.e., the propositions which do not contain *alethic* modalities:

Xp stands for the *permanent* propositions, i.e., those containing the condition "as long as it exists." These propositions are close to the necessary ones, as we will see in the sequel, but they do not contain the word "necessarily", i.e., the alethic modal operator.

Xs stands for the absolute propositions containing the condition "as long as it is S"

Xga stands for the absolute propositions containing the condition "at some time while it exists."

Xm stands for the absolute propositions containing the condition "at some time while S."

In addition, we have the special absolute proposition:

Xspa, which stands for the special absolutes containing the condition "at some times but not permanently."

These special absolute propositions, i.e., those containing "at some times but not permanently" correspond to what Avicenna calls the *wujūdī* sentences in *Manṭiq al-Mashriqiyyīn*, if we take into account his explanations in that treatise. In *al-Najāt*, *al-Ishārāt* and *al-Qiyās*, he also mentions this kind of sentences and provides their contradictories. They are parallel to the bilateral possible propositions.

In addition, we have to note that Avicenna in *Manṭiq al-Mashriqiyyīn* calls the permanent propositions (those named **Xp** above) which contain the condition "as long as it exists" *ḍarūrīyya* (i.e., necessary) (while in other treatises, they are called *dā'ima* propositions). So as noted by one referee, one must distinguish between these **Xp** propositions and the **L** ones. The **L** and **Xp** propositions differ from each other in at least two ways:

1. The **L** propositions contain *explicitly* the necessity operator, while the **Xp** propositions contain no necessity operator but rather the condition "as long as it exists" either explicitly or implicitly.
2. The **L** propositions are always necessary, whether they are universal or particular, affirmative or negative, and the necessity they contain means "not possibly not," while the **Xp** propositions are permanent but their necessity is not obvious in some cases, since Avicenna says (in *al-Qiyās*) that the permanent propositions (**Xp**) are not necessary when they are *particular*. For instance, the sentence "Some humans are white (as long as they exist)" does not express a real necessity, since the property "white" is not essential to humans.

As to the propositions of kind **Xs**, they are clearly absolute, i.e., not necessarily permanent or necessary, given that being S does not always durate the whole existence of the individual. It can durate only some time in his / its life, precisely the time of being S. For instance, when one says "Every runner is moving (as long as he is running)," the time of the running is only one specific time in the existence of these individuals and its duration may be short. So there is no real reason to call these propositions "perpetual" ones, as some authors do[10], despite the fact that "as long as it is S" means "at all times while it is S." Consequently, their permanence or non-permanence, necessity or non-necessity is not determined in advance. Likewise, the propositions containing "at some times" are absolute too, because "some times" means "one time or more," where "more" can include "all times", so that "at some times" does not *exclude* "at all times" although it does *not require* it. However, the proposition containing "as long as it is S" has the advantage of being convertible, unlike the proposition containing "at some times" which is not convertible. This is why it is used by Avicenna in his categorical syllogistic, as we saw above when analyzing the proof of *Cesare*.

With these symbols, we will qualify the propositions used in the different moods provided by Avicenna.

The first mood presented by Avicenna is the following:

Barbara **LsXpL**: When *Barbara* has a necessary major and an absolute minor, it leads to a necessary conclusion. This is expressed thus

"Every C is B by actuality (*ay bi al-iṭlāq*) (**Xp**)
Every B is A by necessity (**Ls**)
Therefore Every C is A by necessity (**L**)" ([34], 125.4–5).

In his argumentation, he gives the following two premises and explains why some people did not consider the conclusion of these two premises to be necessary:

Zayd is white
Every white thing has necessarily a color dispersed to the eye (*mufriqun li-al-baṣar*).

According to Avicenna, these people "did not deduce (*lam yantuj lahum*) [from these two premises] the conclusion 'Zayd has a color dispersed to the eye by necessity' because they thought that a necessary conclusion follows only if 'Zayd is white by necessity'". The conclusion of these two premises should then be absolute, according to them. But this opinion is due to the fact that they did not distinguish clearly between the conditions 'as long as it is S' and 'as long as it exists' and "considered that the necessary, here, is all what is necessary as long as the essence of the subject exists or necessary as long as the subject is described as such, so that if it is said that 'every white thing has necessarily a color dispersed to the eye' ('*kull abyaḍ fa-huwa bi-d-ḍarūra dhū lawnin mufriqin li-al-baṣar*'), they considered that as a real necessary [i.e. as meaning 'as long as it exists']" ([34], 126.6–9). Now the sentences (1) "every white thing has necessarily a colour dispersed to the eye as long as it exists" and (2) "every white thing has a colour dispersed to the eye as long as it is white" differ in their truth values, for the first is false while the second is true. So because they failed to recognize this difference in the truth values, these people

[10]See, for instance, Riccardo Strobino in his recent article ([142], Sect. 3.1).

"did not examine the truth of the [modal] *de omni et de nullo* principle in the right way (*lam yashtaghilū bi-istithbāti haqīqat al-maqūl 'alā al-kull qawlan ḍarūrīyyan*), so that to properly distinguish [in interpreting] the sentence 'every white has necessarily a color dispersed to the eye', between a first meaning, according to which 'everything described as white, no matter how it is described as such, then as long as its essence exists, whether it is white or it is not white, this thing has a color dispersed to the eye, or [a second one] according to which 'every thing described as white no matter how it is, then this thing has a color dispersed to the eye as long as it is white" ([34], 126.14–127.1)

So they failed to recognize that the two interpretations differ in their truth values, since the first one is *false,* while the second is *true* ([34], 127.3), for strictly speaking, one cannot say that "every white thing has necessarily a color dispersed to the eye" ([34], 127.9) as long as it exists, since some white things do not remain white during the whole time of their existence.

If we consider the above example once again, we can show that the conclusion is necessary when the premises are written with the following conditions:

Zayd is white (as long as he exists) (**X**p)
Every white thing has necessarily a color dispersed to the eye (as long as it is white) (**L**s)
Therefore Zayd has necessarily a color dispersed to the eye (as long as he exists) (**L**)

Since if Zayd is permanently white and if the white things are colored as long as they are white, then Zayd (who is white during his whole existence) will be colored necessarily and permanently. The condition "as long as it is white" in the major premise warrants the truth of that premise, for not every white thing remains white as long as it exists, for instance, a skirt can be white for a while but become grey after some time. But Zayd is one of the things that are white permanently as long as they exist; therefore, he also has necessarily a color dispersed to the eye permanently as long as he exists, like all things that are permanently white.

Thus the first *Barbara* should be stated as follows:

Barbara **LsXpL**:
Every C is B (as long as it exists) (**X**p)
Every B is A (as long as it is B) by necessity (**L**s)
Therefore Every C is A by necessity (**L**)

Thus if C (the minor term) enters into B (the middle term), being a subclass of B, then if A is predicated by necessity of B, it is also predicated by necessity of C, because if every element of B is A by necessity, then every element of its subclass C is also A by necessity. Note here that we do not need to add "as long as it exists" in the conclusion, because **L** alone presupposes also "as long as it exists."

In *al-Najāt*, Avicenna says explicitly and simply that with regard to the moods with an absolute premise and a necessary premise, he shares Aristotle's opinion to the effect that the conclusion in that case must be necessary when the necessary premise is the major, and absolute when the necessary premise is the minor, as witnessed by the following quotation:

> "The truth with regard to the mixing between the absolute and the necessary is, as claimed by the first teacher, that the [modality of] the conclusion [is the same as] the [modality of] the major [premise] (*al-'ibra bi al-kubrā*), for if it [= the major] is absolute, then the conclusion is absolute like this premise (*mithlahā*), and if it is necessary, then the conclusion is necessary too." ([32], 37.15–17)

In this quotation, no further precisions about the kind of absolutes contained in the moods are given. So we have to examine the different moods to try to determine what kind of absolutes is used in these moods.

Let us now examine the second of the *Barbara* moods with an absolute premise and a necessary one, i.e., the one where the major is absolute, and the minor is necessary, which, according to Aristotle and Avicenna, leads to an absolute (non-modalized) conclusion. In this case too, we have to determine what kind of **X** propositions is used in this mood, in Avicenna's frame.

This second *Barbara* is expressed as follows:

"Every C is B by necessity
and every B is A by actuality (*bi-l-iṭlāq*),
Therefore every C is A by actuality" ([34], 127.14–15).

What does the condition "by actuality" mean in this mood? Does the major premise here mean "Every B is A (as long as it is B)"? What about the condition in the conclusion? The answer to this question can be found in the following quotation: "This absolute [premise] *should not* be understood as meaning: 'Everything that is B, then only as long as it is B, not permanently, this thing is A by actuality'" ([34], 128.2–3). For some things which are B are permanently B, i.e., are B as long as they exist, not only as long as they are described as B. For instance, some white things are permanently white as long as they exist, not only as long as they are described as white. So the condition "as long as it is S" should not exclude the possibility for the thing to be S permanently, i.e., all the time of its existence. This is why Avicenna says what follows "It is not true, after we made these precisions, to say that all what is described as B is A at some time, and that this time is the one when it is described as B, for some of what is described as B is so described permanently" ([34], 128.5–7). So the actuality must be such that it admits both necessity and non-necessity. In other words "as long as S" accounts for the case where the individual is S at some time, but also where he is S all the time of his existence. In this sense, it accounts for the permanent as well as the non-permanent cases. Thus "if it is true that every B is A by this kind of absoluteness, the conclusion will be *necessary despite its absoluteness* (*kānat al-natīja ma'a annahā muṭlaqa, ḍarūrīyya*), because this conclusion will be absolute like the major, that is, general absolute, for we would have every C is A as long as it is described as B, but *its being described as B will be permanent for it*, therefore its being A will also be permanent for it" ([34], 128.14–129.1, emphasis added).

This *Barbara* mood should then be expressed as follows:

Every C is B by necessity (**L**)
and every B is A (as long as it is B or *all the time it is a B*) (**X**s)
Therefore every C is A (as long as it exists) (**X**p)

This *Barbara* could thus be called *Barbara* **X**s**LX**p, in Avicenna's frame. The concrete example that illustrates it is the following:

"[Every] snow is white by necessity (**L**)
Every colored white dissociates the eye (as long as it is colored white)(**X**s)
Therefore every snow dissociates the eye (permanently)" (**X**p) ([34],129.1–2)

Here the absolute premise is what Avicenna calls "the most general absolute." This absolute is the one containing the condition "as long as it is S," and can be necessary when being S is durable the whole time of the existence of the thing while it is not necessary when being S durates only some time of the existence of the thing.

In commenting this mood, Avicenna says what follows:

"Let those who are astonished by the producing of a necessary conclusion from a minor absolute and a major necessary examine this [mood]. For they [will] find that the necessary follows [too] from a major absolute if the minor is necessary" ([34], 129.3–4).

This comment seems to show that the conclusion in this mood too is considered as necessary, even if the word "necessarily" is not used in it, since Avicenna uses rather the word "permanently" in the conclusion, while he uses the word "necessarily" in the premise.

In *al-Najāt*, *Barbara* **X**s**LX**p is expressed as follows: "Every C is B permanently and every B is A as long as it is B, therefore every C is A permanently" ([32], 38.6–7). However, here, instead of using the word "necessarily", he uses "permanently", while in *al-Qiyās*, it is the *alethic* modality which is explicitly used in the premise. But in *al-Najāt* too, the premise and the conclusion of the mood are stated in the same way, since they both contain the word "permanently" and neither contains the word "necessarily". Should we thus say that "permanently" and "necessarily" amount [almost] to the same? Maybe yes since Avicenna seems to consider the **L** propositions and the **X**p ones as very close, as witnessed by the quotation above where he says explicitly that in both *Barbaras*, the conclusion is "necessary", and opposes this opinion to that of some commentators of Aristotle, who think that this conclusion should be absolute. However, since the word "necessarily" is not explicitly used in the conclusion, whether in *al-Qiyās* or in *al-Najāt*, we will use the symbol **X**p to express the conclusion. This conclusion is thus necessary in the sense of permanent.

Now this syllogism is expressed as follows in W. Hodges and A. Hasnawi's article ([79]):

"*(a-d)* All snow is coloured white throughout its existence.
(a-ℓ) Everything coloured white dissociates the eye so long as it is coloured white.

(a-d) Therefore all snow dissociates the eye throughout its existence (*Qiyās* 129.1f.)" ([79], 62)

Here, what is called (a-*d*) corresponds to the universal affirmative **A** containing the condition "as long as it exists," which is of the kind that we have called **X**p. Likewise, (a-ℓ) is the universal affirmative absolute **A** of the kind **X**s, i.e., containing the condition "as long as it is S." But here, the minor premise is expressed exactly in the same way as the conclusion, since both contain the condition "as long as it exists," while as we have seen, in *al-Qiyās*, it is the *alethic modality* **L** (= necessarily = not possibly not) which is explicitly used. Now according to Wilfrid Hodges, there is a *parallelism* between the *d* propositions, i.e., what we called the **X**p propositions and the necessary propositions in the alethic sense (the **L** propositions). This is what could explain the fact that they seem to be used indistinctly in Avicenna's two texts.

The third mood, i.e., *Celarent* **L**s**X**p**L**, is expressed as follows:

Every C is B [as long as it exists] (**X**p)
No B is A by necessity (**L**s)
Therefore Necessarily No C is A (**L**)

As to the fourth mood, which is *Celarent* **X**s**LX**p, it can be expressed as follows:

Every C is B by necessity (**L**)
and No B is A (as long as it is B) (**X**s)
Therefore No C is A (as long as it exists) (**X**p)

Exactly like for the second *Barbara* mood, for here too "the conclusions follow the major (*wa al-natā'iju tābi'a li al-kubrā*)" ([34], 129.15).

Darii **L**s**X**p**L** which is said to contain "a minor particular affirmative absolute and a major necessary universal affirmative" ([34], 129.11) would be expressed as follows:

Some C is B (as long as it exists) (**X**p)
Every B is A by necessity (**L**)
Therefore Some C is A by necessity (**L**)

While *Darii* **X**s**LX**p would be expressed as follows:

Some C is B by necessity (**L**)
Every B is A (as long as it is B) (**X**s)
Therefore Some C is A (as long as it exists) (**X**p)

The same holds for *Ferio* **L**s**X**p**L** and *Ferio* **X**s**LX**p, which would be stated respectively as follows:

Ferio **LsXpL**

Some C is B (as long as it exists) **Xp**

No B is A by necessity (**Ls**)

Therefore Not every C is A by necessity (**L**)

Ferio **XsLXp**

Some C is B by necessity (**L**)

No B is A (as long as it is B) (**Xs**)

Therefore Not every C is A (as long as they exist)" (**Xp**)

In all these cases, the conclusion, Avicenna adds "as you know (*'alā mā qad 'alimta*)" ([34], 129.7–9) and insists that it follows the major. So **Xp** is considered as an absolute, not as a necessary proposition.

In *al-Ishārāt*, he evokes mainly the **LsXpL** first figure moods and does not enter into more details with regard to those which contain a major absolute. For instance, *Celarent* **LsXpL** is (partly) stated as follows: "Necessarily No B is A [and Every C is B so Necessarily no C is A]" ([36], 390.1). As to *Darii* and *Ferio* **LsXpL**, they are also respectively (partly) stated as follows: "Some C is B, [necessarily every B is A, therefore necessarily some C is A]" and "Some C is B [and necessarily no B is, therefore necessarily some C is not A]" ([36], 390.4–8).

Now in all cases, the absolute premise should be convertible, as we will see in the sequel and as Avicenna says explicitly.

In the second figure, if the two premises are universal but one of them is necessary while the other one is absolute, so that the middle "is said of one of the extremes by necessity, while it is said of the other one by actuality…, so that [the middle] is permanently (*dā'imun lahu*) said of it and not permanently said (*laysa dā'iman lahu*) of the other one,…, whether the judgment is negative or affirmative" ([34], 130.7–9).

The first mood considered is *Cesare* with an absolute premise and a necessary one. This mood can have either a major necessary and a minor absolute or a major absolute and a minor necessary. In the first case, it is stated as follows:

Cesare 1

"Every C is B by actuality

and No A is B by necessity

It follows No C is A by necessity" ([34], 131.7–8).

This is said to be the usually admitted *Cesare* (*al-mashhūr*) with a minor absolute and a major necessary. Avicenna holds it by saying that this mood is undeniable (*lā munāza'a fīhi*). It is proved by the conversion of the necessary proposition, which leads to "No B is A by necessity," and consequently to the following *Celarent* mood of the first figure:

Every C is B by actuality (**Xp**)
and no B is A by necessity (**Ls**)
It follows no C is A by necessity (**L**)

Now in *Celarent* **LsXpL**, "Every C is B by actuality" means "Every C is B (as long as it exists)," i.e., is an **Xp** proposition, as we saw above. So here too, it should be an **Xp** proposition, so that the whole *Cesare* **LsXpL** would be expressed as follows:

Every C is B (as long as it exists) (**Xp**)
No A is B by necessity (**Ls**)
Therefore No C is A by necessity (**L**)

So *Cesare* 1 is *Cesare* **LsXpL**.

As to the second mood mentioned by Avicenna, it is the one where "the negative necessary is the minor" ([34], 131.10). So it is a *Camestres* mood with an absolute major and a necessary minor.

This mood would be expressed as follows:

Camestres **XpLsL**
No A is B by necessity (**Ls**)
Every C is B (as long as it exists) (**Xp**)
Therefore No A is C by necessity (**L**)

This mood is just mentioned, it is not examined in detail. But since Avicenna proves *Camestres* in categorical logic by reducing it to *Celarent* and by using two conversions—on the minor premise and on the conclusion—and since he seems to hold the opinion that in the second figure, the modality of the conclusion follows that of the negative premise, if we take into account the moods presented, then this *Camestres* should be reduced to the first *Celarent*, i.e., *Celarent* **LsXpL**, and should thus be *Camestres* **XpLsL**. The conversions lead from "No A is B (by necessity)" to "No B is A (by necessity)" and from "No A is C (by necessity)" to "No C is A (by necessity) ." The reduction is thus expressed as follows (where the arrows are put for the conversions):

Camestres **XpLsL**		*Celarent* **LsXpL**
No A is B by necessity (**Ls**)	┐	Every C is B (as long as it exists) (**Xp**)
Every C is B (as long as it exists) (**Xp**)	└→	No B is A (by necessity) (**Ls**)
Therefore No A is C by necessity (**L**)	⟶	Therefore No C is A (by necessity) (**L**)

This being so, the absolute premise is more likely to contain the condition "as long as it exists," since the mood *Celarent* of the first figure to which this *Camestres* is reduced contains that condition. And Avicenna does say that in *al-Qiyās* where he claims "…the right combination in that [mood] is that 'every C at all times is described as B permanently as long as it exists, not only as long as it is described as C[11]" ([34], 132.1–2). In *al-Najāt*, Avicenna says explicitly that the conclusion follows the negative premise as witnessed by the following passage: "As to the second figure, what is clear and usual (*al-ḍahir wa al-mashhūr*) is that the negative premise is determining (*al-'ibra li-al-sāliba*), for it becomes the major by conversion or *ekthesis* and the conclusion follows its modality" ([32], 38.9–11). However, this opinion is criticized afterward as follows "and the truth requires (*yūjibu*) what we should not be ashamed of, which is that the conclusion is always necessary" ([32], 38.11–12). We will see in the sequel what are the necessary conclusions that he is talking about in this passage.

The third mood is *Cesare* 2, i.e., *Cesare* with an absolute major and a necessary minor. In this case, the absolute is explicitly said to be "not necessary" ([34], 131.12) and also to contain the condition "as long as it is S," since it must be convertible. The conversion leads to *Celarent* **Xs**L**Xp**. The mood is expressed as follows:

Cesare **Xs**L**Xp**
Every C is B by necessity (**L**)
No A is B (as long as it is A) (**Xs**)
Therefore No C is A by necessity (**Xp**)

Avicenna says explicitly that the absolute in this mood must be convertible and that its conversion leads to the following proposition: "No B is A, as long as it is described as B" ([34], 131.16). Since its other premise is "Every C is B, permanently" ([34], 131.16) [or "by necessity" ([34], 131.11)], what follows is "a necessary proposition" ([34], 131.17). So this *Cesare* seems to be *Cesare* **Xs**L**Xp**. But since the mood to which it is reduced is *Celarent* **Xs**L**Xp**, which contains an **Xp** conclusion, as we saw above, the conclusion of this *Cesare* should be rather necessary in the sense of permanent (see the discussion above about *Barbara* **Xs**L**Xp**). This *Celarent* is thus the following:

Every C is B, by necessity (**L**)
No B is A (as long as it is B) (**Xs**)
Therefore No C is A, permanently (**Xp**)
Which expresses *Celarent* **Xs**L**Xp**.

Once again, we note that the propositions **L** and **Xp** are very close even if they are not exactly the same, for Avicenna uses in many parts of his text the words "permanently" and "necessarily" indistinctly. In any case, if a proposition is necessary, i.e., of kind **L** where necessity means "not possibly not," it surely implies the absolute containing the condition "as long as it exits," which itself involves

[11]Here there is an error in the text, for what is written is "described as *B* only." However the right letter should be C, since C is the subject of the proposition.

some kind of necessity. This is why the conclusion of *Cesare* **XsLXp** can be said to be necessary (in the sense of permanent).

What about the second *Camestres*? This should be the fourth mood, i.e., the *Camestres* mood where the negative minor is absolute and the affirmative major is necessary. In this case, the mood would be *Camestres* **LXsXp** and would be expressed as follows:

Camestres 2 (**LXsXp**)		*Celarent* **XsLXp**
No C is B (as long as it is C) (**Xs**)	→	Every A is B by necessity (**L**)
Every A is B by necessity (**L**)	→	No B is C (as long as it is B) (**Xs**)
Therefore No C is A (by necessity)(**Xp**)	→	Therefore No A is C (as long as it exists)(**Xp**)
		[Then (by conversion) No C is A (as long as it exists)]

since it is proved by the conversion of the negative minor and of the conclusion, which reduces it to *Celarent* **XsLXp**. Avicenna says explicitly that in this mood, the conclusion is necessary, as appears in the following quotation: "and it produces as you know a necessary [conclusion]" ([34], 131.16–17). This confirms once again the closeness of **L** and **Xp** propositions. But Avicenna does not state explicitly the "necessary conclusion" that he is talking about, so he does not explicitly use the necessity operator in the conclusion, while he uses the word "necessarily" (*bi-ḍ-ḍarūra*) in the premise, while first stating the mood ([34], 131.11) and afterward the word "permanently" (*dā'iman*) in the premise in his explanations and comments ([34], 131.16) of this same mood. This is why he seems to consider the **Xp** and the **L** propositions as very close, which explains our use of the symbol **Xp** to account for this kind of necessity (= permanence, here) in the conclusion.

The fifth mood is *Festino* **LsXpL**. It is stated as follows:

Festino **LsXpL**
Some C is B by actuality (**Xp**)
No A is B by necessity (**Ls**)
Therefore Some C is not A by necessity (**L**) ([34], 151.1–2)

This mood is reducible to *Ferio* **LsXpL** by the conversion of the universal negative premise. Since the absolute proposition in *Ferio* is an **Xp** proposition, so will be the absolute proposition in *Festino*, since the particular proposition remains the same in both moods. The reduction is shown in what follows:

Festino **LsXpL**		*Ferio* **LsXpL**
Some C is B by actuality (**Xp**)		Some C is B by actuality (**Xp**)
No A is B by necessity (**Ls**)	→	No B is A by necessity (**Ls**)
Therefore Some C is not A by necessity (**L**)		Therefore Some C is not A by necessity (**L**)

As to *Festino* **Xs****L****Xp**, it is the following:

"Some C is B by necessity (**L**)
No A is B by actuality (**Xs**)
Therefore Some C is not A by necessity (**Xp**)" ([34], 151.3–4)

This is reducible to *Ferio* **XsLXp** by the conversion of the major negative premise, which is an **Xs** proposition in *Ferio*. Here too, the absolute proposition will contain the condition "as long as it is S," and although the conclusion is not stated explicitly in *al-Qiyās*, for he just says that "it produces in the aristotelian tradition (*fī al-mashhūr*) what you know" ([34], 151.4), the whole mood should be *Festino* **XsLXp**, since it is reduced to *Ferio*, which contains an **Xp** conclusion. Now, in *al-Najāt*, he says that all second figure moods produce necessary conclusions ([32], 38).

The reduction can be shown as follows:

Festino **XsLXp**		*Ferio* **XsLXp**
Some C is B by necessity (**L**)		Some C is B by necessity (**L**)
No A is B by actuality (**Xs**)	⟶	No B is A by actuality (**Xs**)
Therefore Some C is not A permanently (**Xp**)		Therefore Some C is not A (permanently) (**Xp**)

Now if we consider that *Festino* is proved by reduction to *Ferio*, the conclusion should be an **Xp** proposition rather, given that it is the same as *Ferio*'s conclusion. So its necessity seems to be a kind of permanence, as in the other moods considered above, since the word "necessarily" is not explicitly used in the conclusion, whether in *al-Qiyās* or in *al-Najāt*. Avicenna's opinion is due to the fact that the **Xp** and the **L** propositions are very close.

As to *Baroco*, it is also expressed in two ways depending on which premise is necessary and which one is absolute. Thus *Baroco* **XpLsL** is stated as follows:

'Not every C is B by necessity
Every A is B by actuality
Therefore Not every C is A by necessity' ([34], 151.5)

It is illustrated by the following concrete example:

'Necessarily not every white is an animal (**Ls**)
Every human is an animal by actuality (**Xp**)
Therefore Not every white is a human by necessity (**L**)' ([34], 152.5–14)

In this mood, the absolute proposition is "Every human is an animal by actuality." This absolute proposition can mean "Every human is an animal (as long as it exists)" or "Every human is an animal (as long as it is human)," since a human remains an animal both as long as he is a human and as long as he exists. But Avicenna says in his justification of this mood that "'Every human is an animal' can be false (*qad yakdhibu*) if the humans are no more existent (*idhā 'adima al-nāsu kulluhum*)" ([34], 152.10–11), so the sentence is true if there are humans, and seems then to contain the condition "as long as it exists" rather than simply the condition "as long as it is (described as) human." The conclusion of this mood is not provided explicitly and its modality is discussed by Avicenna, since he claims that according to some people, the conclusion of this mood should *not* be necessary by saying "and it has been said nevertheless (*ma'a dhālika*) [that] the conclusion is not necessary" ([34], 151.9–10). However, he himself considers it as necessary since he claims what follows: "[You must] know that some people (*ṭā'ifatan min al-muhaṣṣalīn*) have [rightly] considered that the conclusion of this mood is necessary... they proved that the conclusion of this mood is necessary and they demonstrated it by *ecthesis* in a good way (*tabyīnan ḥaqīqīyan*)" ([34], 155.17–156.1). So it seems that according to Avicenna, the conclusion in this mood should be necessary.

As to its proof, it is by *ecthesis* rather than by *reductio ad absurdum*, as Avicenna says in the quotation above. For since the conclusion is necessary, and since its contradictory is a possible proposition, in order to apply the proof by *reductio ad absurdum*, one should already know the moods containing possible propositions, which are not available yet, since they are not yet proved. Now the proof by *ecthesis* used by Avicenna in categorical logic uses both *Camestres* and *Ferio*. In this case, these should be *Ferio* **LsXpL** and *Camestres* **XpLsL**. Let us write down the proof by analogy with the proof provided in categorical logic. Thus let the "some C's" that are "not B" be called "every D", then

1. Not Every C is B by necessity (minor premise)
2. No D is B by necessity (from 1, assumption)
3. Every A is B by actuality (i.e. as long as it exists) (major premise)

Therefore 4. No D is A by necessity (as long as D) (from 2, 3 by *Camestres* **XpLsL**)
and 5. Some C is D by actuality (= as long as they exist) (assumption)
Therefore 6. Not every C is A by necessity (from 4, 5, by *Ferio* **LsXpL**)

The conclusion in this mood is said by Avicenna to be "necessary with regard to the matter" ([34], 152.14). So *Baroco* 1 is *Baroco* **XpLsL.**

What about the second *Baroco*? How is it stated by Avicenna? This *Baroco* contains a major necessary premise together with an absolute minor premise. It is stated as follows:

Baroco **LXsXp**
"Not every C is B by actuality (**Xs**)
Every A is B by necessity (**L**)
Therefore Not every C is A" (**Xp**) ([34], p. 151.5–6)

This mood should also be proved by *ecthesis* as the one above. This proof could then be the following: Let the "some C's" that are "not B's" be called "all D's", then:

1. Not Every C is B by actuality (minor premise)
2. No D is B by actuality (from 1, assumption)
3. Every A is B by necessity (L)

Therefore 4. No D is A by necessity (from 2, 3, by *Camestres* **LXsXp**)

Before continuing with the rest of the proof, let us note that the *Camestres* mood used here should be *Camestres* **LXsXp**, since *Camestres* **XpLsL** is not applicable to the universal propositions used in this proof. This shows that the absolute minor premise of *Baroco* should be an **Xs** proposition, and the universal negative proposition (4) should be an **L** proposition. We can thus rewrite the propositions again and continue with the rest of the proof. We will have the following steps:

1. Not Every C is B (as long as it is C) (**Xs**) (minor premise)
2. No D is B (as long as they are D) (**Xs**) (from 1, assumption)
3. Every A is B by necessity (**L**) (major premise)

Therefore 4. No D is A (as long as they exist) (**Xp**) (from 2, 3, by *Camestres* **LXsXp**)

And 5. Some C is D (as long as they are D) (assumption) (**Xs**)

Therefore 6. Not every C is A (by necessity) (**Xp**) (4, 5 by *Ferio* **XpXsXp**)

However the conditions in (5) and (2) are the same, since it is the same assumption. This proof justifies the almost necessary character of the conclusion, expressed by a permanent proposition, which is claimed by Avicenna against Aristotle.

This mood too is illustrated by a concrete example, which is the following:

'Not every white is an animal (**Xs**)
Every human is an animal [by necessity] (**L**)
Therefore Not every white is human' (**Xp**) ([34], 151.6–7)

To summarize, we could say that the second figure moods with an absolute and a necessary premises held in *al-Qiyās* are the following: *Cesare* **LsXpL**, *Cesare* **XsLXp**, *Camestres* **XpLsL**, *Camestres* **LXsXp**, *Festino* **LsXpL**, *Festino* **XsLXp**, *Baroco* **XpLsL**, and *Baroco* **LXsXp**.

However, in presenting the moods 5–6 (the two *Festinos*), Avicenna says that *Festino* 1 (= **LsXpL**) "leads to what you know" ([34], 151.2) and that the second *Festino*'s "usual (*al-mashhūr*) conclusion is what you know," meaning that this *Festino* is an **XLX** mood in Aristotle's frame. As to the two *Barocos*, they are, respectively, *Baroco* **LXX** and *Baroco* **XLX** in Aristotle's frame as Avicenna notes at page 151 ([34], 151.5–10). But this opinion of Aristotle is discussed to show that their conclusions should be necessary. He illustrates these two *Barocos* as follows:

Baroco 2 [(7) in [34], 151.5]	*Baroco* 1 [(8) in [34], 151.8]
Not every white is an animal	Not every white is an animal (by necessity)
Every human is an animal (by necessity)	Every human is an animal
Therefore not every white is human	[Therefore not every white is human] ([34], 151.5–10)

In his discussion of the two moods, he says, commenting the conclusion of *Baroco* 2 "but they said nevertheless that the conclusion is not necessary (*thumma qīla ma'a dhalika anna al-natīja laysat darūrīyya*)," meaning that he does not share Aristotle's opinion about the non-necessity of the conclusion, for: "when whiteness is always negated from some animals, then it is always negated from some humans" ([34], 152.2–3). Thus the first *Baroco* is **XpLsL**, since the proposition "every human is an animal" means "every human is an animal as long as he exists," which becomes false "when people do not exist any more (*kull insān hayawān, alladhī qad yakdhibu idhā 'adama annāsu kulluhum*)" ([34], 151.10–11). As to the *Baroco* where the affirmative universal is necessary, while the particular negative is an absolute containing the condition "as long as it is S" [since "not every white is an animal" means "not every white is an animal (as long as it is white)"], its conclusion is materially necessary too as the proof by *ecthesis* above shows.

In *al-Najāt*, Avicenna says explicitly that, unlike what Aristotle says, in all these moods, "the conclusion is always necessary" ([32], 38). He even says that "we should not be ashamed to correct this" ([32], 38), meaning that Aristotle's opinion according to which the modality of the conclusion follows that of the negative premise in the second figure should be corrected, because in all cases, the conclusion is necessary. Now what we found in analyzing the second figure moods as they are expressed in *al-Qiyās* shows that the conclusion in *Cesare* 1, *Camestres* 1, *Festino* 1, *Baroco* 1, and *Baroco* 2, the conclusion is indeed necessary, i.e., is an **L** proposition. In the other moods, i.e., in *Cesare* 2, *Camestres* 2, and *Festino* 2, it seems to be an **Xp** proposition, which is not an **L** proposition but has some kind of necessity, since it is a permanent proposition containing "as long as it exists."

In *al-Ishārāt*, which is one of the latest treatises, he says first while discussing *Cesare* that "the conclusion follows the modality of the major (*wa takūnu al-'ibra fī al-jiha li-al kubrā*) ," which would mean that the conclusion in *Cesare* **LXL** is necessary but not in *Cesare* **XLX**. One page later, he says: "The conclusion follows also the *negative* in modal moods (*wa takūnu al-'ibra li-al-sāliba ayḍan fī al-jiha*)" ([36], 411.4, emphasis added). Note, here, the use of the word "also" (*ayḍan*) which suggests that the same idea has already been expressed previously. Finally, some pages later, he claims what follows: "and you know that the conclusion is always necessarily negative (*wa kadhālika ta'lamu anna al-natīja dā'iman takūnu ḍarūrīyyata al-salbi*[12]). And this is what they did not see (*wa hādhā mimmā*

[12]Note here that he does not say "*ḍarūrīyya sāliba*", which would be translated by "*necessary* negative", but rather "*ḍarūrīyyata al-salb*", which I have translated as "*necessarily* negative."

ghafalū 'anhu)" ([36], 422.3–4). Does this mean that the negative conclusion is always necessary in these moods, or rather that in these second figure moods, the conclusion is always negative? The question can be raised because nobody is unaware of the fact that the conclusion is always negative in all second figure moods. So what did these people miss? Maybe the fact that this conclusion is also necessary in these moods, as the discussion of some of Aristotle's interpretations in *al-Qiyās* and *al-Najāt* shows. It seems that despite some ambiguities in his analysis and despite the fact that some proofs are not always stated clearly, the conclusion in all second figure moods should be necessary unlike what we find in Aristotle's analysis, but the necessity in some of these moods can simply have the meaning of permanence.

In the third figure, Avicenna says first, for instance, in *al-Najāt* that "the truth is that the conclusion follows the major (*wa al-ḥaqqu anna an-natīja tatba'u al-kubrā*)" ([32], 39.9). Thus the first mood is *Darapti* **LsXpLst**, which he proves by the conversion of the minor and reduction to *Darii* **LsXgaLst**, itself implied by *Darii* **LsXpL**, which we considered in the first figure. This *Darapti* is stated as follows:

"Every B is C by actuality (**Xp**)
Every B is A by necessity (**Ls**)
Therefore Some C is A by necessity (**Lst**)" ([34], p. 156.7–8)

This mood is proved by the "conversion of the minor," as claimed explicitly by Avicenna in *al-Qiyās* (p. 156.8), which leads to a general absolute particular affirmative proposition, that is, a particular affirmative of the kind **Xga**, expressed by "Some C is B (at some times) ." Thus *Darapti* **LsXpLst** is reduced to *Darii* **LsXgaLst** as follows:

Darapti **LsXpLst**		*Darii* **LsXgaLst**
Every B is C (as long as B)(**Xp**)	⟶	Some C is B (at some times)(**Xga**)
Every B is A by necessity (**Ls**)		Every B is A by necessity (**Ls**)
Therefore Some C is A by necessity (**Lst**)		Therefore Some C is A necessarily at some times (**Lst**)

The conclusion follows by *Darii* **LsXpL**. But the conversion of **A** permanent leads to **I** general absolute. Therefore, the *Darii* to which *Darapti* is reduced could not contain an **Xp** minor premise. Likewise, the conclusion of this *Darii* (to which *Darapti* is reduced) could not be a necessary proposition **L** (where the necessity is permanent), since its minor is a general absolute, but it could be an **Lst**. But since **Xp** implies **Xga**, and **L** implies **Lst** ("st" refers to "Some Times" (see below)), the minor premise and the conclusion of this *Darii* **LsXgaLst** are implied by *Darii* **LsXpL**, the major premise being exactly the same. And since *Darapti* is reduced to this *Darii*, its conclusion should also be the same as that of *Darii*.

This *Darii* could be illustrated by the following concrete example:

Every writer is moving necessarily (as long as he is writing)(**Ls**)
Some humans are writing (at some times) (**Xga**)
Therefore Some humans are moving by necessity (at some times) (**Lst**)

The above *Darapti* would be expressed as follows, if we keep the same terms:

Every writer is moving necessarily (as long as he is writing)(**Ls**)
Every writer is human (as long as he exists) (**Xp**)
Therefore Some humans are moving by necessity (at some times) (**Lst**)

So what the conclusion says is that this moving is necessary for these humans precisely the time while they are writing, not as long as they are human or as long as they exist. However, the proposition in this conclusion seems so unusual that one could ask about its correctness. But if one thinks about the whole mood and all its components, one could understand the fact that if a writer is necessarily moving in so far as he is writing, and if some people are writing at some times, then they are necessarily moving precisely at the times while they are writing. The necessity of this moving seems then quite natural, since it is conditioned by the time of the writing, which is intimately related to it, because there is no writing without moving.

This means also that there is another kind of propositions which is the proposition **L**st, i.e., "A is necessarily **B** (at some times) ," which we did not mention above, but should be added to the whole set of propositions. We could illustrate it, for instance, by the following examples: "Humans are necessarily sleeping (at some times) ," or "Animals are necessarily eating (at some times)" and so on, which seem quite natural.

I have called this **L** proposition, which contains the condition "at some times" just as the general absolute proposition **X**ga, **L**st, because it could not be a general absolute proposition, which is not necessary at all. So if I had called it **L**ga, it would have been strange. Now "st" stands for **S**ome **T**imes and accounts for the fact that the necessity is not durable or permanent; rather it can be present only at some times, whether regular or not. This kind of necessity is more likely to be a physical necessity. As a matter of fact, Allen Bäck evokes this kind of propositions which are necessary without being permanent when he gives the following Avicennan example "The moon has eclipses by necessity" ([39], 221). This proposition is necessary even if "the moon is not eclipsed permanently, but only for short periods of time at great intervals. But, during those periods of eclipse, it is eclipsed necessarily and not merely by accident" ([39], 221) as he notes in the same page. He adds that there is also some kind of necessity that can occur at some times, not necessarily determined as for the eclipse of the moon, for "those periods of time for which the predicate holds are *required* and fixed, but are less definite, as in 'Socrates breathes (inhales): Socrates *must* breathe during a stretch of time, because he is a human being, but when in particular during that stretch of time is left open" ([39], 221, emphasis added). Now this example is precisely one of those that are given by Avicenna to illustrate the general absolute propositions, i.e., those containing the condition "at some times." But when we add the word "necessarily", which is plausible since the predication is at least physically necessary, we get an **L**st proposition, which is comparable to the general absolute one with regard to the condition ("at some times"), but is nevertheless necessary, because the predicate is necessary. The predicate in this case "is not accidental but essential or inseparable attributes (*propria*) of the subject" ([39], 221). So there is some kind of necessary "predication" which "holds discontinuously" ([39], 222) not permanently. Thus the propositions containing "at some times" are not always absolute, and they can be necessary, when they contain the necessity operator. For as anyone can see the two propositions "every human is writing (at some times)" is a general absolute first because it does not contain the word "necessarily" but also because writing is a contingent predicate, while if someone says "every human is necessarily sleeping (at some times)" or "every human is necessarily breathing (at some times)" this proposition is necessary, because of the presence of the necessity operator but also because of the essential character of the predicate, which is the reason why the proposition containing the necessity operator is true.

Now what about the other *Darapti*? This one is said to lead to an absolute proposition unlike what, according to Avicenna, Aristotle claims. Avicenna provides Aristotle's and says that his own one is different and leads to a different conclusion, because the conversion of **I** necessary does not lead according to him to **I** necessary, as we saw above. To explain his own view, Avicenna presents Aristotle's proof and says why he rejects it. According to Avicenna, the usual Aristotelian proof (*al-mashhūr*) proceeds by two conversions and proves *Darapti* **XLL** by reducing it to *Darii* **LXL** as follows:

Darapti **XLL** (Aristotle and his followers)
Every B is C by necessity
Every B is A by actuality
Therefore Some C is A by necessity

The proof runs as follows:
Every B is C by necessity (minor premise)
Some A is B by actuality (by conversion of 'every B is A by actuality')
Therefore Some A is C by necessity (by *Darii* **LXL**)
Then Some C is A by necessity (by conversion of this conclusion).

This last step is what Avicenna finds problematic, since he says that the conversion of **I** necessary "should not lead to **I** necessary," because the converse may very well be simply absolute. According to him, this *Darapti* should lead to a non-necessary proposition as follows:

Darapti **XsLXga** (Avicenna)
"Every B is C by necessity (**L**)
Every B is A by actuality (**Xs**)
Therefore Some C is A by actuality (**Xga**)" ([34], 156.9–13)

Now Avicenna did not provide his own proof, but if we use as above two conversions by applying his own definitions of conversions, we get what follows:

Every B is C by necessity (**L**)
Some A is B (at some times while A) (**Xm**)
Therefore Some A is C at some times (**Xga**) (by *Darii* **XsLXp**, given that **Xs** implies **Xm** and **Xp** implies **Xga**)

Consequently, Some C is A (at some times) (**Xga**) (by the conversion of this conclusion (see Sect. 3.2.2).

The conclusion that we get in this reduction, by the conversion of the particular affirmative proposition deduced by *Darii* is **Iga**. It is a general absolute proposition as shown by the proof above, since the conversion of the **Xs** premise produces an **Xm** proposition and both premises could not lead to an **Xp** proposition or even to an **Xs** one. As to the conversion of the conclusion, it leads to an **Xga** proposition, according to Avicenna's conversion rules which we already stated above, since **Iga** leads to **Iga**.[13]

If we consider the concrete example provided by Avicenna, we would have the following combination, where B = human C = animal and A = breathing:

Every human is an animal by necessity
Every human is breathing (as long as he is human)
Therefore Some animal is breathing (at some times).

Where the condition "at some times" refers precisely to the times where these humans are breathing.

This mood is thus specific to Avicenna since it is proved by his own conversion rules, which are different from those of Aristotle and his followers.

As to the first *Felapton*, it should be *Felapton* **LsXpL**, which is expressed as follows:

"Every B is C by actuality (**Xp**)
No B is A by necessity (**Ls**)
Therefore Not every C is B by necessity (**L**)" ([34], 157.1–2)

This is proved by the conversion of the minor, which leads to "Some C is B by actuality" and reduces the mood to *Ferio* **LsXpL**, which Avicenna admits in the

[13] Even **Ip** (= **I** permanent) leads to **Iga**

first figure. But just as we saw above when analyzing *Darapti* **LsXpL**, the conversion of **A** (**X**p) should lead to **I** (**X**ga) in *Ferio* and consequently the conclusion should be an **L**st proposition in *Ferio*. This *Ferio* is implied by the initial *Ferio* **L**s**X**p**L**.

We would thus have the following reduction:

Felapton **L**s**X**p**L**	*Ferio* **L**s**X**ga**L**st
Every B is C by actuality (**X**p) →	Some C is B at some times (**X**ga)
No B is A by necessity (**L**s)	No B is A by necessity (as long as it is B) (**L**s)
Therefore Not every C is B by necessity (**L**st)	Therefore Not every C is B by necessity (**L**st)

This *Ferio* contains **L**st as a conclusion, which can be explained in the same way as for *Darapti* and *Darii* above.

The second *Felapton* should be *Felapton* **X**s**LX**ga. It is proved by the conversion of the minor, which leads to an absolute proposition and reduces it to the categorical *Ferio*. The conversion of the minor "Every horse is an animal by necessity" leads to "some animal is a horse (at some times)" and the reduction to *Ferio* can be shown as follows:

Felapton **X**s**LX**ga	*Ferio* **X**s**X**ga**X**ga
Every B is C (necessarily)(**L**) ⟶	Some C is B (at some times) (**X**ga)
No B is A (as long as it is B) (**X**s)	No B is A (as long as it is B) (**X**s)
Therefore Not every C is A (at some times)(**X**ga)	Therefore Not every C is A (at some times)(**X**ga)

This *Ferio* is implied by the categorical *Ferio*, which contains as we have seen above (Sect. 3.2.2) only **X**s propositions. This is so because **X**s propositions imply **X**ga propositions. Note that, here too, Avicenna does not evoke possible propositions explicitly when converting the necessary one, presumably because his present analysis involves only absolute and necessary propositions, not possible ones. But since, as shown by W. Hodges, possible propositions (in the unilateral sense of possibility) are parallel to the general possible ones, the conversion above can be admitted.

Avicenna illustrates this mood by the following example (where the terms are horse, animal and sleeping) and where the modalities in the premises are reversed:

Every horse is [necessarily] an animal (**L**)
No horse is sleeping (as long as it is a horse) (**X**s)
Therefore Not every animal is sleeping (at some times) (**X**ga)

The fifth mood is *Datisi* **L**s**X**p**L** and is considered as almost obvious, as appears in the following quotation: "and the fifth [mood] is the combination of a minor particular affirmative absolute, while the major is a necessary universal affirmative.[14] It makes no doubt that the conclusion is necessary (*fa-lā shakka anna al-*

[14]Here, there is an error of editing since what is written is that "the major premise is a universal necessary *negative*" ([34], 157.13, my emphasis). But this could not be so, since the mood *Ferison* is the ninth mood and will be presented and developed by Avicenna at page 158.

natīja ḍarūrīyya)" ([34], 157.12–13). This is so because this *Datisi* **LsXpL** mood is reducible to *Darii* **LsXgaLst** as follows:

Datisi **LsXpL**		*Darii* **LsXgaLst**
Some B is C (as long as it exists)(**Xp**)	→	Some C is B (at some times)(**Xga**)
Every B is A by necessity (**Ls**)		Every B is A by necessity as long as B (**Ls**)
Therefore Some C is A by necessity (**Lst**)		Therefore Some C is A by necessity at some times (**Lst**)

This *Darii* **LsXgaLst** is itself implied by *Darii* **LsXpL**, since **Xp** implies **Xga**. The conversion of the minor premise of *Datisi* leads to an Xga proposition and in turn to *Darii* **LsXgaLst**, since from an **Xga** proposition and an Ls proposition, one gets an Lst proposition as we saw with the moods above.

As to the second *Datisi*, which is the seventh mood in Avicenna's classification, it is expressed at page 158.1–2 as follows: The seventh mood is "Some B is C by necessity and every B is A by actuality, *not by necessity*; this yields an absolute by the conversion of the minor and [by considering] the aforementioned condition (*wa 'alā al-sharṭ al-madhkūr*)" ([34], 158.1–2, emphasis added).

Here he says quite explicitly that the universal affirmative absolute is not necessary, so this absolute proposition should contain the condition "as long as it is S" and the mood should be expressed and reduced to *Darii* **XsXgaXs** as follows:

Datisi **XsLXga**		*Darii* **XsXgaXga**
Some B is C by necessity (**L**)	⟶	Some C is B (at some times)(**Xga**)
Every B is A (as long as it is B)(**Xs**)		Every B is A (as long as B) (**Xs**)
Therefore Some C is A (at some times)(**Xga**)		Therefore Some C is A (at some times) (**Xga**)

The mood *Darii* **XsXgaXga** to which this *Datisi* is reduced is itself implied by the usual categorical *Darii*, which contains only **Xs** propositions, since **Xs** implies **Xga**. This is so because the minor of *Datisi* being "Some B is C by necessity"; its conversion should lead either to a general absolute (**Xga**) or to a possible proposition according to the analysis above (Sect. 4.3.1, p. 143). But the possible propositions are not yet used by Avicenna, and he does not say explicitly that the conversion of this **I** necessary proposition should be a possible proposition in this part of this text. This being so, the conversion seems to lead to an absolute proposition of the broadest sense, that is, to the absolute containing the condition "at some times."

As to *Disamis*, where the universal minor is necessary, Avicenna says that Aristotle expresses it with a necessary conclusion, but he himself considers that the conclusion should not be necessary, rather it should be not necessary. So he admits *Disamis* **XLX** rather than **XLL**, unlike Aristotle, once again (see [113], 117, Table 1) and proves it by *ekthesis* provided that "some B = every D." Aristotle proves it by two conversions and by reducing it to *Darii* **LXL** as follows:

Disamis **XLL** (Aristotle)	[reduction to] *Darii* **LXL**
Some B is A by actuality (X)	Every B is C by necessity (**L**)
Every B is C by necessity (L)	Some A is B by actuality (**X**)(by conversion)
Therefore Some C is A by necessity	Therefore Some A is C by necessity (by Darii **LXL**)
	Consequently Some C is A by necessity (by converting the conclusion)

This proof is not admitted by Avicenna because of the conversion of **I** necessary which does not lead in his frame to a necessary **I**. He also says that the proof he uses for this mood is by *ecthesis* rather than by two conversions ([34], 157.15).

In Avicenna's frame, this *Disamis*, which is the sixth mood, can be stated as follows:

Disamis **XsLsXga**
Every B is C by necessity (**Ls**)
Some B is A by actuality (**Xs**)
Therefore Some C is A by actuality (**Xga**) ([34], 157)

The proof by *ekthesis*, which is only evoked by Avicenna but not provided, could be stated as follows: "let some B = every D" ([34], 157.15–16), then we have

(1) Every D is A by actuality (as long as D) (**Xs**)
And (2) Every D is B (assumption) (**Xp**)
(3) Every B is C by necessity (minor premise of *Disamis*) (**Ls**)
(4) Every D is C by necessity (**L**) (from 2 and 3 by *Barbara* **LsXpL**)
(5) Some C is A by actuality (**Xga**) (from 1 and 4 by *Darapti* **XsLXga**)

This conclusion is a general absolute proposition as Avicenna insists at page 157.14–16.

However, when the major is necessary, the conclusion should be necessary too, as Avicenna says explicitly at page 158.5. So this second *Disamis* should be *Disamis* **LsXsL**, as we will show with the proof below, which uses *ecthesis* too. It is expressed as follows:

"Some B is A by necessity (**Ls**)
Every B is C by actuality (**Xp**)
Therefore Some C is A by necessity (**Lst**)" ([34], 158.3–5)

For he says explaining the necessity of the conclusion, that "*ekthesis* requires the conclusion to be necessary" ([34], 158.5).

The proof should be as follows:

(1) Every D is A by necessity (**L**s)
And (2) Every D is B (assumption) (**L**)
(3) Every B is C by actuality (minor premise of *Disamis*) (**X**s)
(4) Every D is C by necessity (**X**p) (from 2 and 3 by *Barbara* **X**s**LX**p)
(5) Some C is A by necessity (**L**) (from 1 and 4 by *Darapti* **L**s**X**p**L**st)

This conclusion is indeed necessary but it contains the condition "at some times" too, which makes its necessity different from the real one (= the one expressed by **L**), which is also permanent. This conclusion, although necessary, is not permanent.

The ninth mood is *Ferison* **L**s**X**p**L**. It is provable by converting the minor and reducing it to *Ferio* **L**s**X**p**L** as follows:

Ferison **L**s**X**p**L**		*Ferio* **L**s**X**ga**L**st
Some B is C (as long as they exist) (**X**p)	→	Some C is B (at some times) (**X**ga)
No B is A (by necessity) (**L**s)		No B is A (by necessity) (**L**s)
Therefore Not every C is B (by necessity)(**L**)		Therefore Not every C is B (by necessity)(**L**st)

This *Ferio* to which *Ferison* is reduced contains **X**ga in the major premise by the conversion of the **X**p minor premise; since the **X**p propositions convert as **X**ga, it leads then to a proposition **L**st, because the major is necessary but the minor contains the condition at some times. So if some C is B at some times and no B is A as long as it is B, by necessity, then not every C is B at some times by necessity. But since **X**p implies **X**ga and **L** implies **L**st, this *Ferio* is implied by the initial *Ferio* **L**s**X**p**L**.

The second *Ferison* is *Ferison* **X**s**LX**ga, which is expressed as follows:

Ferison **X**s**LX**ga
Some B is C (by necessity) (**L**)
No B is A (as long as it is B) (**X**s)
Therefore Not every C is A (as long as it exists) (**X**p)

It is proved by the conversion of the minor and reduction to Ferio as follows:

Ferison **X**s**LX**ga	*Ferio* **X**s**X**ga**X**ga
Some B is C (by necessity)(**L**)	Some C is B (at some times) (**X**ga)
No B is A (as long as it is B) (**X**s)	No B is A (as long as it is B) (**X**s)
Therefore Not every C is A (as long as it exists)(**X**ga)	Therefore Not every C is A (at some times) (**X**ga)

This *Ferio* to which *Ferison* is reduced is implied by the categorical *Ferio*, which is *Ferio* **X**s**X**s**X**s, since the **X**s propositions imply the **X**ga ones.

The last mood is *Bocardo* **X**s**L**s**X**ga where the conclusion is explicitly said to be absolute ([34], 159.1–2), which is proved by *ekthesis* as in categorical logic (see Sect. 3.2.2). This proof should use *Felapton* **X**s**LX**ga, already admitted.

This first *Bocardo* should be *Bocardo* **X**s**L**s**X**ga, which should be stated as follows:

Every B is C, by necessity (**L**s)
Not every B is A, by actuality (**X**s)
Therefore Not every C is A (**X**ga)

Its proof by *ecthesis* should be the following: Let this "some B" which is not A, be "every D" then we have the following:

1. No D is A (by actuality) (**X**s) (assumption).
2. Every B is C by necessity (**L**s)(minor premise),
3. Every D is B by actuality (**X**p) (assumption),
4. Every D is C by necessity (**L**) (2, 3 by *Barbara* **L**s**X**p**L**).
5. Not every C is A, by actuality (**X**ga) (1, 4, by *Felapton* **X**s**LX**ga).

Thus, this Felapton leads to a general absolute conclusion, which means that Bocardo leads also to a general absolute one and justifies what Avicenna claimed from the start.

As to the second *Bocardo* which contains a necessary particular major premise and an absolute universal minor premise, Avicenna says that its conclusion is *traditionally* (*al-mashhūr*) an absolute proposition ([34], 159.7), so that it would be a *Bocardo* **LXX**. But Avicenna claims that nothing forbids the conclusion to be true if it is necessary, for he says "*fa 'inna hādhā al-'iṭlāqa lā yamna'u ṣidka aḍ-ḍarūrati*" ([34], 159.14). The example he provides, which seems to be classical is the following:

"Every biped is moving, by actuality
Some biped is not a man, by necessity
From which it is usually deduced that
Not every moving [being] is a man, by actuality" ([34], 159.11–12)

In his comment of the conclusion, which is here absolute, he says that it could as well be necessary, given that the proposition "Some of what is moving is not a man by necessity" (or "is necessarily not a man") is also true, since a horse is a moving being but it is necessarily not a man ([34], 159.15–16).

Now how could we prove *Bocardo*, given these details? In categorical logic, this mood is proved by ecthesis in Avicenna's frame. So let us prove it by ecthesis.

If *Bocardo* is stated as follows:

Bocardo **L**s**X**s**L**st

Every B is C, by actuality (**X**s)
Not every B is A, by necessity (**L**s)
Therefore Not every C is A (**L**st)

Its proof by *ecthesis* runs as follows: let this "some B" which is not A, be "every D", then we will have the following:

1. No D is A by necessity (assumption) (**Ls**).
2. Every B is C by actuality (**Xs**).
3. Every D is B by necessity (assumption) (**L**).
4. Every D is C by actuality (**Xp**) (2, 3 by *Barbara* **XsLXp**).
5. Not every C is A by necessity (**Lst**) (1, 4 by *Felapton* **LsXpLst**).

In *al-Najāt*, he states the rules governing the modality of the conclusion in the third figure, and says that "the conclusion follows the major" ([32], 39.8). He recalls the traditional opinion stated above, according to which the modality of the conclusion should be the same as that of the universal premise (the minor one in *Bocardo*) but criticizes it by saying "What is usually held in that figure and the second one is that the conclusion is not necessary [when the universal is absolute] in that case. But the truth is that the conclusion should follow the major even if it is particular, and this is shown by *ekthesis*, for when the major is particular negative, we say that the conclusion is necessary." ([32], 39.11–13). The proof by *ekthesis* of *Bocardo* **LsXsLst** is the following:

Let us call the things that are B's but are not A's, the D's. So Some B = every D. From this assumption it follows that:

(1) Necessarily No D is A as long as it is D (**Ls**) (assumption)
But (2) Every B is C as long as they are B (minor premise of *Bocardo*) (**Xs**)
And (3) Every D is B (assumption) (**L**)
Then (4) Some C's are D (by actuality) (from 2, 3, *Darapti* **XsLXga**)
Therefore (6) Some C's are not A by necessity (from 1, 4, *Ferio* **LsXgaLst**) ([32], 39.14–17, justifications added).

To summarize, if the above analysis is correct, Avicenna admits the following moods in the third figure: *Darapti* **XsLXga**, *Darapti* **LsXpLst**, *Felapton* **XsLXga**, *Felapton* **LsXpLst**, *Datisi* **LsXpLst** and *Datisi* **XsLXga**, *Disamis* **LsXsLst** and *Disamis* **XsLsXga**, *Ferison* **XsLXga** and *Ferison* **LsXpLst** and finally *Bocardo* **XsLsXga** and *Bocardo* **LsXsLst**.

Let us now turn to the moods which contain possible propositions combined either with absolute ones or with necessary ones.

4.3.1.2 The Moods with Possible (and Other) Propositions

What about the moods containing possible premises? Avicenna considers the case where all the premises are possible, the one where only one premise is possible while the other one is absolute and the one where one premise is possible while the other one is necessary. He starts by the first figure moods and holds some opinions which are not all in accordance with Aristotle's ones.

So in the first figure, when the minor premise is possible, the first figure moods "are not perfect" according to Ṭūsī ([36], 391.9) although they are also valid. In that case, the major can be either necessary or actual or possible and the modality of the conclusion is the same as that of the major premise. Consequently, these cases are all **LML** and **MMM** first figure moods. According to Ṭūsī, the last combination "need to be demonstrated" ([36], 391.10), which means that they need some extra assumptions to be admitted as valid. He adds that Avicenna, unlike his predecessors who provide proofs by *reductio ad absurdum*, does not prove them by that way. Rather he uses some principles like the principle according to which "the possibly possible is possible," that Avicenna states in both *al-Ishārāt* ([36], 391.7–8) and *al-Qiyās* ([34], 183.3).

However, Avicenna himself says that *Barbara* **MMM**, which he states in the very beginning of Sect. 4.1 of *al-Qiyās*, is "a perfect syllogism [*fa hādhā qiyāsun kāmilun*]" ([34], 181.9) because when "Every C is B by possibility, and every B is A by possibility, then every C is A by possibility" because "C enters potentially into B, therefore it possesses potentially what B possesses" ([34], 181.8–9). So it is the principle of "what is said of every" (*dictum de omni*) together with the general principle that what is possibly possible (*mā yumkinu 'an yumkina*) is possible stated in *al-Qiyās* ([34], 183.3) and *al-Ishārāt* ([36], 391.7–8) that justifies this mood. As a matter of fact, this last principle is admitted in the system S4 of contemporary modal logic where it is formalized as follows: "$\Diamond\Diamond\alpha \to \Diamond\alpha$" (See [76], Sect. 2). In *al-Ishārāt*, Avicenna says that it "is natural (*qarībun 'inda at-ṭab'i*)" ([36], 391.8). So maybe he considers that the justification of some moods by this natural assumption is comparable to the justification of the first figure (categorical) moods by the "*dictum de omni and de nullo*", which does not make these moods imperfect. Ṭūsī himself adds in the rest of his commentary that Avicenna "tends to consider this combination as perfect and not as in need of a further justification (*ghair muḥtājun 'ilā ziyādati bayānin*)" ([36], Ṭūsī's comment, p. 391, note 8, line 3), which confirms what is said by Avicenna in *al-Qiyās*. The possibility here seems to be understood as potentiality which corresponds to the unilateral possible, for this notion is used in the justification provided by Avicenna. This is also Thom's interpretation of the possibility in this mood, for he states Barbara MMM as follows: "Barbara MMM: if $j_M \subset b_M$ and $b_M \subset a_M$ then $j_M \subset a_M$" and says explicitly that Avicenna "takes the MMM moods in Fig. 1 to be valid" ([143], 364). It seems also to be internal, for the adverb "possibly" is put in the end of each proposition.

But *Barbara* **MMM** is very controversial as many authors after and before Avicenna have shown. Avicenna himself acknowledges that it has been rejected by some people before him ([34], 181.10–12). The principle that what is possibly possible is possible is itself reformulated as follows: "A which is possible to B (itself) possible to C is possible to C" ([34], 183.15). This shows that the possibility puts on the predicate, not on the quantifier. Avicenna compares between what is "possibly possible", what is "necessarily necessary" and what is "actually actual" ([34], 184.3–4). Note that according to the same principle, we could also have *Barbara* **QMQ**, since if B is $\Diamond$A $\wedge$ $\Diamond\sim$ A, and C is $\Diamond$B then C [like B] is $\Diamond$A $\wedge$ $\Diamond\sim$ A. Although Avicenna does not say clearly that one of the premises and the conclusion are narrow-possible, in his proofs of the third figure moods, he says explicitly that these moods contain bilateral possible conclusions and major premises and he proves them by reducing them to first figure moods. These proofs suggest that the other first figure moods can also have the structure **QMQ**.

However one must be careful about the eventual **QMQ** moods containing negative premises, for **E** and **O** bilateral possible are not rightly stated by Avicenna, as we saw above. Thus maybe *Barbara* **QMQ** and *Darii* **QMQ** can be admitted, but *Celarent* **QMQ** and *Ferio* **QMQ** (as well as the third figure moods with the same structure and with a negative premise and a negative conclusion) are much more doubtful.

The second mood, i.e., *Celarent* **MMM,** is stated as follows: "Every C is B by possibility and Possibly no B is A, therefore Possibly no C is A" ([34], 186.7–9). However, here, in the mood as it is stated, the possibility operator is put in the beginning of the second premise and the conclusion, while the first premise contains an internal possibility. Since Avicenna does not provide any explanation of this mood and given the place of the modalities in the second premise and the conclusion, one might think that the second premise and the conclusion contain an external (i.e., a *de dicto*) modality, but this is not the case for later in *al-Qiyās* ([34], 189.7–8), Avicenna says explicitly that when the possibility operator puts on the quantifier, i.e., when the possibility is external, the syllogisms are not conclusive. To illustrate that opinion, he gives the following example: "If we say: It is possible that every human is white, and it is possible that every white is a horse … it follows from these that necessarily no human is a horse," while when the premises are "Possibly every human is white and possibly every white is an animal," what follows is "every human is an animal by necessity." Since these two *Barbaras* lead to two different and contrary conclusions, the mood thus expressed cannot be valid, which means that even in *Barbara*, the possibility operator must be internal. So despite the several formulations of the modal propositions, which tend to make things a little bit confused, one has to interpret the modalities used in the syllogistic moods as internal, i.e., as putting on the predicate rather than on the quantifier.

He also says that the mood where both premises are negative; possible and universal is valid and "produces a negative possible" ([34], 187.11–12). This "mood" is expressed as follows:

"No C is B by possibility (Q)
Every B is A by possibility
Therefore Every C is A by possibility (Q)
Consequently No C is A by possibility (Q)" ([34], 186.10-11. 187.8–9)

It is proved by changing "No C is B by possibility" by "Every C is B by possibility," given the bilateral character of the possibility and the alleged "equivalence" between **A** bilateral possible and **E** bilateral possible. Avicenna says that "all what is not so and so by the real possibility (= the bilateral possibility) is also so and so by the real possibility" ([34], 187.3–4). This justifies, according to him "the shifting (*naql*) of the affirmation to a negation or of the negation to an affirmation" ([34], 187.7). This "shifting" (or "conversion" (*'aks*) as Avicenna will say afterward, p. 198.7) is used in the minor premise to produce an affirmative proposition, and also in the conclusion to produce a negative one.

However, this claim is based on the confusion already mentioned in Sect. 4.1, which is related to the way the negative bilateral possible propositions, where the possibility is internal, are expressed. As we have seen, Avicenna thinks that **A** bilateral possible and **E** bilateral possible amount to the same, since the former would be rendered by **A**$\Diamond \wedge$ **E**$\Diamond$, while the latter would be rendered by **E**$\Diamond \wedge$ **A**$\Diamond$. But the fact is that they are different as we have seen, since **E** bilateral possible contains a negation putting on the whole "$\Diamond \wedge \Diamond\sim$", which means that **E** bilateral possible should be expressed in terms of a disjunction relating a necessary proposition and an impossible one. The same could be said about **O** bilateral possible, when the possibility is internal. So if we take this analysis into account, the mood above is not valid and its proof is not correct.

As a matter of fact, the above mood would have the structure **AEE**, since Avicenna insists on the fact that the minor premise and the conclusion are *negative* ([34], 186.11), but there is no valid **AEE** mood in the first figure. Now, if we consider it as a *Barbara* mood with bilateral possible propositions, then it would be expressed as follows:

Every C is possibly B (and possibly not B) (**AQ**)
Every B is possibly A (and possibly not A) (**AQ**)
Therefore Every C is possibly A (and possibly not A) (**AQ**)

In this case, as we can see, there is a negative side in each premise and in the conclusion, but the whole mood would be a *Barbara* mood with bilateral possible propositions and would not contain **E** propositions. Thus expressed, it would not correspond to what Avicenna is talking about.

As a consequence, there is a problem with this mood and all those which are said to contain negative (bilateral) possible premises, such as the fourth mood mentioned by Avicenna which is supposed to contain "two negative possible premises" and "to produce a negative possible conclusion by shifting the minor to an affirmative [proposition]" ([34], 187.11–12). If we express this "mood" as usual, we have the following:

Every C is $\Diamond\sim$B ($\wedge$ $\Diamond$B) (Q) (**E**?)
Every B is $\Diamond\sim$A ($\wedge$ $\Diamond$A) (Q) (**E**?)
Therefore Every C is $\Diamond\sim$A ($\wedge$ $\Diamond$A) (Q) (**E**?)

As we said above, if the two premises and the conclusion are supposed to express **E** bilateral possible proposition, then they are not rightly expressed and they are not valid, since **E** bilateral possible should not be expressed in that way. If on the contrary, we consider all propositions as being **A** bilateral possible propositions, then they are rightly expressed. However, even in this case, there is a problem related to the middle term which is not exactly the same in both premises.

The middle, here and in the mood that preceded it, is "B" in the major premise and "$\Diamond B \wedge \Diamond \sim B$" in the minor one. This raises a problem because the middle term in all the syllogistic moods should be exactly the same term in both premises.

Avicenna does not seem to pay attention to that particular problem, however, since he expresses the possible propositions explicitly as follows: "Every actual C (*kull mā huwa [C] bi al-fi'li*) is possibly B, and every actual B (*kull mā huwa [B] bi al-fi'li*) is possibly A..." ([34], 183.6–7), which means that in a possible proposition, the possibility operator puts on the predicate but not on the subject, which is said explicitly to be actual. So the middle term is not exactly the same in both premises, as will be the case in Buridan's system, for instance, where the modality operator is put in front of both the subject and the predicate.

What about *Darii* and *Ferio*? Are they valid when their premises are possible? Avicenna evokes them at page 187.13–14, where he just says: "and you can combine yourself these four moods: from a particular minor, and a universal major, whether negative or affirmative, or affirmative and negative, or negative and affirmative." This seems to validate *Darii* **MMM**, *Ferio* **MMM**. In this case too, the moods containing two negative propositions are held by Avicenna ([34], 187.13–14), but they raise the same problems as the ones containing **E** bilateral possible propositions discussed above. Consequently, unlike what Avicenna says, they cannot be considered as valid.

When the major premise is possible while the other one is absolute, the conclusion is possible as shown in *al-Qiyās* and elsewhere. This combination holds for the first figure moods such as *Barbara* **MXsM**, which is stated as follows in *al-Najāt* for instance:

"Every C is B by actuality (**Xs**)
Every B is A by possibility (**M**)
Therefore Every C is A by possibility (**M**)" ([34], 40.6–7).

The other first figure moods **MXsM** are also valid, such as *Darii* **MXsM**, *Celarent* **MXsM** ([32], 41.9–11) and *Ferio* **MXsM**.

In all these moods, when the minor is an absolute proposition not containing any kind of necessity, the conclusion may be a real possible one, that is, a bilateral possible proposition (*al-Najāt*, p. 41.9–11). This seems to allow the first figure moods **QXsQ** to be valid too.

If the major premise is absolute while the minor is possible, this combination is illustrated by the following:

"If every C is B by possibility (**M**)
and every B is A by actuality (**Xs**)
then we say every C is A by possibility (**M**)" ([34], 192.2–3).

In this case, the conclusion is possible too, so that we can say that Avicenna admits *Barbara* **XsMM**. He adds that the possibility is unilateral (*bi-l-'imkān al 'āmm*) as usually admitted ([34], 192. 4).

All these moods are parallel to those containing **Xs** and **Xga** propositions which we found in some of our proofs of the moods containing absolute and necessary propositions (see Sect. 4.3.1.1), since the **Xga** propositions (the general absolutes) are parallel to the **M** ones (the possible ones).

However, Avicenna criticizes Aristotle's proof of this *Barbara* mood by *reductio ad absurdum*, which runs as follows: if the conclusion of *Barbara* **XsMM** stated above ([34], 192. 2–3) is false, this means that the necessary proposition "Some C is not A by necessity" is true. Let us suppose that the original possible premise is absolute (since this hypothesis is not impossible even if the proposition becomes false in that case), we have then the following: "Every C is B by actuality." From these two premises, we can deduce by *Bocardo* **LXL** the following conclusion: "Some B is not A by necessity." We can write this as follows:

Some C is not A by necessity (the contradictory of the conclusion)
Every C is B by actuality (assumption from the possible original premise)
Therefore Some B is not A by necessity (from *Bocardo* **LXL**)

But this conclusion is incompatible with "Every B is A by actuality" which was the major premise of *Barbara*. Therefore it is not acceptable.

According to him, the problem with this proof is the supposition that the possible premise can be replaced by an absolute false proposition. Avicenna says that one can provide a proof without appealing to this supposition, this is why he provides his own proof by *reductio ad absurdum,* which runs as follows: if the conclusion is false, then "Some C is not A by necessity" is true, but we had "Every B is A by actuality," so by *Baroco* **XpLsL**, we can deduce "Some C is not B by necessity." But this contradicts the minor premise "Every C is B by possibility," which is not acceptable ([34], 193.3–5).

In *al-Najāt*, he provides a proof similar to the first one, in that it relies on the same hypothesis, but he also says that the conclusion might express a bilateral possibility provided that the absolute premise contained in the mood is "purely absolute", that is, not necessary at all. In that case, when negating the conclusion, he gets a disjunction and proves that the first side of the disjunction together with one of the premises leads to an incompatibility, and that the same happens with the second side. This means that the conclusion of *Barbara* **XsMM (XsQQ?)** may be a bilateral possibility.

Barbara **XsMM (XsQQ?)** is expressed as follows in *al-Najāt*:

Every C is B by possibility (**Q**?)
Every B is A by actuality (**Xs**)
Therefore Every C is A by possibility (**Q**?) ([32], 40.7–8+ note 2)

If the conclusion expresses a bilateral possibility, then its contradictory is the following disjunction: "Some C is necessarily not A or necessarily A," which is equivalent to "Either some C is necessarily not A or some C is necessarily A." Avicenna reasons as follows: Let the conclusion be false, then its whole negation is true; therefore, either its first side is true or its second side is true. Let us first consider its first negative side, we will thus have

Some C is not A by necessity
and Every C is B (by actuality) (by the same hypothesis as in *al-Qiyās*) (**Xs**)
Therefore Some B is not A by necessity (by *Bocardo* **L**s**X**s**L**st from the third figure).

But this conclusion is incompatible with the major premise "Every B is A by actuality," even if it is not exactly its contradictory, since this premise says that every B is A *as long as it is B*, while the conclusion that we get by *Bocardo* **L**s**X**s**L**st says that "Some B is not A *by necessity at some times*," i.e., there are some times where some B are not A by necessity, which is incompatible with saying that these B's are A *all the time when they are B*.

If the affirmative side is the one which is true, then the proof runs as follows:

Some C is A by necessity
And Every C is B by actuality (by the same hypothesis as above) (**Xs**)
Therefore Some B is A by necessity (by *Disamis* **L**s**X**s**L**st from the third figure)

But this result is said to be "absurd" (*khalf*) ([32], 41.2), for the conclusion is incompatible with the absolute premise "Every B is A," which does not contain any kind of necessity, being purely absolute (*muṭlaqa ṣirfa*) ([32], 40.8), that is, not necessary at all (*lā ḍarūrata fīhā al-battata*) ([32], 40.8).

In *al-Ishārāt*, the same mood is stated differently and it is said to contain a bilateral possible premise and to lead to a general possible one, for Avicenna says "If every C is B by the real narrow possibility (*bi-l-'imkān al-ḥaqīqī al-khāss*) and every B is A by actuality, then it is permissible that every C is A in actuality and it is permissible that it is so in potentiality. And what is common to both must be the general possible" ([36], 392.1–4, part of the translation reported from Thom 2008, in [144], 364). Here the conclusion may be possible, but its possibility is general. Ṭūsī illustrates this mood by the following example: "Every man is a writer by possibility, and every writer uses a pencil by actuality, it does not follow that every man uses a pencil by actuality, rather [every man uses a pencil] by possibility and maybe he could actually (*bi-l-fi'l*) do so" ([36], 392, note 9, lines 8–11). He says explicitly in his commentary of this passage that the mood "produces a possible" ([36], 392, note 9). So the mood here is a *Barbara* **X**s**QM**. But the possibility of the conclusion is general, for what is possible can become actual, given that it is not impossible. We may state it as follows:

"Every human is possibly a writer (and possibly not a writer) (**Q**)
Every writer uses a pencil (**Xs**)
Therefore every human [possibly] uses a pencil (**M**)" ([34],196.15–17)

Does this mean that the conclusion may be absolute? In this case, the mood may be a *Barbara* **X**s**MX**ga (**X**s**QX**ga?). But Avicenna does not seem to admit the combination **X**s**MX**ga (nor **X**s**QX**ga), for he raises that same question "Is this conclusion true when absolute? (*hal taṣduqu muṭlaqa*?)" and answers it by saying "this is not necessary (*lā yajibu dhālika*)" ([34], 196.12). So it seems that the conclusion in this mood is not absolute, but possible. In addition, we note that the examples he gives contain possible conclusions, not absolute ones and even the proofs he uses presuppose that the conclusion is possible, as we saw in the analysis provided above of this particular kind of *Barbara* in *al-Qiyās* and *al-Najāt*.

However, in his analysis of Avicenna's modal syllogistic, Paul Thom says that Avicenna, despite the fact that his text is unclear (i.e., *al-Ishārāt* cited at several places) may have admitted *Barbara* **XMX** as follows: "Barbara XMX: if $j_M \subset b_M$ and $b_M \subset a_m$ then $j_M \subset a_m$" ([144], 364), together with Barbara XMM, which is acceptable too but not stated clearly in Avicenna's text. As a matter of fact, it is Thom's interpretation itself that makes *Barbara* **XMX** "valid, and perfect, on the simple *de re* reading" ([144], 364). For in this interpretation, the formula provided by P. Thom says the following, if we take into account his explanations of the different symbols he uses: "If every possible *j* is a possible *b* and every possible *b* is sometimes *a*, then every possible *j* is sometimes *a*" (See [144], 362, where he says that M = "possible" and m = "at some times"). Thus interpreted, the mood is valid, since it is true by transitivity. But this interpretation does not account for what Avicenna says. For one thing, Avicenna talks in the analysis provided in *al-Ishārāt* about a real (that is, a bilateral) possibility for the possible premise that the mood contains, which is not the case in Thom's formal interpretation. Second, the actual or absolute premise is interpreted by Thom as saying "every possible *b* is sometimes *a*." But this interpretation of the absolute proposition, already analyzed in Sect. 3.2.2, is not the one that is generally used by Avicenna in his syllogistic moods, as we saw in Sect. 3.2.2. This is so because the propositions containing "at some times" *do not convert* as Avicenna insists in several places, in particular in *al-Qiyās*, but also in *al-Ishārāt* ([36], 322.3–9). The convertible propositions used by Avicenna are explicitly said, in *al-Qiyās*, to be those that contain the condition "as long as it is S" or the condition "as long as S exists" as we saw above (Sect. 3.2.2). Third, Paul Thom seems to take the ampliation of the subject in possible and absolute propositions as granted. But this ampliation is not at all obvious in Avicenna's frame, in particular with regard to the absolute propositions and even with the possible ones, first because Avicenna does not explicitly use this way of expressing the modal propositions, second because he gives existential import at least to the affirmative propositions as we already noted above when we evoked the moods containing only possible propositions, which means that their subject must be existent in order for these propositions to be true, when they are absolute, necessary, or possible.[15] This being so, it seems that Thom's interpretation of Avicenna's syllogistic is not entirely convincing and does not conform to the texts.

Now what about the other first figure moods? We could assume that *Darii* functions as *Barbara*, since it contains only affirmative propositions. We might thus

[15]See Saloua Chatti ([55]), where it is shown that the possible propositions too have an existential import.

assume that Avicenna admits also *Darii* **XsMM**, although the text is not very clear ([34], 198.11–12), where he says "If the minor is possible and the major absolute, then the conclusion is as you already know (*'alā mā salafa laka*)". As to *Celarent*, Avicenna talks about it in both *al-Qiyās* and *al-Najāt*. In *al-Qiyās*, he provides the following mood:

"Every C is B by possibility (**M**)
And No B is A by actuality (**Xs**)
Therefore No C is A by (unilateral) possibility (**M**)" ([34], 197.1–2)

This is clearly *Celarent* **XsMM**, since the affirmative premise is the absolute one. He adds that its proof can be made by *reductio ad absurdum* as was the case with *Barbara* **XsMM**. Thus if the conclusion is false, then its contradictory "Some C is A by necessity" will be true. If we add by hypothesis "Every C is B by actuality" (the corresponding absolute of the possible minor premise), we can deduce "Some B is A by necessity," by *Disamis* **LsXsLst** from the third figure. This conclusion is incompatible with the minor premise which says "No B is A by actuality," even if it does not contradict it, which is not acceptable. In the same way, we might admit *Ferio* **XsMM** too, since the only difference with *Celarent* **XsMM** is the particularity of the minor premise. Since what holds for the universal holds for the particular, this validates *Ferio* **XsMM**. However, the "mood" stated just after that, which is the following:

"No C is B by possibility
Every B is A by actuality
Therefore No C is A by possibility" ([34], 198.6–7)

and is said to be provable by "converting (*'aks*) the possible negative premise to an affirmative one" and then by "converting the conclusion to an affirmative one" ([34], 198.7–8) is not valid, unlike what Avicenna says, since the affirmative possible propositions are not equivalent to the negative ones when the modality is internal as we saw above. As to the moods with universaltwo negative premises, Avicenna says that when the "minor is negative and absolute" ([34], 198.8–9), then the mood is not valid.

In addition, if the moods of the first figure contain a necessary premise and a possible one, the conclusion will be possible if the major is possible ([34], 199.1–2), so *Barbara* **MLM** is valid, the proof being made by means of the principle of "what is said of every" (*dictum de omni*) ([34], 199.2), and can be expressed as follows:

Every C is B by necessity (**L**)
Every B is A by possibility (**M**)
Therefore Every C is A by possibility (**M**)

This is so because if every C is by necessity B and if every B is A by possibility, therefore every C, which a subclass of B, will be A by possibility, just as B.

As to the second *Barbara* (*Barbara* **LQM**), where the major is necessary, while the minor is possible, he first provides a proof of it by *reductio ad absurdum* ([34], 199.8–11), which shows that the conclusion is unilateral possible. This mood is expressed as follows:

Barbara **LQM**
Every C is B by possibility (**Q**)
Every B is A by necessity (**L**)
Therefore Every C is A by possibility (**M**)

The conclusion of this mood is said to be possible in the unilateral sense (**M**) "(*mumkina bi al-ma'nā al-'āmm*)" ([34], 199.4). Its proof is by *reductio ad absurdum* and runs as follows:

If the conclusion is false, then "it is not possible that every C is A" (*kānat ghair mumkina an takūna kull C A*) ([34], 199.5), then

Necessarily some C is A (**L**)
and Necessarily every B is A (**L**)
Therefore Necessarily some C is not B (**L**)

But we had "Possibly every C is B (by real [= bilateral] possibility) (**Q**)(*wa kāna bi al-imkān al-ḥaqīqī kull C B*)" ([34], 199.3–7), which is incompatible with this necessary conclusion and is not acceptable.

Likewise, he presents a proof of *Celarent* **LQM** by *reductio ad absurdum* and by using the oppositional relations of the modal propositions. For *Celarent* **LQM** is the following:

Celarent **LQM**
Every C is B by possibility (**Q**)
No B is A by necessity (**L**)
Therefore No C is A by possibility (**M**)

If the conclusion is false, then its contradictory, that is, "Some C is A by necessity" will be true, but we had "'Necessarily no B is A' (*bi-ḍ-ḍarūra lā shay'a min B A*)" ([34], 199.14), from which one can deduce "Some C is not B by necessity," by *Festino* **LLL**. But "Some C is not B by necessity" is incompatible with the minor premise "Every C is B by possibility," which is not acceptable ([34], 199.13–14).

But it seems that these proofs are those provided by Aristotle, for Avicenna says what follows:

"What is said in the first teaching about these moods is a universal opinion (*qawlan kullīyan*): if the necessary major is affirmative, then what follows is only possible, and not an absolute, while if it is negative, it produces a possible [conclusion] and an absolute not necessary [one]" ([34], 199.15–200.2).

As to his own opinion, it seems to be expressed in what follows: "the conclusion in this [mood] and what resembles it is necessary and I say that the affirmative mood and the negative one where the major is necessary produce a necessary conclusion" ([34], 200.3–4).

So, according to Avicenna, and unlike Aristotle, when the major is necessary, the conclusion will be necessary too, whether in *Barbara* or in *Celarent* ([34], 202.3–4). So *Barbara* **LML** is admitted and expressed as follows:

Barbara **LML**
"Every C is B by possibility (**M**)
Every B is A by necessity (**L**)
Therefore Every C is A by necessity (**L**)" ([34], 202.5)

This is also proved by *reductio ad absurdum*, and by first supposing that "the possible [premise] is actual (*mawjūd*)" ([34], 200.6), for then the contradictory of the conclusion is "Some C are not A by possibility," but we had "Every B is A by necessity," so by *Baroco* **LXsXp** we can deduce not only "Some C is not B by possibility" but rather (*bal*) "Not possibly every C is B," which is equivalent to "Necessarily Not every C is B," that is, "Necessarily Some C is not B," which follows "from the doubtful premise" and contradicts the minor premise "Every C is B by possibility" ([34], 202.7–8). But the necessary premise means what follows: "The sentence 'Every C is B by necessity' means: every thing described as C, is described as B, *as long as its essence exists* – even if its description [as C] changes" ([34], 202.12–14, my emphasis). This is why the conclusion that follows means: "Every C is A by necessity as long as its essence exists. So if its essence exists, then it is A by necessity" ([34], 202.12–14). As to the possibility used in this mood, it seems to be a unilateral possibility, for he says "and in general you must know that what can be necessary is permanently necessary and its possibility is the most general (= unilateral) possibility (*al-imkān al-a'amm*)" ([34], 203.4–5).

"Every human is possibly moving (**M**)
Every moving is a body by necessity (**L**)
Therefore evey human is a body by necessity (**L**)" ([34], 203.10–11)

His explanation is the following:

"For when every moving [being], as long as its essence exists – whether it is moving or not – is described as being a body, and when a human when he is moving, is truly said to be a body by necessity, i.e., as long as his essence exists, whatever his situation is, then it follows necessarily – even if he is not moving – that he is a body, because he is a body as long as his essence exists, not only when he is moving" ([34], 203.11–15).

This explanation shows that the necessary proposition is said to be necessary (= **L**) but also parallel to an **Xp** proposition.

The same can be said about *Celarent* **LML**, which also has a necessary conclusion when its major premise is necessary, and about *Ferio* **LML**, which contains the same kinds of propositions ([34], 204.1–2). As to *Darii*, it behaves like *Barbara*, for it contains only affirmative premises, so we can say that *Darii* **LML** as well as *Darii* **MLM** are both admitted. The general rule deduced by Avicenna is then the following: In the first figure, the modality of the conclusion should follow that of the major premise, whether the major is necessary or possible, affirmative or negative ([34], 204.15–16).

The necessary premise, in all these moods, should contain the condition "as long as it exists," as in the following example "Every human is possibly moving, and every moving (thing) is a body by necessity; therefore every human is a body by necessity," that is, as long as he exists.

In *al-Najāt*, he examines once again the first figure **LML** moods and criticizes the proof provided by Aristotle who says that when the major premise is necessary and the minor is possible, the conclusion should be possible, so that the mood is **LMM (LQQ?)**, not **LML**. As it is rendered by Avicenna, Aristotle's proof by *reductio ad absurdum* is the following. *Barbara* **LMM (LQQ**?) is expressed as follows:

Every C is B by possibility (**Q**?)
Every B is A by necessity (**L**)
Therefore Every C is A by possibility (**Q**)

Note that the conclusion expresses a bilateral possibility (*mumkina ḥaqīqīya*) ([32], 41.19), from which we can assume that the premise expresses a real (i.e., a bilateral) possibility too. If the conclusion is false, then its contradictory, that is,

"Some C is not A by necessity" is true.
But we had Every B is A by necessity
Therefore Some C is not B by necessity (by *Baroco* **LLL**)

which contradicts "Every C is B by possibility" (the minor premise). But Avicenna says that this is not a real contradiction ([32], 42.1) because this proposition [Every C is B by possibility] does not express a general possibility, rather it expresses a bilateral possibility, whose contradictory is a disjunction.

He goes on proving the mood once again in order to show that the conclusion should be necessary rather than possible. The mood is expressed as follows:

Every B is A by necessity
Every C is B by possibility
Therefore Every C is A by necessity (as long as it exists) ([32], 42.5–10)

The example that illustrates it is the same as in *al-Qiyās*, and is the following:

"Every human is moving by possibility (**M**)
Every moving (being) is a body by necessity (**L**)
Therefore Every human is a body by necessity (**L**)" ([32], 42.10–11)

The same may be said about *Celarent*, which is also **LML** in Avicenna's frame, while it is **LMM** in Aristotle's one.

Now what about the second and the third figures' moods if one or more premises are possible? According to Avicenna, when the premises are possible, no mood is conclusive in the second figure as appears in the following quotation: "No syllogism is conclusive in the second figure when the premises are possible" ([32], 42.15; [34], 205.4; [36], 404). The same can be said about the moods where one of the premises is possible while the second one is absolute, as Avicenna says in *al-Najāt* ([32], 43.4) and elsewhere ([36], 404), unless the absolute proposition is convertible and can be necessary, that is, "when the absolute is taken to be true when necessary" (*bi ḥaythu taṣuḥḥu ḍarūrīyatan*) ([32], 43.5). He adds that no conclusive mood with two general absolute propositions (*bi-l-iṭlāq al-'āmm*), that is, with absolutes containing the condition "at some times," is held in that figure ([36], 404.6). So the combinations **MM**, **X**ga**X**ga, and **MX**ga do not lead to any conclusion in the second figure ([36], 403–404). But the combination necessary-possible is held in that figure. Likewise, when the absolute "is convertible or has a contradictory of its own kind [i.e. absolute too, like it]" ([36], 408.4), then we could have valid moods in that figure.

When one considers the combination between absolute and possible propositions in the second figure, the absolute must contain conditions which make them behave like the necessary propositions. These conditions may be "as long as it is S" or "as long as S exists." Thus, the first mood containing a possible and an absolute proposition is *Cesare* **X**s**MM**, which is reduced to *Celarent* **X**s**MM** by the conversion of the minor premise. This mood is expressed as follows:

"Every C is B by possibility (**M**)
No A is B by actuality (**X**s)
Therefore No C is A by possibility (**M**)" ([34], 214.2–4)

The conversion of the major absolute leads to "No B is A by actuality," which when added to the minor premise "Every C is B by possibility" leads to the conclusion "No C is A by possibility" by *Celarent* **X**s**MM**.

However, since according to Avicenna, **E** Possible does not convert, when the major premise in this mood is possible while the minor is assertoric, no conversion can be used to reduce the mood to *Celarent* **MX**s**M**, which makes this particular combination (i.e., *Cesare* **MX**s**M**) invalid.

So the only valid moods containing absolute and possible propositions in this figure are those where the absolute or possible premises are convertible, whether they are affirmative or negative.

If one expresses the premises of the second mood (*Camestres* **MX**s**M**) as follows:

No C is B by actuality (convertible) (**Xs**)
Every A is B by possibility (**M**)

What would be the conclusion? It should be "No C is A by possibility" (**M**). But the conversion of the absolute premise, which could reduce this one to *Celarent* **XsMM**, leads to the following:

No B is C by actuality (**Xs**)
Every A is B by possibility (**M**)
Therefore No A is *C* by possibility (by *Celarent* **XsMM**)

Unfortunately "No A is C by possibility" is *not* equivalent to "No C is A by possibility," since **E** Possible does not convert. So it seems that the conversion, here, does not lead to a valid mood, which means that *Camestres* **MXsM** is not valid.

If the major premise is possible (**M**), while the minor is absolute (**X***ℓ*), the conversion of **E** is not even acceptable, so there is no way to reduce it to *Celarent* **MXsM**, as appears in what follows:

No C is B by possibility (**M**)
Every A is B by actuality (**Xs**)
Could this lead to No C is A by possibility (**M**)?

The answer should be: No, since neither "No C is B by possibility" nor the conclusion are convertible, which makes it impossible to reduce this mood to *Celarent* **MXsM**.

What about the other second figure moods, that is, *Festino* **MXsM** + **XsMM** and *Baroco* **MXsM** + **XsMM**?

It is clear that *Festino* **MXsM** is not conclusive, for **E-M** does not convert. But what about *Festino* **XsMM**? Could one reduce it to *Ferio* **XsMM** by conversion?

Festino **XsMM** would be the following:

No C is B by actuality
Some A are B by possibility
Therefore Some A are not C by possibility

The conversion of the first premise leads to

No B is C by actuality (**Xs**)
Some A are B by possibility (**M**)
Therefore Some A are not C by possibility (**M**) (By *Ferio* **XsMM**, which is valid as is *Celarent* **XsMM**).

So we might assume that *Festino* **XsMM** is valid, since it is provable. Avicenna says that explicitly in *al-Ishārāt* where he claims that "if the major is universal negative, and of the mentioned kind of absolute, and the possible is affirmative or negative, then it is reduced by conversion to the first figure, or by *ekthesis* and it produces the conclusion that you know in the first figure" ([36], 414.2–5). This claim holds for the moods whose major premise is an absolute **E** proposition, i.e., *Cesare* and *Festino*, with absolute major premises.

However, we could not say the same about *Festino* **MXsM**, whose major premise is not convertible.

What about *Baroco* with one **M** premise and one **Xs** premise? Are they valid? And what kind of conclusion should they produce? If the major is possible, *Baroco* could be expressed as follows:

Every C is B by possibility (**M**)
Some A are not B by actuality (**Xs**)
Therefore Some A are not C by possibility (**M?**)

Is this conclusive? How could one prove it? We could prove *Baroco* by *ekthesis* or by *reductio ad absurdum*. If one uses *reductio ad absurdum*, one supposes that the conclusion is false, therefore "Every A is C by necessity" will be true. If we add "Every C is B by possibility," then by *Barbara* **MLM**, we would deduce "Every A is B by possibility." Unfortunately, this conclusion does not contradict nor is it incompatible with the minor absolute premise "Some As are not B by actuality," since when one says, for instance, "Some people are not writing by actuality," this is not incompatible with "Every human being is writing by possibility." It follows that *Baroco* **MXs**M is not valid.

What about *Baroco* **XsMM**? This would be expressed as follows:

Every C is B by actuality (**Xs**)
Some A is not B by possibility (**M**)
Therefore Some A are not C by possibility (**M**)

Here too, we could try to prove it by *reductio ad absurdum*. If we consider the conclusion false, then the following is true:

"Every A is C by necessity' (**L**)

Then we have "Every C is B by actuality" (**Xs**), from which we can deduce "Every A is B by actuality (**Xp**)" or 'Every A is B (as long as it exists)', by *Barbara* **XsLXp** (see [32], 38.6–7)

Is this conclusion incompatible with the second premise "Some A is possibly not B"? The answer is "no" if we consider that when "Every A is B (as long as it exists)" is true, "Some A is possibly not B" or more precisely "Not (Every A is B by necessity)" since **O**$\Diamond$ does not have an import, is undetermined as can be shown by the formula below (where **O**$\Diamond$ is formalized as $\sim$ **A**$\Box$):

$$[(\exists x)Ax \wedge (\forall x)(Ax \supset Bx)] \;\underline{\vee}?\; \sim [(\exists x)Ax \wedge (\forall x)(Ax \supset \Box Bx)$$

If we consider just one element in the universe (x_1) we have the following lines of the table:

$[Ax_1 \wedge (Ax_1 \supset Bx_1)]$?	$\sim[Ax_1 \wedge (Ax_1 \supset \Box Bx_1)$
1 1 1 1 1	0? 1 ? 1 ? ? 1
1 0 1 0 0	1 1 0 1 0 0 0

In the first line of this table, we show that when **A** is true, **O◇**, that is ~ **A□**, is undetermined, i.e., neither false nor true. So there is no real contradiction, even if when **A** is false, **O◇** (= ~ **A□**) is true. In the second line, we show that both propositions cannot be false together, so there is no contrariety between them.

Now what about the combinations between possible and necessary premises in the second figure moods?

In *al-Qiyās*, he expresses the second figure mood *Cesare* in the following way:

"Every C is B by possibility
No A is B by necessity
Therefore No C is possibly A" ([34], 216.2–4 formulation slightly modified)

Avicenna's own formulation of the conclusion is not very clear, however, for he says "it follows by general possibility *and* by necessity that No C is A" ([34], 216.3, my emphasis). But things become clearer once one recalls the equivalence between "No C is possibly A" and "Every C is necessarily not A," which means that the conclusion is necessary. The mood is then *Cesare* **LML**.

This interpretation is confirmed by the proof by *reductio ad absurdum* provided by Avicenna in the sequel, for he says that if the conclusion is false, then

"Some C are possibly A" (unilateral possibility)

and "No A is B by necessity," from which we can deduce:
"Some C are not B by necessity" by *Ferio* **LML**.

This contradicts the minor premise "Every C is B by possibility," and is not therefore acceptable.

As to *Camestres*, where the negative minor is necessary, it could "be proved by two conversions, for the conclusion is necessary and its converse is necessary too" ([34], 216.11–12). So here we would have a *Camestres* **MLL**, which would be expressed as follows:

No C is B by necessity
Every A is B by possibility (bilateral possibility)
Therefore No C is A by necessity

Avicenna proves it by *reductio ad absurdum* in the following way: if the conclusion is false, then

"Some C are possibly A (unilateral possibility)",
but we had "Every A is B by possibility (bilateral possibility)"
Therefore "Some C is B by possibility (bilateral possibility)" (by *Darii* **QMQ?**) ([34], 217.2–5)

But this contradicts "No C is B by necessity," which is not acceptable.

If the affirmative premise in *Camestres* is necessary, the conclusion is also necessary, for in all the second figure moods where one of the premises is necessary and the other one possible, the conclusion is necessary.

Camestres **LML** would then be expressed as follows:

No C is B by possibility
Every A is B by necessity
Therefore No C is A by necessity

For if the conclusion is false, then,

Some C is A by possibility
but we had Every A is B by necessity
Therefore Some C is B by necessity (by *Darii* **LML**)

In the same way, in the other second figure moods containing one possible and one necessary, the conclusion is necessary. So *Festino* **LML** as well as **MLL**, and *Baroco* **LML** as well as **MLL** should be admitted.

Festino **LML** could be expressed and proved thus as follows:

Some C is B by possibility (**M**)
No A is B by necessity (**L**)
Therefore Some C is not A by necessity (**L**)
Otherwise Every C is A by possibility
But we had No A is B by possibility,
Therefore No C is B by necessity (by *Celarent* **LML**)
This contradicts Some C is B by possibility, which is not acceptable.

As to *Festino* **MLL**, it could be expressed and proved as follows:

Some C is B by necessity (**L**)
No A is B by possibility (**M**)
Therefore Some C is not A by necessity (**L**)
Otherwise Every C is A by possibility (**M**)
But we had No A is B by possibility (**M**)
Therefore No C is B by possibility (**M**) (by *Celarent* **MMM**)
This contradicts "Some C is B by necessity", the minor premise, which is not acceptable.

Baroco **LML** is provable in the same way as follows:

Every C is B by necessity (**L**)
Some As are not B by possibility (**M**)
Therefore Some As are not C by necessity (**L**)
Otherwise Every A is C by possibility
But we had Every C is B by necessity
Therefore Every A is B by necessity (by *Barbara* **LML**)
This contradicts "Some As are not B by possibility", the minor premise, which is not acceptable.

Baroco **MLL** can also be expressed and proved in the same way as follows:

Every C is B by possibility (**M**)
Some As are not B by necessity (**L**)
Therefore Some As are not C by necessity (**L**)
Otherwise Every A is C by possibility
But we had Every C is B by possibility
Therefore Every A is B by possibility (by *Barbara* **MMM**)
This contradicts "Some As are not B by necessity", the minor premise, which is not acceptable.

This shows that all second figure moods contain necessary conclusions when one of the premises is possible while the other one is necessary.

What about the third figure moods?

According to Avicenna, the third figure moods with possible premises are conclusive provided that "one of the premises is universal and the minor premise is negative" ([32], 43.14–15). In that case the conclusion expresses a bilateral possibility (*mumkina ḥaqīqīya*) ([32], 43.15).

In *al-Qiyās*, he presents and proves these moods in more detail. For instance, *Darapti* **QMQ** is expressed and proved as follows:

"Every B is C by possibility
And Every B is A by possibility (bilateral: *ḥaqīqīya*) (Q)
Therefore Some C is A by possibility (bilateral: *ḥaqīqīya*)" (Q) ([34], 223.6–7).

In this mood, the major and the conclusion express a bilateral possibility, for he says "for the minor converts by general possibility, and the major is a real possible, so the conclusion is a real possible" ([34], 223.7–8). The proof is made by the conversion of the minor, which leads to the following *Darii* **QMQ** from the first figure:

Some C is B by possibility (**M**)
Every B is A by possibility (**Q**)
Therefore Some C is A by possibility (**Q**)

In the same way, *Felapton* **MMM** is expressed and proved by conversion as follows by taking into account Avicenna's remarks:

Every B is C by possibility (**M**)
No B is A by possibility (**M**)
Therefore Some C is not A by possibility (bilateral: *ḥaqīqīya*) (**Q?**) ([34], 223.5–6).

Here, Avicenna says that the conclusion expresses a bilateral possibility, so that one should use **Q** to label it. However, the negative **Q** propositions are not expressed correctly by Avicenna as we already noted so that what he thinks to be an **O** bilateral proposition is just in fact an **I** bilateral one. This is why it is preferable to state the mood with unilateral possible negative propositions instead.

The proof is made by the conversion of the minor and reduction to *Ferio* **MMM**, from the first figure. It runs as follows:

Some C is B by possibility (**M**) (by conversion)
No B is A by possibility (**M**)
Therefore Some C is not A by possibility (bilateral: *ḥaqīqīya*) (by *Ferio* **MMM**)

Datisi **QMQ** is reduced to *Darii* **QMQ** by the same method, while *Ferison* **MMM** is reduced to *Ferio* **MMM** by the conversion of the particular affirmative, since the negative proposition is not convertible. In all cases, the conclusion is a bilateral possible.

As to *Disamis* **QMQ**, which is the following:

Every C is B by possibility (**M**)
Some C is A by possibility (**Q**)
Therefore Some B is A by possibility (**Q**)

Avicenna provides a proof by *ecthesis* which could be stated as follows:

Let us suppose that some C = Every D, then

(1) Every D is B by possibility (**M**) (assumption)
And (2) Some C is D by possibility (assumption)
Then (3) Some B is D by possibility (**M**) (by *Darii*)
Then we have (4) Every D is A by possibility (**Q**) (since every D = Some C)
Therefore (5) Some B is A by possibility (**Q**) (3, 4, by *Darii* **QMQ**) ([34], 223.12–224.1–3).

As to *Bocardo* **MMM**, which could be stated as follows:

Every C is B by possibility (**M**)
Some C is not A by possibility (**M**)
Therefore Some B is not A by possibility (**M**)

It is also proved by ecthesis, which proceeds by supposing that some C = every D. So

(1) Every D is B by possibility (**M**) (assumption)
And (2) No D is A by possibility (**M**) (assumption)
Then (3) Some B is D by possibility (**M**) (from (1) by conversion)
Therefore (4) Some B is not A by possibility (M) (from (2) and (3) by *Ferio* **MMM**)

Now, what about the moods mixing actuality with possibility? According to Avicenna, when one of the premises is actual (or absolute) the conclusion may be absolute too. It does not always express possibility, nor real (bilateral) possibility. So he does not agree in this respect with Aristotle, whose theory is stated but criticized afterward, for instance, at page 224, where he says that the usual theory (*al-mashhūr*) assumes that "all the conclusions [in these moods] are possible, not actual" ([34], 224.5–6), "But the truth (*wa 'amma al ḥaqqu*) is that it does not need to be a real possible, so that it does not allow the absolute to be true (*wa 'an lā yasduqa ma'ahā al-iṭlāqu*)" ([34], 225.2–3). For the bilateral possible affirmative propositions, when converted become general possible, and when one adds an absolute premise, the conclusion may be absolute too.[16] As an example, he provides the following concrete syllogism, which is a *Darapti* **X**s**QX**ga:

[16] Here there is an error in the text, not noted by the editor, for what is written is that the minor premise is absolute, but this could not be so, since the conversion leads to a possible minor premise, which with the absolute major leads to an absolute conclusion. Furthermore, the case of the mood with an absolute minor is considered just after the first one, in the sequel, where Avicenna says "If the minor is absolute, then the conclusion is a bilateral possible" ([34], 225.7)

Every man is writing (by possibility) (bilateral) (**Q**)
Every man is breathing by actuality (**Xs**)
Therefore Some writing (being) is breathing by actuality (**Xga**) ([34], 225.5–7).

However, Avicenna's opinion is not very clear here, because if one proves this mood by reducing it to *Darii*, the reduction should lead to *Darii* **XsMXga**, and not to *Darii* **XsMM**. But we saw above that Avicenna admits *Darii* **XsMM** (or even **XsQM**), rather than *Darii* **XsMXga**, for all the first figure moods with an absolute major and a possible minor contain a possible conclusion in his frame. But if we consider that the propositions **Xga** and **M** are logically comparable as shown by Wilfrid Hodges, we could say that the conclusion **Xga** is acceptable. So maybe the idea he stresses in his defense of this kind of third figure moods is simply that the conclusion should not in any case be a *real* (i.e., bilateral) possible as it is in Aristotle's frame; rather, its possibility should be unilateral, so that it allows the absolute to be true ([34], 225.5).

Now, if the minor is absolute, while the major is possible, the conclusion will be a (bilateral) possible (*mumkina ḥaqīqīyya*) ([34], 225.7), when the major is a real possible too. This is so because the mood is reducible to *Darii* **QXsQ** by conversion as follows:

Every C is B by actuality (**Xs**)
Every C is A by possibility (**Q**)
Therefore Some B is A by possibility (**Q**)

By the conversion of the absolute premise, we obtain

Some B is C by actuality (**Xs**) (by conversion)
Every C is A by possibility (**Q?**)
Therefore Some B is A by possibility (**Q**) (by *Darii* **QXsQ**)

As to *Bocardo* and *Ferison* whose major premises are negative, when their major premises are absolute while their minors are possible, they produce a possible conclusion in the usual interpretation. For instance, *Bocardo* **XsMM** is expressed as follows:

Every B is C by possibility (**M**)
And Some B is not A by actuality (**Xs**)
Then Some C is not A by possibility (**M**)

This is *usually* (*al-mashhūr*) proved by *reductio ad absurdum* as follows: if the conclusion is false, then, "Every C is A by necessity" ([34], 226.1),

but we had "Not every B is A by actuality" (major premise)

Therefore "Some B is not C by necessity" (by *Baroco* **LXsL**) ([34], 226.1–2). But this contradicts "Every B is C by possibility," the minor premise of *Bocardo*, which is not acceptable.

But this proof is criticized by Avicenna, who says that if the conclusion of *Bocardo* **XsMM** is a *real* (bilateral) possible, its contradictory is not only "Every C is A by necessity," but a disjunction containing that proposition "Every C is A by necessity" and a negative one expressed as follows: "No C is A by necessity" ([34], 226.4), in which case the whole proof needs to be modified or supplemented.

However, he does not provide another proof in his text, for he goes on talking about another kind of moods, which is the one containing a possible major and an absolute minor, that is, the **MXsQ** moods, where the conclusion can express a *real* (bilateral) possibility ([34], 226.5–6), according to him.

In *al-Najāt*, his opinion about the third figure moods where one premise is absolute and the other one is possible is expressed more clearly, although very briefly in the following quotation:

> "But the truth is that all the conclusions are possible; so if the absolute is pure, then the conclusion is a real (*ḥaqīqīya*) possible and if it is not pure, then the possible [conclusion] is unilateral. And this is proved either by one conversion or by *ekthesis* otherwise" ([32], 44.10–12)

So, it seems that in all cases, the conclusion is possible. But the possibility may be unilateral or bilateral, depending on the kind of absolute the mood contains. The question is rather: what does he mean by a pure absolute and a "not pure" one? A plausible answer would be the following: the pure absolute is a proposition of kind **Xs**, while a "not pure" one would be of kind **Xp**, since **Xp** may be in some cases necessary and its logical behavior is comparable to that of **L**. One is then entitled to say that in Avicenna's frame, this kind of third figure moods should contain the following combinations: **XsMM** and **MXpM** just as **MLM**.

Now what about the third figure moods containing a possible premise and a necessary one? These are examined in pages 226–227 of *al-Qiyās*.

If the minor is particular, as in *Datisi* and *Ferison* and if this minor is possible, the major will be necessary, so that the premises are **LM**, then the conclusion is necessary ([34], 226.11–13), whether the major is affirmative or negative. So *Datisi* **LML** and *Ferison* **LML** are hold valid. If the major is a particular affirmative necessary, the conclusion is necessary too ([34], 226.16–17). So *Disamis* **LML** is hold valid. If the negative particular premise (as in *Bocardo*, for instance) is possible, then the conclusion is possible ([34], 226.14–15). So it seems that *Bocardo* **MLM** is admitted. In the same way, *Disamis* **MLM** is also held valid ([34], 226.18–227.1). If the major is negative and necessary as in *Felapton* and *Bocardo*, the conclusion is necessary ([34], 227.3). So *Felapton* **LML** and *Bocardo* **LML** are also hold valid.

In *al-Najāt*, he examines these combinations in one paragraph and says simply that in that case "the conclusion follows the major" ([32], 44.13). So we can say that the moods held are the following: *Darapti* **LML**, *Felapton* **LML**, *Datisi* **LML**, *Disamis* **LML**, *Ferison* **LML** and *Bocardo* **LML** on the one hand and *Darapti* **MLM**, *Felapton* **MLM**, *Datisi* **MLM**, *Disamis* **MLM**, *Ferison* **MLM**, and *Bocardo* **MLM** on the other.

4.3.2 The Modal Syllogistic in Averroes' System

As to Averroes, the main thing that characterizes his attitude in general and with regard to modal syllogistic in particular is that he aims at being as faithful as possible to Aristotle unlike Avicenna who criticizes Aristotle in several places and does not hold valid the same modal moods. This attitude has been stressed by many of his commentators such as Abdelali Elamrani-Jamal, for instance, who says that Averroes' "fundamental project, identical for Logic as it was for Physics and Metaphysics" is "to restore the authentic doctrine of Aristotle" ([69], 51, my translation). This being so, it is not estonishing to find some differences between Avicenna's views and Averroes' ones in categorical as well as in modal logic. In modal logic, as we will see below, these differences are particularly noticeable.

Averroes presents his analysis of the modal moods in his *Talkhīs al-Qiyās*, where he states the moods that he considers as valid and proves them.

His analysis starts with the conversions of the modal propositions, that is, **E**□, **A**□, **I**□, as well as **E**◊, **A**◊ and **I**◊. Unlike Avicenna, Averroes holds that **A** Necessary converts to **I** necessary, **E** necessary converts to **E** necessary and **I** necessary converts to **I** necessary ([26], 147–148). All the modalities are internal in his frame, as they are in Aristotle's one.[17]

He proves **A** necessary conversion in the following way: The conversion holds that "Every B is A by necessity" leads to "Some A is B by necessity." To prove this, Averroes proceeds as follows:

"If "Some A is B by necessity" is not the admitted converse, then this converse would be "Some A are possibly B (not necessarily)". But if "Some A are possibly B (not necessarily)", then "Some B are possibly A (not necessarily)" as has been shown by *ekthesis* with regard to the categorical propositions (*fī-al-wujūdīya*). For if we suppose that some B, which exists in A by possibility, is a sensitive thing, this thing would be some A and some B; therefore "Some B is A by possibility". But we have assumed that "Every B is A necessarily", and this is unacceptable [for it is incompatible with "Some B are not necessarily A"]. Therefore 'Every B is A necessarily' converts to 'Some A are B necessarily'" ([26], 148).

This proof resembles a proof by *reductio ad absurdum*, because it shows an incompatibility in the end, which leads to the conclusion searched for. But it somehow differs from the proof by *reductio ad absurdum*, because it does not start by the contradictory of the conclusion (the contradictory of **I** necessary). For Averroes seems to start from an alternative: either the converse of **A** necessary is a *necessary particular* or it is a *possible particular*. So if it is not necessary, it must be possible (but in all cases, it has to be particular), and he continues to prove that if we suppose it possible, this leads to an incompatibility, given that we would have

[17]With regard to Aristotle's opinion about the modalities see, for instance, Marko Malink who says in his article "A reconstruction of Aristotle's Modal Syllogistic" ([113]) the following: "In Aristotle's modal syllogistic, however, there are no iterable modal sentential operators and, as a consequence, no *de dicto* modalities." ([113], 96)

"Some B is A by possibility," that is, "Some B are not necessarily A," given that the possibility is considered as bilateral. But this contradicts **A** necessary which says "Every B is A by necessity." Therefore, the possible converse is incompatible with **A** necessary, which means that this converse must be necessary. This is why Averroes says that the converse of **A** necessary must be **I** necessary.

He thus differs from Avicenna in this respect, for according to Avicenna, the possibility used in the modal syllogistic is not always the bilateral possibility; it may be unilateral too. And the converse of **A** necessary is precisely **I** possible, because here the possibility is understood and considered in its *unilateral* meaning. Avicenna says precisely *when* one should use the bilateral meaning of possibility and *when* the unilateral meaning is more adequate in all the rules and moods he provides in his modal syllogistic. But Averroes seems to privilege, *from the start*, the bilateral meaning over the unilateral one in the modal syllogistic he presents.

What about **E** necessary conversion? This leads from "No B is A by necessity" to "No A is B by necessity" and is proved as follows by Averroes:

> "If "No A is B by necessity" is not true, then its contradictory will be true, and this [contradictory] is : either the affirmative particular in the possible matter, which is contrary to the necessary matter, or the necessary particular affirmative, for here there is no other matter than these two ones; the absolute has the nature of the possible and the impossibility which results from supposing them is the same…If we suppose the necessary particular [**I** necessary], the above proof of the conversion of the universal negative absolute shows that it leads to an impossibility (*muḥāl*), and if we suppose (*anzalnāhā*) the particular possible as when we say "Some As are B by possibility", then it is clear that if we suppose "Some As are B by actuality" (*bi-l-fi'l*) no impossibility can be deduced from that, but when we say "Some As are B by actuality", then "Some Bs are A by actuality", because the particular affirmative has been shown to convert. However we have assumed that "No B is A by necessity" and this is a contradiction that cannot be accepted. For the existential is of the nature of the possible and the possible is contrary to the necessary, and if the particular affirmatives necessary and possible are false, then the universal necessary negative must be true, because what does not exist by possibility or by necessity, is negated (*maslūbun*) necessarily." ([26], 147.5–19, my translation).

This proof seems rather bizarre, although it arrives at a real incompatibility. For one thing, Averroes seems not to know exactly what is the real *contradictory* of the universal necessary negative proposition, since he talks about *two* kinds of modal propositions: the absolute particular affirmative and the necessary particular affirmative. The former is given the status of a possible proposition since its matter is possible. This introduces some confusion with regard to the nature of each kind of propositions. Second, he wants to rely on the previous proofs admitted in categorical logic. This is why he uses the conversions of the absolute propositions rather than the possible ones. But what he shows is just an incompatibility (a contrariety); it is not a real contradiction, for **I** absolute is *not* the contradictory of **E** necessary; rather it is one of its contraries, since they cannot be true together. As to **I** necessary, it is not needed in the proof, for it is *not* the contradictory of **E** necessary either and it does not add anything to the incompatibility already shown by assuming **I** absolute.

Note that this proof is not equivalent to Avicenna's one (see above, Sect. 4.3.1) which relies on some principles explicitly claimed by Avicenna, despite the fact that both proofs rely on the conversions already proved in categorical logic. Avicenna's proof seems more convincing and less confused than this one, given that the modal oppositions are very clear in Avicenna's logic, and no confusion is made by him about the real contradictory of **E** necessary.

I necessary is shown to convert as itself, i.e., as **I** necessary because "if 'Some Bs are A by necessity', then it must be the case that some As are necessarily B, otherwise, no B is A by necessity" ([26], 148.7–9).

As to possible propositions, he says that **A** possible converts to **I** possible, **I** possible converts to **I** possible, given that possibility here has its bilateral meaning as explicitly claimed by Averroes, when he says "I mean those that are really possible (*al-mumkin bi-l-ḥaqīqati*), i.e. those that can exist and can not exist in the future" ([26], 148.14–15). This is proved as follows: "If "Every A is B by possibility" or "Some As are B by possibility," then I say that "B is A by possibility," because if it is not possible but necessary, then "A is B by necessity" as we showed above; but we assumed that "Every A is B by possibility," this is absurd (*khalf*) and cannot be accepted" ([26], 148). This proof relies also on the incompatibility between the bilateral possible and the necessary. So the converse of **I** possible and **A** possible is also a bilateral possible. This conversion is thus different from that admitted by Avicenna, for as we saw above, Avicenna distinguishes between both kinds of possibility and says that **A** (bilateral) possible converts to **I** (unilateral) possible. But here, both the initial proposition and its converse are said to express a bilateral possibility.

As to **E** possible, it does not convert to **E** possible because "it is not really a negative proposition" ([26], 148.25), since it is expressed thus: "Every S is possibly not P." Consequently, its converse should be particular, as is the case with **A** propositions. But Averroes does not provide that converse in this part of his text. He says that he will return to the subject in his syllogistic ([26], 148.23).

4.3.2.1 The Moods with Assertoric and Necessary Propositions

Now, given these rules and definitions, what are the moods considered as valid in Averroes' modal syllogistic?

These moods and their proofs are presented in Sects. 8–12 for the apodeictic moods and in Sects. 13–21 for the problematic moods, where the possible propositions are mixed with the assertoric ones first and with the necessary ones afterward in the three figures. The assertoric propositions are called "*wujūdī*" in the title of the first paragraph of Sect. 9 ([26], 177) but it is called "*muṭlaq*" in the title of the newt paragraph of the same section. Now in the main text, Averroes uses the word "*muṭlaqa*" ([26], 179.7) to qualify these non-modal propositions. But as we will see in the sequel, it does not mean the same as in Avicenna's frame, for Averroes does not add the several conditions that Avicenna adds in his propositions. Rather, he sticks to Aristotle's definitions and rules and does not change the formulations of the *muṭlaqa* propositions. He states these "*muṭlaqa*' propositions by using the expression "*bi-al-fi'l*", when the proposition is not modalized and he opposes it with "*bi-l-imkān*" (by possibility) ([26], 177.19) and "*bi-ḍ-ḍarūra*" (by necessity) ([26], 177.14). For instance, he says "Everything that is C is actually (*bi-al-fi'l*) A" ([26], 177.14). So the non-modal proposition says simply what is actually (*bi-al-fi'l*) the case (or is not actually the case), without adding "necessarily" or "possibly". It asserts something about real facts, not about what *could have been* the fact or about what *must be* a fact. This is why we could call them assertoric, since they are neither problematic (possible) nor apodeictic (necessary). Averroes follows then Aristotle in his distinction between apodeictic, i.e., which "necessarily belongs," problematic, which "may belong," and assertoric, which simply "belongs" ([24], I, 29b29). According to Aristotle, assertoric propositions say "what is," while problematic propositions say "what is possible," and apodeictic propositions say what "is necessary" ([24], I, 29b29–29b35).

In the first figure, when the premises are necessary and assertoric, i.e., not containing any alethic modality, the modality of the conclusion follows that of the major premise. So Averroes admits the **LXL** and the **XLX** first figure moods. He justifies this opinion by means of several proofs, rules, and arguments. He also criticizes the peripateticians such as "Theophrastus and Eudemus, among the ancient peripateticians, and Themistius and his followers among the later ones" ([26], 179.8–9) who hold the opinion according to which "the modality of the conclusion follows the least modality" ([26], 179.14–15) in the first figure moods. We will return to these criticisms below.

For one thing, he proves the **LXL** moods by using the *dictum de omni and de nullo*, which, according to him, applies to the moods with necessary and assertoric premises, but not to those which contain a possible premise. More precisely, the way this *dictum* is stated is different when possible propositions are considered. For when the necessary or assertoric propositions are considered, the *dictum de omni and de nullo* requires that what is said of every or of none "is said affirmatively or negatively of what is *actually* (*bi-l-fi'li*) B" ([26], 176.10), while when the possible premises are considered this condition (of effectiveness) might not be respected, given that the subject of a possible proposition may be *potential* rather that *real*.

Now how does Averroes prove *Barbara* and *Celarent* **LXL**? The proof uses the general principle that "If something is predicated of the whole, then it must be predicated of the part, with the modality by which it has been predicated of the whole" ([26], 178.1–2), which justifies the necessity of the conclusion in case the major premise is necessary.

So *Barbara* **LXL** is the following:

"Every C is B by actuality (**X**)
Every B is A by necessity (**L**)
Therefore Every C is A by necessity (**L**)" ([26], 177.14–15)

While *Celarent* **LXL** is the following:

"Every C is B by actuality (**X**)
Every B is not A by necessity (**L**)
Therefore No C is A by necessity (**L**)" ([26], 177.14–16)

Because in both cases, what is necessarily true of B (the whole) is also necessarily true of C (the part), whether affirmatively or negatively, the part and the whole being, respectively, the subject and the predicate of the *minor premise*, which for this reason, must always be affirmative in the first figure, in order for the *dictum* to hold.

When the major premise is assertoric while the minor premise is necessary, Averroes provides other kinds of proofs to validate both *Barbara* **XLX** and *Celarent* **XLX**.

He first uses a proof by *reductio ad absurdum*, second by considering the relation between the part (C) and the whole (B), and finally by providing a counterexample which illustrates the case where the conclusion is assertoric rather than necessary.

The proof by *reductio ad absurdum* is the following, for *Barbara* **XLX** and *Celarent* **XLX**.

Barbara **XLX**, for instance, is stated as follows:

Every C is B by necessity (L)
Every B is A by actuality (X)
Therefore Every C is A by actuality (X)

If we suppose that the conclusion is necessary, then it would say what follows:

Every C is A by necessity

but we have assumed that

Every C is B by necessity

So by *Darapti* (which is also valid when the premises are both necessary), we may deduce the following:

Some B are A by necessity

"But we have assumed that every B is A without necessity, which is absurd (*khalf*) and impossible (*lā yumkinu*)" ([26], 178.9–10). This shows that the conclusion cannot be necessary; therefore it has to be assertoric.

So according to him the universal assertoric affirmative proposition is incompatible with the necessary particular affirmative one, even if he does not say explicitly that they are contrary, for instance. His rejection of the necessary conclusion is justified by this incompatibility. As a matter of fact, the universal affirmative assertoric and the necessary particular affirmative may both be false as is the case in the following examples: "Every table is round" and "Some tables are necessarily round."

The second proof is the following: C is part of B, therefore "if A is predicated of all (what is part of) B, which is the whole, without necessity, then it must be predicated of C without necessity too, for C is part of B" ([26], 178.11–13). This is clear from the *dictum* which says that what is true of the whole is true of the part, in the same way, that is, with the same modality.

The third justification is provided by a counterexample which shows that some concrete illustrations of *Barbara* **XLX** can be easily provided, such as the following:

"Every man is living by necessity (L)
Every living (being) is moving by actuality (i.e. without necessity)(X)
Therefore Every man is moving by actuality (i.e. without necessity)(X)" ([26], 178.14–16)

This shows that the conclusion is not necessary in some cases.

The same kind of proofs can be applied to validate *Celarent* **XLX**, which is expressed as follows:

"Every C is B by necessity
No B is A by actuality
Therefore No C is A by actuality" ([26],178. 4–5)

As to *Darii* and *Ferio* **LXL** as well as *Darii* **XLX** and *Ferio* **XLX**, they are said to be provable by the *dictum de omni and de nullo*, by *reductio ad absurdum* and by providing counterexamples showing in the case of the moods **XLX** that the conclusion is not necessary, but assertoric, i.e., non-modal. The proofs are not provided but they are said to be the same as those of the moods containing universal premises such as *Barbara*, which use the "*dictum de omni et de nullo*". A concrete example of *Darii* **XLX** is then provided, which is the following: "Some white (beings) are living by necessity, every living (being) is moving not by necessity, therefore some white (beings) are moving not by necessity" ([26], 179.2–3), where "not by necessity" means "by actuality."

This opinion is intended to be a strong and strict defense of Aristotle's one against some of his commentators who endorsed different positions. Averroes states these positions endorsed, for instance, by Theophrastus, Eudemus, and Themistius, whom he evokes explicitly ([26], 179.8–9) and discusses their arguments which, according to him, are not convincing.

According to these commentators, the conclusion of the moods containing one assertoric premise and one necessary premise should always be assertoric, for the modality of the conclusion should always be the least modality (in this case, it should be assertoric, given that actuality is weaker than necessity), exactly like the quality of the conclusion is thc least quality (if one of the premises is negative, the conclusion must be negative). Their argument uses also the relation between the whole and the part, but in a non-standard way which is criticized by Averroes.

According to them too, what is true of the whole is true of the part, but the whole and the part are not always the same, depending on the modalities of the premises. As it is summarized by Averroes, their opinion is the following:

> "If the necessary premise is the *minor*, then... the middle term is the whole, and the minor term is the part; therefore if something is predicated with some modality (*bi jihatin mā*) of the whole which is the middle term, then this same modality must be predicated of the part which is the minor term." ([26], 179.19–23).

This justifies the **XLX** moods of the first figure, for in that moods, (A) the major term is predicated of the whole, that is, the middle term (B) by actuality; therefore, it has to be predicated of the part, i.e., the minor term (C) by actuality.

Now when the necessary premise is the *major* (rather than the minor), the whole becomes "the major term," while the part is "the middle term." Therefore

> "when the part, which is the middle term, is predicated of the minor term with a certain modality, then this same modality must be the one used when predicating the whole, which is the major term, of the minor term." ([26], 179.24–180.1–2).

But this last opinion contains, according to Averroes, some confusion. For he notices that the whole and the part should always be the same terms, if one applies the *dictum de omni et de nullo* ([26], 180.5–6), namely, the terms contained in the *minor premise* ([26], 180.10), which are the middle and the minor terms. So the change of the wholes and the parts does not respect this *dictum*, which is the main argument validating the first figure moods. When one respects the *dictum de omni et de nullo*, then one finds that the modality of the conclusion, in the first figure, should be that of the major premise, as is the case in Aristotle's modal logic ([26], 180.16). If one takes into account the terms of the major premise, without considering those of the minor premise, then according to Averroes, the validity of the mood is no more warranted, for the truth of the conclusion would be "accidental" ([26], 180.16), given that the major term is not necessarily true of the minor, if we consider that what is true of the major (the whole) is true of the middle (the part), for that particular condition does not warrant the truth of the conclusion (which contains the *minor* and the *major*).

His second criticism stresses the illegitimacy of the comparison between the quality and the modality. For, according to him, the rule saying that the conclusion should be negative when one of the premises is negative, which is correct, does not rely on the *weakness* of the negation with regard to affirmation. For it is *not because* the negation is *weaker* than the affirmation that this rule holds, rather it is the very specificity of the negation that validates it. Consequently, if one says that the *weakest* modality is the one that should be present in the conclusion, because he compares between the *quality* and the *modality*, one is wrong, because this particular comparison or analogy is not correct ([26], 180.20–24).

The third criticism considers the case of some concrete syllogisms, which are given to illustrate the thesis that the conclusion is sometimes assertoric, even when the major is necessary. The syllogism provided by these authors is the following:

"Every man is walking by actuality
Every walker is moving by necessity
Therefore Every man is moving by actuality" ([26], 180.25–26)

According to him, in this syllogism, the major is necessary only if one says that "Every walker is moving by necessity as long as he is walking," for if one considers the walker only "as a man," then, it is not necessary ([26], 181.1–3), because "Every walker is moving as long as he exists" is not necessary. So if one adds the same condition in the conclusion, then it is itself necessary because "Every man is moving by necessity, as long as he is walking" ([26], 181.3).

However, this answer is not very convincing because the conclusion provided by Averroes contains three terms (man, moving and walking), while it should contain only two terms. As a matter of fact, the conclusion of the example is not even true, because the minor premise is not true, given that not every man is actually walking, consequently, not every man is actually moving. So the example relies on premises which cannot anyway be necessary, given that they are not even true, the only true (and necessary) premise being the major one, provided the condition "as long as he is walking" is added.

Now how could one express the modal *dictum de omni et de nullo*? Should it apply only to what is *actually* said to be so-and-so (in the minor premise) or to what is either potentially so-and-so or actually so-and-so (in the same minor premise)? Averroes notes that if something is true of what is potentially so-and-so, it is also true of what is actually so-and-so, but the converse is not valid ([26], 181.19–21) given that what is true of a man, for instance, is not necessarily true of a potential man (or a fetus, to recall al-Fārābī's example). This is why "the *dictum de omni et de nullo* in the necessary and the assertoric requires the major term to be predicated of all what satisfies the middle term *actually*, that is, all of what the middle term is predicated to *actually* (*kull mā yuḥmalu 'alayhi al-ḥadd al-awsaṭ bi-al-f'il*), not [only] by possibility" ([26], 181.22–24, emphasis added). For this reason, if this *dictum* applies only to what satisfies *actually* the minor term, it does not validate the moods containing *possible* premises ([26], 181.24–25); it only validates those containing necessary and assertoric premises.

Here, he evokes two possible interpretations of the modal *dictum de omni* (*et de nullo*), the first one is endorsed by al-Fārābī and the second one by Alexander of Aphrodisias. According to al-Fārābī, the *dictum* applies to all kinds of premises, including the possible ones ([26], 182.11–12), for he considers that the *dictum* says that "A is predicated affirmatively or negatively, in whatever modality, of all what is posited as B or described as B, be this description of whatever modality" ([30], 112). This definition, however, is not satisfying for Averroes, because it is not sufficient to validate the moods with necessary and assertoric premises. While

Alexander of Aphrodisias's interpretation, according to which the *dictum*, in all cases, applies only to what is "actually B", does not validate the moods where both premises are possible, for if this were true "nothing would follow from two possible premises by means of the *dictum*" ([26], 182.16–17) and is not satisfying for this reason, according to Averroes. This would mean that the figure moods MMM would not be perfect any more as A. Elamrani-Jamal notes when discussing this option ([69], 61) and would be therefore contrary to Aristotle's opinion according to which these moods are perfect. This is why this solution too is rejected by Averroes, who wants to defend Aristotle's position in that treatise.

So unlike both al-Fārābī and Alexander of Aphrodisias, Averroes considers that the *dictum* requires the subject to be actually B, when it concerns the assertoric and necessary premises, but it does not require the same thing when possible propositions are concerned, for in that case, it says that "'what is B is A by possibility' is true of what is *potentially or actually* B" ([26], 182.8–9, my emphasis). This is the opinion defended in the *Middle Commentary on the Prior Analytics* (See [28]) *(Talkhīs al-Qiyās)*.

The discussion of the *dictum de omni et de nullo*, which reveals three possible interpretations of Aristotle's text, and whose aim is to solve the general character of the *dictum*, given that it should be a universal rule validating all first figure moods, is examined by Abdelali Elamrani-Jamal, who analyzes the evolution of Averroes' thought in ([69]) and says that Averroes criticizes both al-Fārābī's interpretation and Alexander's one because "Applied according to Alexander's interpretation, the principle of universal attribution is valid only for modal syllogisms one of whose premises is necessary and the other assertoric; according to al-Fārābī's interpretation, it is verified only when the minor premise is possible" ([69], abstract, 51). This author notices that Averroes changed his mind several times because of the difficulty to find a formulation of the principle that preserves its general character. He counts at least three different positions related to this *dictum*, starting from an early writing of Averroes entitled *al-mukhtaṭar* or *al-ḍarūrī fī al manṭiq* (554H/1159) ([69], 51) to the treatise entitled *maqālāt fī al-manṭiq wa al-'ilm at-tabī'ī* (591H/1194) ([69], 51) including the treatise that we are presently commenting, i.e., *Talkhīs al-Qiyās*, which is in between and is called the Middle Commentary. He shows in his article the evolution of Averroes' thought from one treatise to another, starting from the earliest one.

According to A. Elamrani-Jamal, Averroes did not, from the start, i.e., in his early treatise "*al-mukhtaṣar fī al-manṭiq*", defend scrupulously Aristotle's position, for he says that in that treatise, he did not, strictly speaking, comment on Aristotle's text; rather he presented his logic as a summary and followed the "tradition of the peripateticians" ([69], 52) without really asking about the legitimacy and the correct formulation of the *dictum de omni et de nullo* as we can read in what follows: "the *Abrégé* [*al-mukhtaṣar*] does not make any room to the problems raised by the modal syllogistic, which will occupy Ibn Rushd during his writing of the *Middle Commentary of the APr.* and subsequently" ([69], 55, my translation). So, in that treatise, Averroes defends a position close to that of the peripateticians and does not raise the problem of the *dictum de omni et de nullo* and of its right formulation. It is from the middle commentary (i.e., the *Talkhīs al-Qiyās*) that Averroes started to raise this problem, according to what A. Elamrani-Jamal says, for it is in that treatise, as we saw above, that he criticizes al-Fārābī's and Alexander's formulations and offers his own one, which is a strict defense of Aristotle's view. A. Elamrani-Jamal notes that Averroes' view is a "reaction" against al-Fārābī, whom he was following before, starting from that treatise ([69], 55). In that commentary, he considers two main positions: those attributed to Theophrastus and Eudemus, who "were opposed to Aristotle" ([69], 59) and those of al-Fārābī and Alexander of Aphrodisias, who "interpreted Aristotle's text (*APr.*, I, 9, 30a15–20) to make his position coherent" ([69], 59). But even the latter positions, as we saw above, are not satisfying according to Averroes, because they are not in agreement with *all what* Aristotle said, since as it is interpreted by al-Fārābī, the principle does not apply to the moods with assertoric and necessary premises, while as it is interpreted by Alexander it does not apply to those containing possible premises. As a result he himself searched another solution which A. Elamrani-Jamal summarizes as follows: "First solution: The condition of application of the principle of universal attribution varies depending on the modalities of the premises in the *Prior Analytics*" ([69], 63). This solution is said to have been defended in both "the *Middle Commentary* and the Opuscule of 567H/1171" ([69], 63). It combines between al-Fārābī's and Alexander's formulations by considering that, when the premises are assertoric or necessary the principle says that "A is attributed affirmatively or negatively to all what is actually B and not to what is potentially B" ([69], 64) while with possible premises it says that "A is attributed, possibly, affirmatively or negatively, to all what B is attributed to affirmatively, whether in a possible way or in an assertoric way or in a necessary way" ([69], 64, my translation). This solution according to Averroes and as noted A. Elamrani-Jamal validates the following positions: "in the moods with assertoric and necessary premises, the conclusion follows the major" ([69], 64) and "the mixed syllogisms where the major is assertoric or necessary and the minor possible are imperfect syllogisms" ([69], 64). Both positions are in accordance with Aristotle. However, as it is stated by Averroes, the principle is not "sufficiently general and formal" as noted by A. Elamrani-Jamal at page 64, which made Averroes reconsider it and search for another solution.

Then different solutions are stated by Averroes and reported by A. Elamrani-Jamal such as the following "Second solution: The principle of universal attribution used in producing the syllogism is different from the one that determines the mode of the conclusion" ([69], 64) and the following: "Third solution: The miscomprehension of Aristotle's claims is due to the equivocal character of the expression 'necessary conclusion'" ([69], 69). But both are criticized, since the former lacks generality while the latter, which defends the idea that when Aristotle said that the conclusion follows the major he meant that it follows the major "by its modality" ([69], 69) not by its matter, is not sufficiently justified by Averroes, since as noted by A. Elamrani-Jamal, Averroes does not provide "a mixed syllogism composed of a necessary major and an assertoric minor which would produce a necessary conclusion by its modality, assertoric by its matter." ([69], 70). As a result, another (final) solution is stated by Averroes as follows: "Fourth solution: The mode of the conclusions of the mixed syllogisms is conditioned by the modality of the premises, which is determined by the properties of their terms" ([69], 71). In this case, the terms are said to be either assertoric or possible or necessary, which leads, according to A. Elamrani-Jamal to a distinction between necessary premises and four kinds of assertoric premises; The necessary premise contains an essential term or a proper one, which would be unseparable from the subject, while the assertoric premises would contain, in the first kind, an accidental subject and predicate (e.g., "every walking thing is moving"). The assertoric premise of the second kind contains a substantial subject and an accidental predicate (e.g., "every walking being is an animal"); that of the third kind contains "a necessary subject and an accidental predicate," while the fourth kind would be a temporal assertoric, e.g., "every (actually) moving is a human" This solution seems to solve the problem and to validate Aristotle's claims since "only kind 2 of assertoric premises" is the one that validates the **LXL** first figure moods ([69], 73), and the *de omni et de nullo* can be stated in a uniform and general way.

Now what about the moods of the second and third figures, which contain necessary and assertoric premises?

According to Averroes, in the second figure, the modality of the conclusion follows that of the *negative* premise, so that if the negative premise is necessary, the conclusion will be necessary too ([26], 183.2–3). This rule validates the following *Cesare* **LXL**:

Every C is B by actuality (*bi al-fi'li*) (X)
No A is B by necessity (L)
Therefore No C is A by necessity (L)

This mood is proved by the conversion of the major premise which reduces it to *Celarent* **LXL** of the first figure ([26], 183.7–8).

As to *Cesare* **XLX**, it is the following:

Every C is B by necessity
No A is B by actuality
Therefore No C is A by actuality

The proof runs as follows: if the conclusion were necessary, that is, if we had

(1) No C is A by necessity,

Then, since the minor premise says that

(2) Every C is B by necessity

Which, when converted becomes

(3) Some B is C by necessity

We can deduce

(4) Some B is not A by necessity (By *Ferio* **LLL** from (1) and (3)).

But since "No A is B by actuality" converts to "No B is A by actuality," which implies "Every B is possibly not A," itself compatible with "Every B is possibly A," "because the assertoric has the nature of the possible (*al muṭlaq min ṭabī'ati al-mumkin*)" ([26], 184.7), this leads to a contradiction between this last proposition and "Some B is not A by necessity" ([26], 184.1–8). Because of this absurdity, the conclusion cannot be necessary; it is therefore assertoric.

The same kind of proofs is said to validate the moods *Camestres* **LXX** and *Baroco* **LXX**, that is, those where "the necessary affirmative is the major and the negative assertoric is the minor" ([26], 184.8–9).

The same rule is said to validate *Festino* **LXL**, *Festino* **XLX,** and *Baroco* **XLL,** which are the moods where the minor premises are particular ([26], 184. 15–18).

In the third figure, the modality of the conclusion is said to be the same as that of the major premise, for "the major [premise] in the third figure is the same as in the first figure, and the minor [premise] is the one which is convertible in it" ([26], 185.10–11). Thus when the major premise is necessary, then the conclusion is necessary too, in that figure. Thus *Darapti* **LXL** as well as *Darapti* **XLX** is admitted as valid.

But if one of the "universal premises is negative" ([26], 186.5) as in *Felapton*, then the modality of the conclusion follows that of the negative premise, because the conversion will concern the affirmative premise, so that the mood will be reduced to *Ferio* of the first figure. This validates *Felapton* **LXL** and *Felapton* **XLX**.

If the mood contains a particular premise, then if both are affirmative as in *Datisi* and *Disamis*, then "the conclusion follows the modality of the universal premise, because it is the one that does not convert in that figure, for if it were to convert, the two premises would be particular" ([26], 186.11–12). This validates *Datisi* **LXL**, *Datisi* **XLX**, *Disamis* **LXX**, and *Disamis* **XLL**.

If "one of the premises is affirmative while the other one is negative" ([26], 186.16) as in *Ferison* and *Bocardo*, then "the modality of the conclusion follows that of the negative premise" ([26], 186. 15–16). For *Ferison* is reducible to *Ferio*, while *Bocardo* is provable by *ecthesis*, which posits a universal premise inside the proof making it reducible to moods of the first figure ([26], 186.17–20). This validates *Ferison* **LXL**, *Ferison* **XLX**, *Bocardo* **LXL** and *Bocardo* **XLX**.

4.3.2.2 The Moods with Possible (and Other) Propositions

As to the moods containing possible premises, they are analyzed starting from the first figure, where according to Averroes, the moods with two possible premises are conclusive when both premises are universal, that is, in *Barbara* and *Celarent* **MMM**. For *Barbara* **MMM** says what follows:

"Every C is B by possibility (M)
Every B is A by possibility (M)
Therefore Every C is A by possibility (M)" ([26], 191.6–7)

This mood is proved by means of the modal *dictum de omni*, for what "Every B is A by possibility" means is that "Everything described as B by possibility or by actuality, that is, everything that is B actually or potentially, this thing is A by possibility, that is, A is predicated of it by possibility" ([26], 191.8–11). So given that C is potentially B and that B is potentially A, it follows that C is potentially A. Here, the kind of possibility considered seems to be unilateral. It is presumably the possibility understood as likeliness (*'alā al 'akthar*), although Averroes does not explicitly say that in this part of his text. However, he does evoke this kind of possibility in the rest of his analysis, for instance, in page 192, while discussing the moods whose premises are both negative.

Celarent **MMM** is stated as follows:

"Every C is B by possibility
No B is A by possibility
Therefore No C is A by possibility"

Which follows from the modal *dictum de nullo*, given that "C is potentially part of the Bs (*juz'un bi 'imkānin li [B]*)" ([26], 191.13), so if B is not A by possibility, C itself will not be A by possibility.

To these two moods, which are considered as perfect because they are valid by means of the *dictum de omni et de nullo*, he adds two other moods which he considers as conclusive but not perfect, for their conclusiveness is only due to their reduction by conversion to the conclusive moods. These two other moods may be those which contain two negative premises which may either both be converted to affirmative ones and reduced by this way to *Barbara* **MMM** or reduced to *Celarent* **MMM** by converting just the minor premise to an affirmative one. These conversions rely on the equivalences we examined above (Sect. 4.1), according to which "unlikely $\sim \alpha$ = likely α." We would thus have a mood which would look like the following:

Every C is unlikely (to be) ~B
Every B is unlikely (to be) ~A
Therefore Every C is unlikely (to be) ~A

Given the equivalence above between "unlikely not" and "likely", this "mood" would be reducible to *Barbara* by replacing in each premise "unlikely not" by "likely", which would lead to the following:

Every C is likely (to be) B
Every B is likely (to be) A
Therefore Every C is likely (to be) A.

This seems to be what is meant by Averroes in his defense of this kind of moods, for he says explicitly what follows:

"And we may have an imperfect syllogism if we convert the two negatives to the two affirmatives which are implied by them, or if we convert the minor negative to the affirmative that is implied by it. And this syllogistic mood is useful mainly when the negative possible are those which express *unlikeliness*; for this kind of moods is useful in dialectic, and they are good ruses (*ḥīla*) in that art (*ṣināʻa*)." ([26], 192.10–13, emphasis added)

The same may be said about the fourth mood, which is proved by the conversion of only one premise, the minor negative one. We could express it as follows:

Every C is unlikely (to be) ~B
No B is A by possibility (understood as likeliness)
Therefore No B is C by possibility (understood as likeliness)

If the conversion is made only on the premise which expresses unlikeliness, while the two other premises express likeliness and remain negative, then we could reduce this mood to *Celarent* **MMM**, for it would say what follows:

Every C is likely (to be) B
No B is A by possibility (understood as likeliness)
Therefore No B is C by possibility (understood as likeliness)

This seems to be Averroes' interpretation of Aristotle's following passage:

"Consequently if B is possible for every C, and A is possible for every B, the same deduction again results. Similarly if in both propositions the negative is joined with 'it is possible': e.g. if A may belong to no B and B to no C. No deduction results from the assumed propositions, but if they are converted we shall have the same deduction as before. It is clear then that if the negation relates either to the minor extreme or to both the propositions, *either no deduction results, or if one does it is not perfect.* For the necessity results from the conversion" ([24], I, 14, 32b38–33a20, emphasis added).

But it is clear also from this passage that Aristotle is much less tempted than Averroes to admit these moods as conclusive, for the text seems to express some doubt about their conclusiveness, as the sentence emphasized shows clearly. In addition, the reformulation of the likely propositions in terms of unlikely ones seems to be merely verbal, so that there is no significant change in the propositions or the moods stated.

Anyway, Averroes himself says that these moods are more useful in dialectic (*jadal*) or rhetoric than they are in logic *stricto* sensu, which means that their importance is not obvious from a strictly logical point of view.

As to the moods containing two possible premises and where the minor premise is particular, that is, *Darii* **MMM** and *Ferio* **MMM**, they are considered as valid and perfect (*qiyās tāmm*) ([26], 192.17), for they are proved by means of the modal *dictum de omni et de nullo* too. But here too, two more moods are added, which contain two negative premises. As for the moods above, one could either convert both premises to affirmative ones, in which case one could obtain *Darii* **MMM** or convert just the minor one to an affirmative premise, which would lead to *Ferio* **MMM**. These moods are not perfect, but reducible to perfect moods. This is so because the negative modal propositions behave as affirmative ones given that their modality is internal and puts on the predicate.

He thus finds *eight* valid moods ([26], 193.25), four of which are perfect while the other four are imperfect but useful in rhetorical and dialectical reasonings. The four perfect ones are *Barbara* **MMM**, *Celarent* **MMM**, *Darii* **MMM** and *Ferio* **MMM**, while the four imperfect ones contain negative possible premises (where the possibility is understood as unlikeliness) and may be reduced to the perfect ones either by the conversion of the two premises into affirmative ones or by the conversion of the minor premise into an affirmative one. He thus follows Aristotle in assuming the validity of these moods, but unlike Aristotle, he does not seem to use the bilateral kind of possibility in these moods, given that, according to him, with the possible as likely and the possible as unlikely the affirmative implies the negative and vice versa, provided that the two kinds of possibility are permuted when passing from the affirmative to the negative or vice versa, as we saw above.

By his defense of the latter four moods containing negative premises, he criticizes overtly Themistius, evoked explicitly in the text ([26], 194.1), who is said to have expressed some doubts about these moods by considering that they are unuseful in case the negative premises express likeliness, for then their affirmative converses would express unlikeliness and would be uninteresting and not useful in any kind of science. Averroes' answer to Themistius is that when the negatives express unlikeliness, they are useful as we showed above, and even when they express likeliness so that their conversion would lead to affirmatives expressing unlikeliness, they could also be useful in some domains, in particular in theology (*al 'ilm al-ilāhī*).

Whether adequate or not, this answer shows what Averroes understands from the Aristotelian text and his will to defend Aristotle's text as much as possible. In this context, he also states some moods that he does not consider as conclusive such as those containing "a particular major premise and a universal minor one" ([26], 192.20), whether affirmative or negative, for in these cases, the *dictum de omni et de nullo* does not apply either directly or indirectly. The invalidity of such moods is illustrated by means of concrete examples, which show that the conclusion may be negative in some cases and affirmative in other ones, that is, there is no unique conclusion falling from a unique set of premises. For instance, when the terms are the following: "man, white and living," the conclusion is affirmative, for from

"Every man is white by possibility, and some white (things) are living by possibility, one can deduce every men is living by necessity" ([26], 193.16–17), while if the terms are the following: "clothe, white and living," the syllogism is the following: "Every clothe is possibly white, and some white is living by possibility, therefore no clothe is living by necessity" ([26], 193.17–18). Since the conclusion is negative in this last case, while it was affirmative in the first one, this shows that such a mood is not conclusive. Its invalidity is also due to the fact that in both examples, the conclusion is necessary, while the premises are possible, which is an incongruity.

What about the mixed moods with assertoric and possible premises?

These are examined in Sect. 14 of *Talkhīṣ al-Qiyās*, where Averroes says that the first figure moods containing a major possible and a minor assertoric are perfect, because they are valid by means of the *dictum de omni et de nullo*, and produce a "bilateral possible conclusion (*wa takūnu natā'ijuhā mumkina ḥaqīqīya*)" ([26], 195.6).

The first figure mood where the major is possible are thus *Barbara* **QXQ**, where **Q** expresses the bilateral or two-sided possibility, *Celarent* **QXQ**, *Darii* **QXQ,** and *Ferio* **QXQ.** *Barbara* **QXQ** is the following:

Every C is B by actuality (*bi al-fi'l*) (X)
Every B is A by possibility (Q)
Therefore Every C is A by possibility (Q)

The conclusion is explicitly said to express a bilateral possibility ([26], 195.6). So the major possible premise should express a bilateral possibility too, for this is indeed the case in Aristotle's frame too (See, for instance [113], Table 2, p. 118).

This could be illustrated by the following concrete example, where the major premise could very well express a bilateral possibility:

Every Greek is human by actuality (X)
Every human is walking by possibility (Q)
Therefore Every Greek is walking by possibility (Q)

We could generalize this opinion to *Darii* and even to the moods containing negative premises such as *Ferio* and *Celarent*, for Averroes says that the conclusion is a real possible one, when "it produces the universal affirmative, the universal negative, the particular affirmative and the particular negative" ([26], 195.5–6).

So *Celarent*, *Darii,* and *Ferio* should contain a bilateral possible conclusion and also a bilateral possible major premise. We would thus have *Celarent* **QXQ**, *Darii* **QXQ**, and *Ferio* **QXQ**.

We can illustrate these moods by the following examples:

Every Greek is human by actuality (**AX**)
Every human is not writing by possibility (= can not write and can write) (EQ?)
Every Greek is not writing by possibility (= can not write and can write) (**EQ?**)

However, as we saw above when analyzing Avicenna's interpretation of the bilateral possible in the context of quantified propositions, this way of expressing **E**-bilateral possible (and likewise **O** bilateral possible) is not correct, given that the negation should put on the whole conjunction "possibly P and possibly not P" and not on each part separately. This criticism applies to Averroes' analysis too and shows that since the moods containing negative bilateral possible propositions are not correctly stated, the correct *Celarent* **QXQ** and *Ferio* **QXQ** are not really the moods stated by Averroes, who confuses **E** bilateral possible and **O** bilateral possible with **A** bilateral possible and **I** bilateral possible, respectively. For here, **E** bilateral possible seems to be rather **A** bilateral possible and the whole mood seems to be *Barbara* **QXQ** rather than *Celarent* **QXQ**. The same can be said about *Ferio* **QXQ**, where the negative premise and conclusion seem rather to be affirmative bilateral possible ones. This confusion makes it equivalent to *Darii* **QXQ.** So we can say that he is just stating *Barbara* **QXQ** and *Darii* **QXQ** in other ways (i.e., by starting with the negative side of the bilateral possible proposition), he does not really provide *Celarent* **QXQ** or *Ferio* **QXQ**.

Darii **QXQ** may be illustrated by the following concrete example:

Every Greek is human by actuality
Some humans are possibly sitting and possibly not sitting
Therefore Some Greeks are possibly sitting and possibly not sitting.

According to Averroes, the moods **QXQ** with one particular premise are perfect if the major is possible and the minor assertoric. For in this case "… the modality of the conclusion is the same as that of the conclusion [?] [it should be 'premise' rather]" (*wa takūnu jihatu al-natījati hīa jihatu tilka al-natīja [al-muqqadima?] bi 'aynihā*) ([26], 203.20). Here despite the error made by the editor who writes "conclusion" instead of "premise" in the end of the quotation, the text is clear with regard to the modality of the conclusion. But as we just saw, *Ferio* **QXQ** contains one negative bilateral possible premise and one negative bilateral possible conclusion, which are not correctly stated in Averroes' frame.

Note, however, that if one states the real *Celarent* **QXQ** and *Ferio* **QXQ**, one finds that they are indeed valid, for *Celarent* **QXQ** is the following:

Every C is B (X)
Every B is $\Box{\sim}A \vee \Box A$ (**EQ**)
Therefore Every C is $\Box{\sim}A \vee \Box A$ (**EQ**)

While *Ferio* **QXQ** is the following:

Some C is B (X)
Every B is □~A ∨ □A (**EQ**)

Therefore Some C is □~A ∨ □A (**EQ**)

Both are valid as the reader can check. But these formulations are not the ones provided by Averroes (nor are they provided by Avicenna either), which means that we cannot credit him with them.

He also says that these moods are provable by means of the *dictum de omni et de nullo*, which applies to the possible matter and says that "all what is potentially or really B is A by possibility" ([26], 195.15), where B is the middle term and A the major one.

However, if the major is assertoric and the minor possible, then according to him, "the moods are not perfect, and their conclusion is possible if it is affirmative, while it is either possible or necessary when it is negative, whether it is universal or particular" ([26], 195.7–10). These moods are said to be "imperfect" because the *dictum* does not apply to them. Their conclusion is said to be possible when it is affirmative, and "either possible or necessary" when it is negative ([26], 195).

Consider the following example:

Every walking is moving by actuality (X)
Every man is possibly walking (and possibly not walking) (Q)
Therefore Every man is moving (by possibility?) (M?)

Should the conclusion contain a bilateral or a unilateral possibility? Some authors consider that according to Aristotle, the minor premise contains a bilateral possibility while the conclusion contains a unilateral possibility, for in Table 2 provided by M. Malink in the above-cited article, we find some **XQM** moods such as *Barbara* **XQM**, *Celarent* **XQM**, *Darii* **XQM**, and so on ([113], Table 2, p. 118).

As to Averroes, he says that in these moods the conclusion should express a unilateral possibility too, for he provides the following proof by *reductio ad absurdum* of the above mood expressed as follows:

Every C is B by possibility
Every B is A by actuality
Therefore Every C is A by possibility

The proof supposes that the conclusion is false; therefore, its contradictory ""“it is impossible that Every C is A" (*laysa yumkinu an yakūna kull [C] [B]*)" ([26], 197.22) will be true. This contradictory, as it is formulated expresses impossibility, so what it negates is itself a unilateral possibility, given that the bilateral possibility is contradicted by a disjunction whose elements are a necessary proposition and an impossible one.

Then from the contradictory of the conclusion:

"Not possibly every C is A" (= "Some C is not A by necessity" ([26], 197.26)

and if the possible premise

"Every C is B by possibility"

is replaced by an assertoric one, that is, by:

"Every C is B by actuality" ([26], 197.23)

which would be in this case "false but not impossible (*kadhibun ghair muḥāl*)" ([26], 197.23–24), then one can deduce by *Bocardo* **LXL** from the third figure: "Some B is not A by necessity," which, according to Averroes, contradicts (*naqīḍu*) the major premise "Every B is A by actuality" ([26], 198.3) and is therefore absurd. This proves the validity of the mood.

However, this proof contains some confusion. For one thing, as it is stated, the premise expressing an impossibility "Not possibly every C is A" is *de dicto*, for the operator "not possibly" is in the beginning of the proposition, while the necessity contained in what is said to be its equivalent is *de re*, for it puts on the predicate. This proposition containing a *de re* necessity itself should be universal if the first one had contained a *de re* impossibility instead of a *de dicto* one. For if we formalize the proposition containing a *de dicto* modality, we obtain the following: $\sim \Diamond(x)(Cx \supset Ax)$, which is equivalent to $\Box \sim (x)(Cx \supset Ax)$, itself equivalent to $\Box(\exists x)(Cx \supset \sim Ax)$, which is a particular necessity **O**, as Averroes says in his text, but expresses a *de dicto* necessity, not a *de re* one, unlike what Averroes suggests by his formulation. While if the proposition expressing an impossibility were *de re*, then its formalization would have been the following: $(x)(Cx \supset \sim \Diamond Ax)$, which is equivalent to $(x)(Cx \supset \Box \sim Ax)$. This last proposition is necessary *and universal*. So it seems that Averroes does not distinguish adequately between *de dicto* and *de re* formulations, and also that he makes a logical error when he states the equivalence between the proposition containing an impossibility and the particular necessary one which he considers to be its equivalent, given that this last proposition should express a *de dicto* necessity too, not a *de re* one, if it is rightly formulated. Second, the whole syllogistic, as we saw above should contain only *de re* modalities, since Averroes is very faithful to Aristotle and Aristotle himself does not use *de dicto* formulations in his syllogistic. Third, the conclusion at which he arrives by applying *Bocardo* **LXL** (= **O** necessary) is *not* the contradictory of the major premise (= **A** assertoric) unlike what Averroes says; rather, it is only one of its *contraries*. Even if this does not mean that the proof is not conclusive, there is nevertheless some confusion between the contrary and the contradictory of the modal proposition. Finally, one may ask about the possible premise, which is modified and changed into an assertoric premise. Does it express a bilateral possibility or a unilateral one? Averroes does not say explicitly what kind of possibility is expressed in that premise, but it seems that Aristotle treats it as a bilateral one as assumed by M. Malink, for instance. If we consider that Averroes follows Aristotle very faithfully, we should assume that this mood is not a *Barbara* **XMM**, but rather a *Barbara* **XQM**, as is the case in Aristotle's system, if we follow M. Malink's interpretation ([113), 118, Table 2). To understand this, let us first return back to Aristotle's text itself. Aristotle expresses (and proves) this mood as follows in his *Prior Analytics*:

> "Since we have clarified these points, let *A* belong to every *B*, and *B* be possible for every C: it is necessary then that *A* should possibly belong to every *C*. Suppose that that it is not possible, but assume that *B* belongs to every *C*: this is false but not impossible. If then *A* is not possible for every *C* but *B* belongs to every *C*, then *A* is not possible for every *B*; for a deduction is formed in the third figure. But it was assumed that *A* possibly belonged to every *B*. It is necessary then that *A* is possible for every *C*. For though the assumption we made is false and not impossible, the conclusion is impossible." ([24], I, 34a34–34b6)

In this passage, the proof provided is very similar to that given by Averroes, but there is no confusion in Aristotle's text between *de dicto* and *de re* modalities, for the way he expresses things shows that the modalities are all *de re*, given that he starts by the predicate and puts the modal word in from of the copula ("belongs to"). For instance, when he says "*A* is not possible for every *B*," this does not mean "It is impossible for every B to be A" (or "Not possibly every B is A"); rather it means "Not every B is possibly A," that is, "Some B is necessarily not A," for this is the only way, first, to understand the use of *Bocardo* **LXL** afterward, second, to express things in terms of *de re* modalities, given the permutation of the subject and the predicate.

However, when Averroes followed this mood and its proof and expressed them in his own way, he seems to not have paid attention to this permutation of the subject and the predicate and to the use of the copula "belongs to," which made his own proof less convincing than Aristotle's one, because he did not express the propositions in the right ways and mixed the modalities *de re* and *de dicto* in his formulations.

Note that in the Arabic translation of Aristotle (made by Tadhārī), which has been published by Badawi, there is no confusion either between *de re* and *de dicto* modalities, for the text follows Aristotle by starting with the predicate and using a copula like "belongs to," since the mood is expressed as follows: "Let A belong to every B (*li-takun A mawjūda fī kulli B*) and B possible in every C (*wa [B] mumkina fī kulli [C]*), then necessarily A is possible for every C (*fa idhan bi-ḍ-ḍarūrati [A] mumkina fī kulli [C]*)" (*Analīṭīqā al-'Ūlā*, in [40], 182.14–15).

But Aristotle does not indicate explicitly, in his formulation of the mood, what kind of possibility he was using in the premise. We can only understand that the conclusion expresses a unilateral (one-sided) possibility, because its contradictory is said to be impossible, as is the case in Averroes' proof. But what about the possible premise? Nothing in the text indicates that it should express a bilateral (two-sided) possibility, as assumed by M. Malink. So if it does express a two-sided possibility, this may be due to other reasons, for instance, to the assumption that the real possible is the bilateral one. So the premise should be a bilateral possible in all cases, but the conclusion cannot express a bilateral possibility when the second premise is assertoric (or necessary). Nevertheless, we can note that Aristotle too changes the possible premise into a (false) assertoric one. This could mean that this change is not incompatible with the bilateral kind of possibility in Aristotle's frame; consequently, there should not be an incompatibility of that kind in Averroes' frame too.

As a consequence, we could assume that the moods containing a possible premise and an assertoric one are of the form **XQM,** just because the possible premises are all supposed from the start to express a bilateral kind of possibility, while the conclusion, in this kind of moods, cannot be a bilateral possible proposition because of the assertoric premise. However, the moods containing negative bilateral possible premises raise the same problem that we evoked above.

In the rest of his text, Averroes gives some precisions about the kind of assertorics that should be used in these syllogisms. For according to him, the assertoric should express facts that happen "all the time or most of the time" ([26], 199.10), for instance, when someone says "every snow is white" or "every raven is black" ([26], 199.12) but not those that happen rarely as when someone says "every moving (thing) is a man," which could be true only in very rare cases ([26], 199.15): with this latter kind, no syllogism is possible. If this latter assertoric is used and mixed with a possible premise, it can produce different conclusions because "it is true [only] in the present time" ([26], 199.19), while the possible premise is more likely to be true in the future ([26], 200); for this reason, the syllogism would not be conclusive.

This interpretation of Aristotle's text is said to be opposed to al-Fārābī's interpretation which, according to Averroes, uses the *dictum de omni* to explain Aristotle's position. According to Averroes, the use of the *dictum* is not adequate in this case, because it is not compatible with what Aristotle says about this kind of syllogisms, in particular with the fact that he proves them by *reductio ad absurdum* ([26], 200.11), which shows that they are not perfect.

As an illustration of the plurality of conclusions, Averroes gives the following examples: Let the three terms be the following: human (as the minor), moving (as the middle), and horse (as the major). This combination gives the following mood:

Every human is possibly moving
Every moving (thing) is a horse (at a certain time)
Therefore No human is a horse

In this case, the conclusion is negative.

While with the following three other terms: man, moving, and living, we can get the following mood, where the conclusion is affirmative:

Every human is possibly moving
Every moving (thing) is living
Therefore Every human is living. ([26], 201.10–11)

When the major premise is assertoric and negative, and the minor is possible, then the conclusion is negative and assertoric ["*bi ishtirāk al-'ism*" ([26], 201.16)], that is, of the kind of those that may be either possible or necessary. In this case, the conclusion is negative and possible but compatible with necessity. Averroes gives the following formulation:

Every C is B by possibility (**Q**)
No B is A by actuality (**X**)
Therefore No C is A by possibility (**M**) (= *Celarent* **XQM**)

In this case, however, we may have a necessary conclusion, for the conclusion may be in certain cases "No C is A by necessity" ([26], 201.22). Despite this difference, Averroes considers this kind of moods as conclusive, because the compatibility with necessity is inherent to the unilateral kind of possibility.

The proof he provides, which is by *reductio ad absurdum*, assumes nevertheless the conclusion to be possible, for it starts by the contradictory of the possible conclusion, which is "Not possibly No C is A" ("*laysa yumkinu wa lā shay'a min [C] huwa [A]*" ([26], 201.24). However, here too, "Not possibly" is put in the very beginning of the proposition, which may suggest that possibilitythe is *de dicto*. Thus the proposition is said to be equivalent to the following particular necessary affirmative one: "Some C is A by necessity," and if one analyzes the two propositions by using the modern formalism, one gets the following proposition: $\sim\Diamond$ No C is A = $\Box \sim$ No C is A = $\Box$ Some C is A [since $\sim$ No C is A = Some C is A].

While if one uses a *de re* kind of possibility, one should formalize the proposition in the following way: No C is $\Diamond$A = $(x)(Cx \supset \sim\Diamond Ax) = (x)(Cx \supset \Box \sim Ax)$ which, as we can see, is a universal negative and necessary proposition (**E**$\Box$).

This shows that the distinction *de re* / *de dicto* is not made at all by Averroes, who expresses things in the vernacular usual language without paying attention to the differences of scopes regarding the modal operator.

Averroes provides a proof by *reductio ad absurdum* and two concrete illustrations of that mood. He also stresses the difference between the assertoric and the necessary in Aristotle's frame especially when he criticizes other people's interpretations such as al-Fārābī and the Greek commentators Alexander of Aphrodisias, Themistius, and Theophrastus who, according to him, do not sufficiently clarify that difference and interpret the assertoric in a confused way.

When the major premise is assertoric and the minor is possible then, according to Averroes, these two moods are conclusive but not perfect ([26], 203.24). He says that these moods could be proved by *reductio ad absurdum* "when the premises are universal" or by conversion when one of the premises is particular ([26], 204.1–2). However, if the minor premise is "a negative assertoric," the mood is not conclusive. This opinion is the same as that defended by Aristotle, for M. Malink says in his article that the moods **QXQ** where the assertoric premise **X** is negative are not conclusive in Aristotle's frame (for instance in *Pr.Ana*, 35b11) as appears in Table 2 ([113], 118).

Averroes shows the invalidity of these moods by providing concrete counterexamples where the conclusion is sometimes affirmative and sometimes negative depending on the terms chosen. These examples are the following: when the terms are "snow", "living", and "white", we may have the following syllogism, where the conclusion is affirmative despite the negative character of one of the premises:

Some snow is not living (**OX**)
Every living (being) is possibly white (AQ?)
Therefore Some snow is white (IX?) ([26], 204.4–5)

The case where the conclusion is negative is illustrated by the terms "stable", "living" and "white" and the mood that may be constructed with them:

Some stable (things) are not living
Every living (being) is possibly white
Therefore Some stable (things) are not white ([26], 204.6–7).

What about the moods mixing between a possible premise and a necessary one? Averroes analyzes these moods in chapter 15 of his *Talkhīs al-Qiyās*. These moods function exactly like the ones containing an assertoric premise and a possible one as shown in the following quotation: "If one of the premises is possible while the second one is necessary, then the kinds (*'anwā'*) of moods considered as conclusive are the same in number as those that contain a possible premise and an assertoric one, whether they are perfect or not" ([26], 205.2–4). He adds that the "perfect" moods are those where the "major" is possible while the minor is necessary, and the imperfect ones are those where the major is assertoric while the minor is possible. Thus he admits as Aristotle whose opinion is summarized by M. Malink in his Table 2 ([113], Table 2, p. 118), the moods **QLQ**, and **LQM**.

For *Barbara* **LQM**, for instance, is expressed as follows:

Every C is B by possibility
Every B is A by necessity
Therefore Every C is A by possibility

This mood is proved by *reductio ad absurdum* as follows:

If the conclusion is false, this means that its contradictory, which is
"Some C is not A by necessity" is true

if we add the possible premise, after changing it into an assertoric one, we get:
"Every C is B by actuality"

From which we may deduce
"Some B is not A by necessity"
By *Bocardo* **LXL**, already admitted above.

This conclusion, although it is not, as Averroes claims, the contradictory (*naqīdu*) ([26], 206.7) of the major premise, is nevertheless, contrary to that universal premise, which means that if it is true, the universal will be false. And this is enough to validate the proof and consequently the mood. The proof shows also that the conclusion expresses a unilateral kind of possibility given its contradictory, which is a necessary particular negative.

Note here that Averroes uses *Bocardo* **LXL**, which he admits as conclusive, while in Aristotle's system, this mood is not considered as conclusive if we take into account the summary and the reconstruction made by M. Malink in his 2006 article cited above. For this author presents in his Table 1 *Bocardo* **LXL** as an invalid mood in Aristotle's frame ([113], Table 1, p. 117). However, despite the invalidity of *Bocardo* **LXL**, Aristotle admits *Barbara* **LQM** proved above, according to the same author ([113], 118), which means that he may have proved it by other ways.

As to the moods where the possible premise is the major and the necessary one the minor, Averroes says that they are "perfect" ([26], 206.12) because one can apply the *dictum de omni* to prove them and does not need other kinds of proofs. This proves, according to him, *Barbara* **QLQ** as well as *Celarent* **QLQ** ([26], 206.11–13). However, this claim does not take into account the problem that the negative bilateral possible propositions raise. Here too, it seems that he just reformulates *Barbara* **QLQ** by starting by the negative side of **A** bilateral possible instead of presenting what really corresponds to *Celarent* **QLQ**.

Now, if the necessary is the major premise and if it is negative, as in *Celarent* **LQM**, then the mood is not perfect and the conclusion does not express a bilateral kind of possibility, which Averroes expresses by saying that this conclusion might be "either actual or possible" ([26], 206.19–20). This means that from:

Every C is B by possibility
and No B is C by necessity
One may deduce
No C is A "by actuality or by possibility" ([18], 206.19–20)

The proof by *reductio ad absurdum* provided assumes that this conclusion expresses a unilateral kind of possibility, for its contradictory is said to be the following:

"Some C is A by necessity" ([26], 206.22).

If one adds the negative premise

"No B is A by necessity"

One may deduce

"Some C is not B by necessity" by *Festino* **LLL** from the second figure, which contradicts "Every C is B by possibility" the minor premise ([26], 206.21–25) and is absurd.

This proof validates *Celarent* **LQM**, but since the proof is made by *reductio ad absurdum*, the mood is not perfect.

He also admits the first figure moods containing a particular premise, such as *Darii* **LQM** and *Ferio* **LQM** as well as *Darii* **QLQ** and *Ferio* **QLQ** ([26], 206.16–25–207.1–5)). According to him, the latter two moods are perfect unlike the former ones which are not. But *Ferio* **QLQ** is not stated correctly and raises the same problem as *Ferio* **QXQ** above.

This means that the valid moods containing a necessary premise and a possible one are the same as the valid moods containing an assertoric premise and a possible one as Averroes notes ([26], 208.24–26), for there is an exact parallelism between both kinds of moods, given that when a mood is valid or invalid or perfect or imperfect in the assertoric case, it is valid or invalid or perfect or imperfect in the necessary case too.

Now, according to M. Malink, Aristotle adds some other moods which have the structure **LQX**, that is, where the conclusion is assertoric while the first premise is necessary and the second one possible. In Aristotle's frame, these are *Celarent* **LQX** (but *not Barbara* **LQX**, which is not said to produce an assertoric conclusion in that case) and *Ferio* **LQX** essentially, together with some moods with two negative premises (see [113], 118, Table 2, column 6). Can we say the same about Averroes? Let us see what he says about these moods where the major is a negative necessary universal (**E** necessary) in both cases. At page 206, he says explicitly that "if one of the universal premises is affirmative and the other one negative, and the negative is necessary and is the major one while the minor is possible, then the mood is conclusive but not perfect and it produces two conclusions: one of them is assertoric and the other one is possible" ([26], 206.13–15). So he seems to admit *Celarent* **LQX** too in this passage, together with *Celarent* **LQM**. But the proof he provides is the one already analyzed above, which is by *reductio ad absurdum* and starts with what is said to be the contradictory of the conclusion, i.e., a necessary particular premise (which contradicts the *unilateral possible* conclusion). He does not provide another proof for the mood whose conclusion is said to be assertoric, for in that case, he would have started with an assertoric contradictory of the conclusion, which he did not.

As to *Ferio* **LQX**, he says what follows about it: "If one of the premises is particular and the major is a necessary negative, then the conclusion is a negative assertoric and a negative possible" ([26], 207.17). Here too, he seems to admit *Ferio* **LQX** too as was the case with Aristotle.

Let us now examine the second figure moods containing either only possible premises or a possible premise mixed with an assertoric or a necessary premise.

As in Aristotle's frame (see [113], Table 3, p. 119), Averroes says that no mood is valid in the second figure if both premises are possible ([26], 215.2). If one of the premises is possible while the other one is assertoric, then the mood is not conclusive when "the affirmative is the assertoric and the negative is the possible" ([26], 215.5–6). So *Cesare* **QXM**, *Camestres* **XQM**, *Festino* **QXM** and *Baroco* **XQM** are *not* valid. In this respect, he is completely in accordance with Aristotle who does not validate these moods either (see [113], 119, Table 3, columns 2 and 3).

If the assertoric premise is the negative one and if it is universal, then the mood is conclusive, according to Averroes ([26], 215.6). This validates *Cesare* **XQM**, *Camestres* **QXM** and *Festino* **XQM** and is also in accordance with Aristotle who validates these moods according to the analysis provided by M. Malink ([113], 119, Table 3, columns 2 and 3).

Averroes adds that the same may be said about the moods containing a necessary premise instead of an assertoric one, together with a possible premise. He thus validates *Cesare* **LQM**, *Camestres* **QLM**, and *Festino* **LQM**. In this respect, he agrees with Aristotle too, whose valid second figure moods are the same whether the negative premise is assertoric or necessary ([113], 119, Table 3, columns 2, 3, 4, and 5).

The validation or invalidation of these different moods seem to follow some rules stated in the very beginning of the chapter, according to which when the negative premise is possible, no second figure mood is valid while when it is assertoric or necessary, the moods are valid provided that this negative premise is universal.

The rest of the chapter justifies these "rules" by analyzing the conversion of the possible negative premises. These possible negative premises do not convert as possible, according to Averroes, for when one starts with a proposition like "Every C is possibly not A" (that is, "No C is A by possibility": **E**◊), one cannot deduce from it the following: "Every A is possibly not C" ([26], 215.11–14). For instance, from "Every man is possibly not white," one cannot deduce "Every white is possibly not a man," because "Some white (things, e.g. the snow) is necessarily not a man" ([26], 216.5). Since this last necessary negative sentence is true, the possible negative one is not true, according to Averroes ([26], 216.5–7). This seems to suggest that the possibility here is taken in its unilateral meaning, for in that case, it is opposed clearly to a necessary proposition.

As a matter of fact, Averroes provides a whole discussion ([26], 216–217) where he tries to argue that the proof by *reductio ad absurdum* for **E**◊ is not really conclusive. This proof shows that "Every A is ◊ not B" implies "Every B is ◊ not A" by *reductio ad absurdum*. It runs as follows:

If "Every B is ◊ not A" is false, then its contradictory, that is, according to Averroes, "Every B is not ◊ not A" will be true. From this contradictory, one may deduce "Some B is □A" [since "not ◊ not" = "□" and the particular is implied by the universal]. But then, by the conversion of **I**□, we have "Some A is □B." But we had "Every A is ◊ not B" which was supposed to be true and which contradicts "Some A is □B." This is absurd and proves the validity of the conversion.

But Averroes objects that the proposition "Every B is ◊ not A" is implied by "Every B is ◊A," and is contradicted by "Some B is □A," while the first one is contradicted by "Some B is □ not A." This means that the proposition "Every B is not ◊ not A" may imply both "Some B is □A" (= **I**□) and "Some B is □ not A" (= **O**□) ([26], 217.1–3), in which case there is no contradiction between the proposition deduced (the particular necessary) and the antecedent of the conversion (the universal possible), thus no absurdity is deduced in the above proof, which means that the proof is not conclusive.

This objection seems to reject the proof because Averroes interprets the possibility of **E** in the bilateral sense, which is the reason why he says that "Every B is ◊ not A" is implied by "Every B is ◊A." But this implication is far from obvious, for as we saw above, the formulation of **E** bilateral possible is not correct in Averroes' frame. As a consequence, the proof above is not convincing.

Second, he assumes that the contradictory of "Every B is ◇ not A" is the following: "Every B is not ◇ not A." Unfortunately, this assumption is wrong, first because "Every B is not ◇ not A" is equivalent to "Every B is □A," given that "not ◇ not" = "□", i.e., to **A□**, which is *not* the contradictory of **E◇**, and does not imply **O□**, even if it does imply **I□**, second because if the possibility is unilateral, the contradictory of **E** possible is the following: "Not (Every B is ◇ not A) ," given that the negation should be put in the very beginning of the proposition to get the real contradictory. This contradictory is thus a particular proposition, and not a universal one, given that $\sim (x)(Bx \supset \Diamond \sim Ax) = (\exists x)(Bx \wedge \sim \Diamond \sim Ax) = (\exists x)(Bx \wedge \Box Ax) =$ **I□**. If **E** is a bilateral possible proposition, its contradictory is the bilateral possible **I,** i.e., "Some S is $\Diamond P \wedge \Diamond \sim P$," given that **E** bilateral possible itself is the following proposition: "Every S is $\Box \sim P \vee \Box P$," as we saw above.

Now, the result of his criticism is that the universal possible negative does not convert as itself ([26], 217.7). Given the invalidity of that conversion, he considers that the following premises do not validly lead to a conclusion:

Every C is B by possibility
and Every A is possibly not B

This means that in the second figure, the mood containing two possible universal premises is not valid, because of the invalidity of **E** possible conversion ([26], 217.10).

However, given the proof above and its criticism by Averroes, which uses the two meanings of possibility, we do not know exactly if **E** possible conversion is invalid when the possibility is bilateral or when it is unilateral too. So it seems that there is some confusion between the two meanings which are not separated sufficiently clearly.

Anyway, Averroes provides also some counterexamples which show that such moods are not valid, since they may produce different propositions. The first example is the following: Take the terms "man", "white", and "horse", then we have the following "mood":

Every human is possibly white
Every horse is possibly not white
Therefore No human is a horse

The conclusion in this mood is negative and at least materially necessary, since it is not possible for any man to be a horse.

But if one changes the terms, then the conclusion will be different in its modality and its quality, as in the following example:

Every human is possibly white
Every living (being) is possibly not white
Therefore Every human is living
where the conclusion is affirmative ([26], 218.3).

This shows that the mood is not conclusive since its conclusion depends on the terms chosen and is not determined by the structure of the mood.

In the same way, if the negative premise is the minor one or if the two premises are either both affirmative or both negative, the mood is not conclusive ([26], 218.8–9) for the same reasons.

In chapter 17 of the same treatise, he examines the combination between the assertoric and the possible in the second figure. He notes first that when the negative premise is the possible one, the mood is not conclusive for the reasons provided above ([26], 219.2).

Now if the negative premise is the assertoric one, then since **E**-assertoric conversion is valid, the mood may be valid too, since it would be reducible to a first figure mood, whether the negative premise is the major one or the minor one. Thus, *Cesare* **XQM** and *Camestres* **QXM** should be valid, because they are reducible to *Celarent* **XQM** of the first figure.

If *Celarent* **XQM** is:
No B is A
Every C is possibly B
Therefore No C is A by possibility
Then *Cesare* **XQM** is:
No A is B (by conversion)
Every C is possibly B
Therefore No C is A by possibility
And *Camestres* **QXM** is:
Every C is possibly B
No A is B (by conversion)
Therefore No A is C by possibility (by the conversion of "No C is A by possibility)

However, since *Camestres* is proved by two conversions even in categorical logic (see Sect. 3.2.2, p. 74 in the case of Avicenna and 3.3.3, p. 87 in the case of Averroes), it should also use two conversions in the modal case, which raises a problem here, given what has been said about **E** possible conversion above. Does this mean that **E** possible conversion is admitted after all when the possibility is unilateral? Since Averroes criticizes the usual proof of **E** possible conversion and says explicitly that **E** possible "does not convert" ([26], 217.7), and since he does not state the conversion of **E** possible when the possibility is unilateral, for instance, at p. 148 of *Talkhīṣ al-Qiyās*, where he evokes the conversions of the modal propositions, the proof provided above which relies on the conversion of the conclusion of *Camestres* **QXM** is not in accordance with the rest of the text and the rules evoked. In other words, there is some confusion in this proof and in the definition of possibility in general, which oscillates between the bilateral and the unilateral meanings. This confusion extends to the rules related to the notion of possibility, which are not stated sufficiently clearly.

If one of the premises is particular, as in *Baroco* and *Festino*, then according to Averroes, when the assertoric is the affirmative one, whether it is particular or universal, the mood is not valid. Thus *Baroco* **XQM** and *Festino* **QXM** are *not* valid. As to *Baroco* **QXM**, it is not admitted too (together with an eventual **EOO** mood), for "if the negative assertoric is particular then there is no [conclusive] mood, whether the other premise is affirmative or negative" ([26], 220.12–13). This is in accordance with Aristotle, who does not validate *Baroco* **QXM** ([113], 119, Table 3).

If the universal negative is assertoric, then the mood is valid and may be proved by conversion and by reduction to the corresponding first figure mood. This means that *Festino* **XQM** is valid and may be reduced to *Ferio* **XQM** by the conversion of the major negative universal premise. The same may be said about the mood where the two premises are negative and one of them is possible. In that case, the possible negative premise (**E**-possible) may become an affirmative one by conversion and lead to a valid mood.

When the combination involves one necessary premise with one possible one, the moods admitted are valid if the negative universal premise is the necessary one. So *Cesare* **LQM**, *Camestres* **QLM,** and *Festino* **LQM** are valid since they can be proved by the conversion of **E** necessary and reduction to their correspondent first figure moods. These moods produce "negative assertorics and possible negatives" according to Averroes. This validates *Cesare* **LQX**, *Camestres* **QLX,** and *Festino* **LQX** too, which, as a matter of fact, are admitted by Aristotle too, if we follow the analysis of M. Malink ([113], 119, Table 3).

But these moods together with their analogues that contain a possible conclusion raise a problem, given that the conclusion is sometimes assertoric and sometimes possible. So one can ask: what is the criterion that makes the conclusion possible in some cases and assertoric in other ones? Given that there are two conclusions resulting from the same two premises, can one consider these moods as valid? It is clear that such moods cannot be *formally* valid, because their conclusive character is related to the nature of the terms that they contain. The question thus is the following: why are they admitted both by Aristotle and Averroes? This question is legitimate if we consider that the plurality of conclusions is one of the reasons that make both authors *reject* some moods. Maybe one could answer by saying that these moods lead first to an assertoric conclusion, which in turn implies a unilateral possible one. But this is not explicitly stated by Averroes, although it would be a plausible reason to accept the two conclusions.

But if the necessary premise is the affirmative one, then the mood is *not* valid, for "if the affirmative is necessary, then there is no [conclusive] mood at all" ([26], 222.20–21). Thus *Festino* **QLX** and *Festino* **QLM** are not admitted. The same may be said about *Baroco* **LQM** and *Baroco* **LQX**, *Cesare* **QLM**, *Cesare* **QLX**, *Camestres* **LQM,** and *Camestres* **LQX** which should not be admitted too, given that the necessary premise is the affirmative one in all these moods.

What about the third figure moods?

According to Averroes, in the third figure moods, when both premises are possible and universal, the conclusion is possible too, but particular and the mood is conclusive ([26], 225.4–5). Thus it seems that according to him, *Darapti* **QQQ** and *Felapton* **QQQ** are both valid. He adds that the proofs of these moods are the same as those of their corresponding assertoric moods. This is in accordance with M. Malink's results about the third figure moods in Aristotle's logic (see [113], 120, Table 4). But as we saw above *Felapton* **QQQ** raises a problem, because both **E** and **O** bilateral possible are not well expressed.

When both premises are negative, then the moods are conclusive but not perfect because one of the negative premises would have to be converted to an affirmative one in order for the conclusion to be deduced ([26], 225.7–9). Thus the moods containing two negative possible premises, whether universal or particular, lead to a particular negative possible conclusion, which is also in accordance with Aristotle's results ([113], 120). Here too, the validity of these moods can be seen as problematic for the same reason.

However, when the premises are both particular and possible, no mood is conclusive, because in that case, one may obtain different conclusions depending on the terms used. For instance, if the terms are: man, white, and living, the conclusion is affirmative if the mood is stated as follows:

Some white (being) is possibly human
The white (beings) are possibly living
Therefore Humans are necessarily living ([26], 226.1–2)

But if the terms are man, white, and horse, then the mood is stated as follows:

White (beings) are possibly human
White (beings) are possibly horses
Therefore No human is a horse ([26], 226)

and produces a negative conclusion. This difference in the conclusions shows that the mood is not valid. Note that the indefinites in both examples are treated as particulars.

If one of the premises is possible while the other one is assertoric, then if both of them are affirmative and universal, then the conclusion is particular, possible and affirmative. Its proof is made by the conversion of the minor premise ([26], 226). So if the minor premise is the possible one, then the mood is *Darapti* **XQM** which is proved by conversion to *Darii* **XQM**. If the minor premise is the assertoric one, then the conversion reduces *Darapti* **QXQ** to *Darii* **QXQ**.

As to *Felapton*, if its negative premise is possible while its affirmative one is assertoric, then "the conclusion is a real possible one (*mumkina ḥaqīqīya*)" ([26], 226.13). So *Felapton* **QXQ** is admitted. But if the negative premise is assertoric while the affirmative one is possible, then the conclusion is possible but its possibility is not bilateral. This probably means that *Felapton* **XQM** is admitted as it is in Aristotle's frame ([113], 120). But the way Averroes expresses the nature of the conclusion is confusing, for he says "I mean that [this mood] produces two conclusions: a negative necessary and a negative possible" ([26], 227.14–15). We might say that the necessity evoked here is a kind of matter necessity and depends on the nature of the terms. So maybe in some cases, the terms are such that the conclusion is materially necessary, while in some other cases, it is only possible. Anyway, the possibility of the conclusion, in this combination, cannot be bilateral, it is thus unilateral. But *Felapton* **QXQ** is problematic because of the fact that neither **E** bilateral possible nor **O** bilateral possible are correctly expressed, as we saw above.

Now in case the two premises are affirmative, but one of them is particular [as in *Datisi* and *Disamis*], or if the universal premise is negative and is the major [as in *Ferison*], then the conclusion is particular [and possible] and "the proof will be made by the conversion of the particular proposition" ([26], 228.3–6), which reduces these moods to their corresponding first figure moods. So it seems that he admits *Datisi* **QXQ** and **XQM** together with *Disamis* **QXQ** and **XQM** and *Ferison* **QXQ** and **XQM**, by reducing them, respectively, to *Darii* **QXQ** and **XQM**, and *Ferio* **QXQ** and **XQM**. But once again, the validity of both *Ferison* **QXQ** and *Ferio* **QXQ** (as they are stated by Averroes) is problematic for the reasons evoked above.

As to *Bocardo* **MXM**, which is stated as follows:

> Every B is C, by actuality (X)
> Some B is not A by possibility (M or Q?)
> Some C are possibly not A (M)

it is proved by ***reductio ad absurdum***, for assuming that the contradictory of the conclusion, that is, "Every C is A by necessity" is true, and given that "Every B is C, by actuality", we can deduce from both propositions that "Every B is A by necessity", *Barbara* **LXL**, which contradicts the major premise of *Bocardo* above "Some B is not A by possibility" and is not acceptable ([26], 228.11–15).

From this proof, we can note that the conclusion of *Bocardo* above is assumed to express a unilateral possibility, given what Averroes considers explicitly as being its contradictory ([26], 228.13), namely, the necessary proposition "Every C is A by necessity." This proposition is not a disjunction between a necessary one and an impossible one, as it would be if were to contradict a bilateral possible proposition. The result of the proof, which is a universal necessary proposition is also incompatible with the major premise "Some B is not A by possibility." If this premise expresses a unilateral possibility, then the necessary proposition is its contradictory, if it expresses a bilateral possibility, the necessary proposition is its contrary. So both cases are admissible, given that Averroes does not specify what kind of possibility the major premise expresses. As a result, the mood could be a *Bocardo* **QXM** or a *Bocardo* **MXM**. In Aristotle's frame, the mood admitted is a *Bocardo* **QXM**, as shown by M. Malink ([113], 120, Table 4).

Averroes admits also *Bocardo* **XQM**, where the major is assertoric while the minor is possible. He says that it can be proved by *ekthesis*. In categorical logic, the proof by ekthesis reduces *Bocardo* to *Felapton* by using *Barbara* in the deduction. As we saw above, Averroes' proof by *ekthesis* for *Bocardo* in categorical logic is not very different from that provided by both al-Farabi and Avicenna. But if *Bocardo* **XQM** is the following mood:

Every B is ◊C (and ◊ not C?)
Some B is not A
Therefore Some C is ◊ not A

the proof by *ekthesis* would run as follows: Let the thing which is B and is not A, be called D, then we have:

Every D is ◊C (and ◊ not C?)
No D is A
Therefore Some C is ◊ not A (by *Felapton* **XQM**)

Barbara **QXQ** would be used to obtain "Every D is ◊C (and ◊ not C?)", starting from "Every B is ◊C (and ◊ not C?)" (Q) and "Every D is B" (X) (assumption).

Now, what about the moods containing one necessary premises and one assertoric premise in this figure three?

Averroes says the following: "If both premises are universal and one of them is necessary while the other one is possible, and both are affirmative, then the mood is conclusive and produces a possible conclusion and this can be shown by the reduction to the first figure." ([26], 229.4–6). This does not indicate which kind of possibility is used, but the "reduction to the first figure" mentioned in the quotation could make things clearer.

Darapti **QLQ** may be proved by the conversion of the minor and the reduction to *Darii* **QLQ** as follows:

Darapti **QLQ**		*Darii* **QLQ**
Every B is C by necessity (L)	→	Some C is B by necessity (L)
Every B is A by possibility (Q)		Every B is A by possibility (Q)
Therefore Some C is A by possibility (Q?)		Some C is A by possibility (Q)

If the minor premise is the possible one, then we can have the following reduction to *Darii* **LQM** by the conversion of the minor to a possible proposition:

Darapti **LQM**		*Darii* **LQM**
Every B is C by possibility (Q)	→	Some C is B by possibility (Q)
Every B is A by necessity (L)		Every B is A by necessity (L)
Therefore Some C is A by possibility (M)		Some C is A by possibility (M)

As to *Felapton* **QLQ?**, it may be reduced to *Ferio* **QLQ** when the major is possible and the minor necessary by the conversion of the necessary premise as follows:

Felapton **QLQ?**		*Ferio* **QLQ**
Every B is C by necessity (L)	→	Some C is B by necessity (L)
No B is A by possibility (Q)		No B is A by possibility (Q)
Therefore Some C is not A by possibility (Q?)		Some C is not A by possibility (Q)

Note that in Aristotle's frame, as shown by M. Malink, the conclusion of both moods expresses a unilateral possibility and the moods are said to be **QNM** (**N** being the symbol used by Malink for necessity) ([113], 120, Table 4). But Averroes does not say explicitly what kind of possibility the conclusion expresses. He only says that the moods can be reduced respectively to *Darii* and *Ferio* of the first figure by the conversion of the necessary premise. Since *Darii* and *Ferio* contain a bilateral possible conclusion, there is no reason to change it to a proposition expressing a unilateral possibility. However, the conclusiveness of these two moods is problematic given that **E** bilateral possible and **O** bilateral possible are not correctly expressed.

If the major premise is the necessary one while the minor is possible, *Felapton* would be expressed thus and reduced to *Ferio* **LQM** by the conversion of the possible premise:

Felapton **LQM**		*Ferio* **LQM**
Every B is C by possibility (Q)	→	Some C is B by possibility (Q)
No B is A by necessity (L)		No B is A by necessity (L)
Therefore Some C is not A by possibility (M)		Some C is not A by possibility (M)

The moods where the minor premise is negative and necessary are said to be non-conclusive, while those where this same negative minor is possible are conclusive only in case the negative premise is converted to an affirmative one ([26], 229.13–15).

As to *Datisi* **QLQ + LQM** and *Disamis* **QLQ + LQM**, they are said to be reducible to *Darii* **QLQ + LQM,** respectively, by conversion. In this case, he says explicitly that the conclusion "is possible as it is in the moods of the first figure to which they are reduced" ([26], 230.3–4). So, since the conclusion of *Darii* **QLQ** is a bilateral possible proposition, so will be the conclusions of both *Datisi* **QLQ** and *Disamis* **QLQ**. However, in Aristotle's frame, these conclusions of the third figure moods express a unilateral possibility if we consider what says M. Malink in his article ([113], 120, Table 4).

Ferison **QLQ + LQM** are reduced, respectively, to *Ferio* **QLQ + LQM** by the conversion of the particular premise. Here too, he says explicitly that the conclusion is "a real possible one (*mumkina ḥaqīqīya*)" ([26], 230.10), that is, a bilateral possible proposition when the "negative premise is possible," i.e., in *Ferison* **QLQ**, as it is in *Ferio* **QLQ**. So, if we follow M. Malink in his analysis of Aristotle's moods, who considers that these third figure moods contain a unilateral possible conclusion, it seems that Averroes departs from Aristotle in this respect or interprets Aristotle's text in a way different from his modern commentators.

Bocardo is proved by *ekthesis* as in categorical logic. It is reducible to first figure moods and contains the same kinds of conclusions. So we may say that Averroes admits *Bocardo* **QLQ** together with *Bocardo* **LQM**, since as we saw above, *Bocardo* is reduced to *Felapton* via *Barbara*.

However, the problem raised by **E** and **O** bilateral possible makes the validity of the **QLQ** moods containing negative propositions problematic.

4.4 Conclusion

This analysis of the modal syllogistics of both Avicenna and Averroes shows that the two systems are very different since they do not admit the same rules or the same valid moods. Averroes is much closer to Aristotle than Avicenna is, and the differences between his system and Aristotle's one are very seldom. As to Avicenna, he defends a different system and uses very rarely the bilateral possible proposition in his valid moods, which contain mostly unilateral possible premises and conclusions. The rules of conversion are not the same in both systems, since in Avicenna's frame, **A** necessary and **I** necessary both convert to **I** possible (similarly **A** permanent and **I** permanent convert to **I** general absolute), while in Averroes' frame, they convert to necessary propositions as it is the case in Aristotle's system. As to **A** possible and **I** possible, they convert to **I** unilateral possible in Avicenna's frame while they convert to **I** bilateral possible in Averroes' one. This explains many of the differences between the two systems.

Summarizing the results obtained in the above analysis, we can provide the lists below that show the differences between both authors.

First Figure valid moods

Premises	Avicenna's valid moods	Averroes' valid moods
LX	**LsXpL** and **XsLXp**	**LXL** and **XLX**
MM	**MMM**	**MMM**
QX	**QXsQ, XsQQ** (*al-Najāt*), **XsQM**	**QXQ** and **XQM**
MX	**MXsM** and **XsMM**	
LM	**MLM** and **LML**	
LQ	**LQL?** (*al-Najāt*)	**LQM, LQX, QLQ**

The differences between both authors concern first the way they express the assertoric propositions, for Avicenna seems to privilege the propositions containing the condition "as long as it is S," while Averroes expresses them without any condition, second the way they express the possible premises which are interpreted in a bilateral way in Averroes' frame while they are mostly interpreted in a unilateral way in Avicenna's one. Note that **LQX** does not hold for *Barbara* in Averroes' frame.

Second Figure valid moods

Premises	Avicenna's valid moods	Averroes' valid moods
LX	**LsXpL** (eae, eio), **XsLXp** (eae, eio) **LXsXp** (aee, aoo), **XpLsL** (aee, aoo)	**LXL** (eae, eio) **XLX** (eae, eio) **LXX** (aee, aoo)
MX	**XsMM** (eae, eio, aoo), **MXsM**	
QX		**XQM**
LQ		**LQM, QLM, LQX, QLX**
LM	**LML, MLL**	

Third Figure valid moods

Premises	Avicenna's valid moods	Averroes' valid moods
LX	**LsXpLst** (aai, eao, aii), **XsLXga** (aai, eao, eio), **LsXsLst**(iai, oao), **XsLsXga** (iai, oao), **LsXpLst** (eio, aii, eao)	**LXL, XLX, LXX, XLL**
MM	**MMM**	
QQ (QM)	**QMQ**	**QQQ**
QX	**QXsQ, XsQM**	**QXQ, XQM**
MX	**XsMXga, (XsMM?), MXsQ**	**MXM** (oao)
LQ		**LQM, QLQ**
LM	**LML, MLM**	

These tables show that Avicenna's frame is simpler than Averroes' one since it admits less moods with two conclusions. But the propositions in Avicenna's frame are much more complex and sophisticated than they are in Averroes' frame, given the conditions added which give rise to very specific moods that we don't find in Aristotle's or Averroes' frames.

References

2. Al-Fārābī, Abū Naṣr. 1960. *Ṣharh al-Fārābī li kitāb Arisṭūṭālīs fī al-'Ibāra*, ed. Wilhelm Kutch and Stanley Marrow, 2nd edn. Beirut: Dar el-Mashriq.
8. Al-Fārābī, Abū Naṣr. 1986. *Kitāb al-'Ibāra*. In *al-Manṭiq 'inda al-Fārābī*, vol. 1, ed. Rafik Al Ajam, pp. 133–64. Beirut: Dar el Machriq.
14. Al-Fārābī, Abū Naṣr. 1988. *al-Qawl fī al-'Ibāra*, in *al-Manṭiqiyāt li-al-Fārābi*, vol. 1, texts published by Mohamed Teki Danesh Pazuh, Edition Qom, pp. 83–114.
12. Al-Fārābī, Abū Naṣr. 1988. *Sharh al-Ibāra*. In *al-Manṭiqiyāt li-al-Farābi*, vol. 2, texts published by Mohamed Teki Danesh Pazuh, Edition Qom, 1409 of Hegira.
18. Al-Fārābī, Abū Naṣr. 1988. *Mā yanbaghī an yuqaddama qabla ta'allum al-falsafa*. In *al-Mantiqiyyāt li-al-Fārābī*, texts published by Mohamed Teki Danesh Pazuh, Edition Qom, pp. 1–10.
24. Aristotle. 1991. Prior analytics. In *The Complete Works of Aristotle*, ed. Jonathan Barnes, vol. 1, the Revised Oxford edn.
26. Averroes. 1982. *Talkhīṣ Manṭiq Arisṭu* (*Paraphrase de la logique d'Aristote*), volume 1: *Kitāb Al-Maqūlāt* (pp. 3–77), *Kitāb al-'Ibāra* (pp. 81–141), *Kitāb al-Qiyās* (pp. 143–366), ed. Gérard Jehamy, Manshūrāt al-Jāmi a al-lubnānīya, al-Maktaba al-sharqiyya, Beirut.
28. Averroes. 1983. *Middle commentary on Aristotle's prior analytics,* Critical edition by M. Kassem, completed, revised and annotated by C. E. Butterworth, and A. Abd al-Magid Haridi, Cairo.
29. Averroes. 1983. *Maqālāt fī al-manṭiq wa-al-'ilm al-ṭabī'ī* [Essays on logic and natural science], ed. Jamāl al-Dīn al-'Alawī. Casablanca.
30. Averroes. 2001. *Kitāb al-Muqaddamāt fi al-Falsafa, al-Masā'il fi-al Manṭiq wa al 'ilm al-ṭabī'ī wa al-Ṭibb*, ed. Assad Jemaa. Tunis: Markez al-Nashr al-Jāmi'ī.
32. Avicenna. 1938. *al-Najāt*, Muḥyi al-Dīn Sabrī al-Kurdī, 2nd edn. Cairo: Library Mustapha al Bab al Hilbi.
34. Avicenna. 1964. *al-Shifā', al-Manṭiq* 4: *al-Qiyās*, ed. S. Zayed, rev. and intr. by I. Madkour, Cairo.
35. Avicenna. 1970. *al-Shifā', al-Manṭiq 3*: *al-'Ibāra*, ed. M. El Khodeiri, rev. and intr. by I. Madkour, Cairo.
36. Avicenna. 1971. *Al-Ishārāt wa l–tanbīhāt, with the commentary of N. Ṭūsi,* intr by Dr. Seliman Donya, Part 1, 3rd edn. Cairo: Dar al Ma'arif.
39. Bäck, Allan. 1992. Avicenna's conception of the modalities. *Vivarium* XXX (2): 217–255.
40. Badawi, Abderrahman. 1980. *Manṭiq Arisṭu*, vol. 1 and 2. Beirut: Dar al Kalam.
47. Buridan, Jean. 1985. *Jean Buridan's logic, the treatise on supposition, the treatise on consequences*, trans. P. King. Dordrecht and Holland: D. Reidel.
49. Carnap, Rudolf. 1988. *Meaning and necessity, a study in semantics and modal logic*, 2nd edn. Chicago and London: Midway Reprint Edition, The University of Chicago Press.
53. Chatti, Saloua. 2014. Avicenna on possibility and necessity. *History and Philosophy of Logic* 35 (4): 332–353.
55. Chatti, Saloua. 2016. Existential import in Avicenna's modal logic. *Arabic Sciences and Philosophy* 26 (1): 45–71. Cambridge University Press.
56. Chatti, Saloua. 2016. Les oppositions modales dans la logique d'al-Fārābī. In *Soyons logiques / Let's be Logical*, ed. Amirouche Moktefi, Alessio Moretti, and Fabien Schang. Collection: *Cahiers de logique et d'épistémologie* (ed. Shahid Rahman and Dov Gabbay). London: College Publications.
65. Dadkhah, Gholamreza, and Asadollah Fallahi. 2018. *Logic in 6th/12th century Iran (Arabic texts with Persian notes and English introduction)*. Tehran: Iranian Institute of Philosophy.
69. Elamrani-Jamal, Abdelali. 1995. Ibn Rušd et les Premiers Analytiques d'Aristote: Aperçu sur un problème de syllogistique modale. *Arabic Sciences and Philosophy* 5: 51–74.
76. Garson, James. 2018. Modal logic. In *Stanford encyclopedia of philosophy*, ed. Edward N. Zalta. http://plato.stanford.edu/entries/logic-modal/.

79. Hasnawi, Ahmed, and Wilfrid Hodges. 2016. Arabic logic up to Avicenna. In *The Cambridge Companion to Medieval Logic*, ed. Catarina Dutilh Novaes and Stephen Read, pp. 45–66. Cambridge: Cambridge University Press.
94. Hodges, Wilfrid. *Ibn Sina: Qiyas* ii.3, translation based on the Cairo text, ed. Ibrahim Madhkur et al. Draft. http://wilfridhodges.co.uk/arabic21.pdf.
96. Hughes, George Edward, and M.J. Cresswell. 1972. *An introduction to modal logic*. London: Methuen and Co. Ltd.
101. King, Peter. 1985. Introduction to Jean Buridan's logic. In *Jean Buridan's logic*. D. Reidel: Dordrecht.
103. Klima, Gyula, *John Buridan: His nominalist logic, metaphysics, and epistemology*. Great Medieval Thinkers. Oxford University Press. http://www.phil-inst.hu/~gyula/FILES/John-Buridan.pdf.
104. Knuutila, Simo. 2008. Medieval theories of modalities. In *Stanford Encyclopedia of Philosophy*, ed. Edward N. Zalta. http://plato.stanford.edu/entries/modality-medieval/.
106. Lagerlund, Henrik. 2010. Medieval theories of the syllogism. In *Stanford Encyclopedia of Philosophy*. http://plato.standord.edu/entries/medieval-syllogism/.
113. Malink, Marko. 2006. A reconstruction of Aristotle's modal syllogistic. *History and Philosophy of Logic* 27 (2): 95–141.
117. Moktefi, A., A. Moretti, and F. Schang (eds.). 2016. *Let's be logical*. London: College Publications.
120. Movahed, Zia. 2010. De re and de dicto modality in the Islamic traditional logic. *Sophia Perennis* 2 (2): 5–19.
118. Movahed, Zia. 2017. Ibn-Sīnā's anticipation of the formulas of Buridan and Barcan. In *Logic in Tehran*, ed. Ali Enayat et al., pp. 248–255. Wellesley, MA: Association for Symbolic Logic and A. K. Peters. First edition (2003).
124. Read, Stephen. 2012. The medieval theory of consequence. *Synthese* 187: 899–912.
140. Street, Tony. 2014. Afḍal al-Dīn al-Khunājī (d. 1248) on the conversion of modal propositions. *ORIENS* 42: 454–513.
142. Strobino, Riccardo. 2018. Ibn Sīnā's logic. In *The Stanford Encyclopedia of Philosophy*. https://plato.stanford.edu/archives/fall2018/entries/ibn-sina-logic.
143. Thom, Paul. 2008. al-Fārābī on indefinite and privative names. *Arabic Sciences and Philosophy* 18 (2): 193–209.
144. Thom, Paul. 2008. Logic and metaphysics in Avicenna's modal syllogistic. In *The Unity of Science in the Arabic tradition*, ed. Shahid Rahman, Tony Street, and Hassen Tahiri, pp. 361–376. Dordrecht.

Chapter 5
The Hypothetical Logic

The Arabic logicians include the study of the hypothetical syllogisms in their counterparts of the *Prior Analytics*, that is, in the treatises called *al-Qiyās*. In al-Fārābī's frame, they are also evoked in *al-Maqūlāt* (the counterpart of the *Categories*). The hypothetical logic is the kind of logic that we find in the Stoïc system. So the fact that it is included in the Arabic authors' systems means that their logics have some non-Aristotelian features due to the fact that they take into account the post-Aristotelian studies, together with the Aristotelian ones.

However, the importance and the interest given to the hypothetical logic and the syllogistic moods that it may contain are not the same in the three frames. For while Avicenna constructs a whole hypothetical system parallel to the categorical syllogistic, al-Fārābī studies the Stoïc syllogistic moods and provides some variants, and Averroes devotes only a few pages ([26], 234–237) to the study of the hypothetical syllogisms and does not even consider the various variants that al-Fārābī has presented. But he does use the hypothetical syllogisms in the *reductio ad absurdum* proof, which is a reason to give them some interest and examine them even in Averroes' frame, where they are not very much developed.

My aim, in the following, is to examine the analysis provided by the three authors of these syllogisms and to compare them with each other and with the Stoïc syllogisms, in order to find out the differences between all these treatments and their common points. I will try to determine, as far as possible, the historical sources of this kind of logic, and to compare between the three Arabic logicians and some ancient authors that they might have followed or that presented systems that were close to what we can find in their own systems. For instance, Avicenna's quantified hypothetical logic might be seen as relatively close to some developments introduced by Theophrastus. We will thus determine to what extent these two systems are close to each other, although at first sight, as we will see below, Avicenna's hypothetical system is far more developed than any other system in traditional logic and highly original.

Furthermore, since the hypothetical syllogisms rely on the definitions and the meanings of the so-called logical constants, I will also try to clarify these meanings

S. Chatti, *Arabic Logic from al-Fārābī to Averroes*, Studies in Universal Logic,
https://doi.org/10.1007/978-3-030-27466-5_5

in the three frames and to compare them with the Stoïc definitions and conceptions of the same connectives. This examination will take into account the rules hold and the syllogisms considered as valid in that domain, together with the reasons that make the different authors validate these particular rules and syllogistic moods.

5.1 The Hypothetical Syllogisms

Before analyzing the hypothetical syllogisms endorsed by the Arabic logicians, let us first examine the Stoïc syllogisms as a starting point, for it is only by considering these Stoïc syllogisms that one can compare between them and those provided by the Arabic logicians. This will be the subject of the next subsection.

5.1.1 The Stoic Hypothetical Syllogisms

According to Łukasiewicz ([111], French translation in [109]), the Stoics present their syllogisms as inferences, unlike Aristotle. They use the following connectives: Negation, disjunction, implication, and conjunction. The negation, in their frame, is classical since their logic is bivalent.

The implication is the material philonian implication: it is false only in case the antecedent is true and the consequent false, and is true in the three other cases.

According to Susanne Bobzien, in the Stoic logic, the implication is expressed in terms of a conjunction as follows: "If p then q = Not (p and not q)." This is why it is truth-functional, since the conjunction is truth-functional. This opinion is endorsed by Łukasiewicz too, who says that the Stoïc implication is the philonian one in his 1934 paper ([111], in [109], 15). Since the philonian implication is false only when the antecedent is true and the consequent false, and true in all other cases, it is without any doubt truth-functional.

The conjunction is true when the two conjuncts are true, false otherwise. Since its truth conditions are sufficiently clear, it is without any doubt truth-functional.

However, the disjunction, unlike the conditional and the conjunction, is not truth-functional in the Stoïc logic, according to Susanne Bobzien ([45], Sect. 5.2), given that its truth conditions are not all determined clearly. The Stoic disjunction is interpreted as an exclusive disjunction, that is, it is true only when the two propositions do not have the same truth value, but its truth value is not clearly determined if we suppose that the two elements of the disjunction are both true. In fact, this supposition is not even made by the Stoics, since all the examples they give and the cases they consider involve propositions which are incompatible, that is, never true together. This is why the disjunction is intensional and non-truth-functional. Its truth value depends on the meanings of its elements and not only on their truth values.

The Stoics present their hypothetical syllogisms by using variables; each variable stands for a whole proposition. The propositional variables are expressed as follows: the first, the second, etc.

Consequently, the first hypothetical syllogism (I) is stated as follows:

- If the first then the second
- But the first
- Therefore the second ([111], in [109], 12, my translation).

In Latin, it says: "si primum, secundum; atqui primum; secundum igitur" (Apuleius, *De inter*, 279, cited by [111] in [109], 12, note 5).

This is the syllogism known as the *Modus Ponens*. It may be reformulated by using the modern propositional letters p, q, and so on. And this is the way Łukasiewicz uses to express the Stoic syllogisms in his 1934 paper. These syllogisms, called "indemonstrables", are formulated as follows by Łukasiewicz:

"I. If p, then q; but p; therefore q
II. If p, then q; but not q; therefore not p
III. Not both p and q; but p; therefore not q
IV. p or else q; but p; therefore not q
V. p or else q; but not q; therefore p" ([111] in [109], 16).

The mood II is called *Modus Tollens*. It involves also the material or philonian implication.

The mood IV shows that the disjunction is exclusive, for it would not be valid in the inclusive meaning, given that when "or" is inclusive, it is true when both propositions are true; so one cannot deduce "not q" from "p" because the truth of p does not exclude the truth of q.

As to the mood III, it is considered as a negated conjunction rather than a disjunction. But we will see that in this particular point, the Arabic logicians defend another view, since they consider this kind of syllogisms as disjunctive. It states that if two propositions are incompatible and one of them is true, then the other one is false.

The mood V is the so-called disjunctive syllogism and is valid whether the disjunction is inclusive or exclusive.

From these indemonstrables, the Stoics deduce many complex syllogisms, such as the following: "If p and q, then r; but not r; but p; therefore not q" ([111] in [109], 16).

This syllogism may be reduced to the indemonstrables by the following steps:

- If p and q, then r
- but not r

therefore - not both p and q (by II)

but - p

therefore - not q (by III) ([111] in [109], 16).

As we can see, the complex syllogisms combine between simple ones by using successively the indemonstrables.

According to Łukasiewicz, this very complex syllogism has been used (implicitly) by Aristotle himself in his proof of *Baroco* (by reduction to *Barbara*), because this reduction assumes that "if (if p and q, then r), then (if p and not r, then not q)" ([111] in [109], 19).

As a matter of fact, Aristotle says what follows:

> "If N belongs to every O [q] and M is predicated also of every N [p], M must belong to every O [r]; but we assumed that M does not belong to some O [not r] and if M belongs to every N [p] but not to every O [not r], we shall conclude that N does not belong to every O [not q]" ([24], I, 5, 27a35–27b5, symbols inside brackets added).

This shows that the inferences used by the Stoic logicians were used by Aristotle (although not explicitly expressed) in his proofs of his own moods, and means consequently that the hypothetical logic, which is a kind of propositional logic, is more fundamental than the categorical syllogistic, which could be seen as a part of monadic predicate logic, because it is at the heart of the proofs themselves. But this opinion was not widely shared by the traditional logicians, including the Arabic ones, who tended to privilege Aristotle's logic over the Stoic logic. Even Avicenna, who gave much attention to the hypothetical logic and presented a whole system much more elaborated than al-Fārābī's and Averroes'rules and syllogisms, tended to consider the hypothetical logic as secondary with regard to the categorical Aristotelian logic, since the latter was the basis of the former in his frame.

These basic rules and syllogisms attributed to the Stoics were known by the Arabic logicians who present them and some of their variants in their correspondents of the *Prior Analytics*. But the systems they develop and present are not exactly the same as that of the Stoics, for the definitions of the connectives are not exactly similar and the inferences presented are less numerous and less elaborated in al-Fārābī's and Averroe' systems and for many of them very different from that of the Stoics in Avicenna's one.

This is why we have to examine their treatment of this subject in order to determine their respective contributions and the differences between their respective systems on the one hand and between them and the Stoics on the other.

5.1.2 The Hypothetical Syllogisms in Arabic Logic

5.1.2.1 The Hypothetical Syllogisms in al-Fārābī's System

Let us start by al-Fārābī. In *al-Qiyās*, al-Fārābī says that the syllogisms may be either predicative (*ḥamli*) or hypothetical (*sharṭī*) ([15], 117; [10], 20). Among the hypothetical syllogisms, he distinguishes between the conjunctive ones (*'ittiṣālī*) and the disjunctive ones (*'infiṣālī*). In the conjunctive syllogism, the first premise contains an implication, while in the disjunctive syllogism, it contains a disjunction. The implication may be complete or incomplete: when it is complete, it is an equivalence (mutual implication); when it is incomplete, it does not convert ([13], 78). The syllogism is presented as an inference as it is in the Stoic logic. Note that even the categorical syllogisms are presented as inferences in his frame, since he deduces the conclusion by using the word "therefore", which makes a difference with Aristotle's presentation. This practice might not be new, since we can find it in other post-Aristotelian texts, as one reviewer suggests by saying that "Paul the Persian gives examples for all moods of predicative premise pairs, with a conclusion if the mood is productive and with pseudoconclusions if it is not. He regularly puts 'therefore' before the conclusion in the productive case, but never for the unproductive."

There are two conjunctive syllogisms and three disjunctive syllogisms.

In *al-Qiyās*, the first conjunctive syllogism is expressed as follows:

"- If this visible thing is a man, then it is an animal
- But it is a man
- Therefore, it is an animal" ([15], 137; [10], 32).

As we can see, this is the indemonstrable I of the Stoïcs, that is, the *Modus Ponens*; but here, it is expressed with concrete terms, not with variables. However, we will see that al-Fārābī does use variables in other contexts. We can also note that the propositions involved are predicative as the ones used in syllogistic and the syllogism looks like a *Barbara* with a singular term. The first premise contains an antecedent (*muqaddim*) and a consequent (*tābi'*). The link between them is expressed by the Arabic words *"in...fa"* (= if...then) which introduce a condition (= *sharṭ*). The word "in" (followed by "fa" = then) may be replaced by *"iḍhā kāna"*, *"iḍhā"*, and *"law kāna"*, which are almost its synonyms.

He also gives variants of this first kind of hypothetical syllogisms, which are the following:

I/ " - If the sun does not rise [not p], then it is not daytime [not q]
- But the sun does not rise [not p]
-Therefore it is not daytime [not q]" ([16], 167, symbols added).

In fact, only the first premise is given in the text. But al-Fārābī certainly holds it for he says: the first hypothetical syllogism contains as a second premise the antecedent itself (*bi 'aynihi*), thus deduces the consequent itself (*bi 'aynihi*)" ([16], 167).

This means that the syllogism may be expressed by means of a negative antecedent and a negative consequent, since its validity is not due to the quality of the propositions involved, rather it is due to the fact that the antecedent of the first premise is the second premise and the consequent of the first premise is the conclusion of the whole inference.

So we may consider that he provides the structure of the syllogism, given that he talks about "the antecedent" and "the consequent", whatever content and quality they may have. We can then assume that, like the Stoïcs, he uses variables.

The other variant of the first hypothetical syllogism is the following:

II/ "-If it is not night-time [not p], it is daytime [q]
-But it is not night-time [not p]
-Therefore, it is daytime [q]" ([16], 167 symbols added).

Here too, only the first premise is given, but the whole syllogism can be constructed by applying the general rule given above.

The second hypothetical syllogism is the following:

"- If this visible thing is a man [p], then it is an animal [q]
- But it is not an animal [not q]
- Therefore it is not a man [not p]" ([15], 138, symbols added).

This is the indemonstrable II of the Stoics (the *Modus Tollens*), but expressed with concrete terms and predicative singular propositions.

Al-Fārābī adds here an interesting remark, that is, "if the second premise was the opposite of the antecedent, then the syllogism would not be conclusive" ([15], 138). He thus rightly rejects "If p then q; but not p, therefore not q," which is not valid. This means that he admits the principle of contraposition, that is, $(p \supset q) \equiv (\sim q \supset \sim p)$.

He also gives a variant of this kind of syllogisms, that is,

"- If God is not unique [not p], then the world is not ordered [not q]
- But the world is ordered [q]
- Therefore God is unique [p]" ([16], 167, symbols added).

This variant has the following structure:

- If not p then not q; but q, therefore p.

What about the implication used in the first premise of these syllogisms? Al-Fārābī distinguishes between three kinds of conditional propositions, which are the following. The first one is called "accidental" (*bi-l-'araḍi*), because the antecedent and the consequent may be related only a few times or even only once as when one says: "If Zayd comes, 'Amr leaves" ([13], 78). Here, one cannot say that the consequent really follows or depends on the antecedent unless "following from" means just a succession in time. The second kind is called "essential" and is divided into two main sub-kinds: In the first sub-kind, the consequent follows from the antecedent "most of the time" ([13], 78), while in the second sub-kind, the consequent always follows the antecedent. This first sub-kind is illustrated by the following example: "When Sirius rises in the morning, the heat will be severe and the rains will cease" (translation W. Hodges in [92], 226).

The second sub-kind is also subdivided into two groups: the complete conditional and the incomplete one. The complete one is a biconditional (or an equivalence) and means what follows: necessarily when the antecedent holds, the consequent holds, and conversely. It is illustrated by the classical Stoic example: "If the sun rises, it is daytime" [and vice versa].

This complete conditional admits some further syllogisms, such as the following:

- 'p iff q; but q; therefore p'
- 'p iff q; but $\sim$p; therefore $\sim$q' ([13], 79, my formalization).

given that they are convertible and behave like biconditionals or equivalences. As to the incomplete one, it is a single conditional which does not convert, for instance, when one says: "If this is a man, it is an animal" [but not conversely] ([13], 78).

This second example involves the notion of inclusion, for the antecedent is clearly included inside the consequent, but not conversely; while in the first kind, there is some reciprocity, for the rising of the sun and the daytime always occur together. In both cases, however, there is either a semantic or a causal relation between the elements and this relation is necessary, unlike what happens in all the previous kinds considered by al-Fārābī.

However, one might ask about the conditional which is said to be essential but not necessary, that is, the one that holds "most of the time" but not always. For if it is essential, why is it not necessary? What does it mean exactly to say that this conditional is "essential"? If the relation is essential, why does it hold only in most cases, and not in all of them? Wilfrid Hodges, for instance, finds this essential character rather strange, for he says: "Curiously he allows that some 'essential' relations hold only for the most part" ([92], 227). Maybe we could say that "essential", here, means simply "natural" as opposed to the conventional, voluntary, or even arbitrary character of the events involved in the first kind of conditionals.[1] For in the latter case, Zayd and 'Amr may behave as they want, and eventually change their behavior, while the natural events involved in the example illustrating the second kind of conditionals, which have to do with the planets and their effects on nature could not change as easily and are neither artificial or conventional nor arbitrary although they are not observable at all times.

Note that the rules provided by al-Fārābī are mainly applied to the last kind of conditionals, which are said to be necessary and to always relate the antecedent and the consequent.

Al-Fārābī also considers more variants with complex premises:

"- If the infinite body exists [p], then it is either simple (*basīṭ*) [q] or complex (*murakkab*) [r]
- But the infinite body is neither simple [not q] nor complex [not r]
- Therefore the infinite body does not exist [not p]" ([16], 167, symbols added).

If we consider the above structure of this kind of syllogism, we would say that premise 2, i.e., ($\sim$q $\wedge$ $\sim$r), is the negation of the consequent of premise 1, that is, (q or r). But generally speaking, a conjunction of two negated propositions "$\sim$p $\wedge$ $\sim$q" is the negation of an inclusive disjunction "p $\vee$ q" and not of an exclusive one "p $\underline{\vee}$ q"; so the disjunction, here, should be inclusive. But al-Fārābī does not

[1] See [59] for a fuller analysis of the conditional in al-Fārābī's theory.

talk about an inclusive disjunction "p ∨ q", for he only mentions two kinds of disjunctions: "p $\veebar$ q" and "∼(p ∧ q)" as we will see below.

In addition, the inference is valid even when the disjunction is exclusive, because when (∼p ∧ ∼q) is true, ∼(p $\veebar$ q) is true too, even if with the exclusive disjunction "p $\veebar$ q", there is a case where ∼(p $\veebar$ q) is true while (∼p ∧ ∼q) is false (the case where p and q are both true). So given the validity of the above inference even when "or" is exclusive, we could not credit al-Fārābī of the use of the inclusive disjunction.

The second kind is called the "disjunctive syllogism" (*'infiṣālī*). Its first premise is disjunctive, and its second one is "predicative" ([15], 138). The disjunction may contain two elements or more. It can be complete or incomplete. The complete disjunction is the exclusive disjunction. It is expressed by "or"; the incomplete disjunction is rather expressed by "∼(p ∧ q)" ([15], 139). It is equivalent to "p ⊃ ∼q", and also to "q ⊃ ∼p" ([13], 79). The word "incomplete" suggests that the disjunction is not a real alternative, for the propositions could be both false and there might be an indeterminate number of disjuncts ([15], 138).

In *al-Maqūlāt*, al-Fārābī says that the complete disjunction expresses a strong conflict (*'inād*), while the incomplete disjunction expresses an incomplete conflict ([13], 79). In the complete disjunction, which is the exclusive one, if one disjunct (whatever it may be) is true, the other one is false, and if one of them is false, the other one is true. The incomplete disjunction is defined as follows: if one of the disjuncts (whatever it may be) is false, the other one is not necessarily true (idem, p. 79), and if one of them is true, the other one must be false. In both cases, the disjunction involves incompatible elements. This incompatibility is semantic, which means that the truth value of the disjunction does not only depend on the truth value of its elements but rather on their semantic and intensional incompatibility. It is thus not truth-functional, as it is in the Stoic logic.

As to the incomplete disjunction, it may have an indeterminate number of elements ([15], 138), for instance, when one says: "this color is either white or red or pink, or yellow, or …."

In the incomplete disjunction, the incompatibility may be either essential (or natural: *bi-ṭ-ṭab'i*) as in "this color is either black or white" or accidental as in "[it is] not [the case that] (Zayd is present and 'Amr is talking)" ([15], 138). Note, here, that there is no real semantic incompatibility in the second example.

According to him, the incomplete disjunction is not usually expressed by the word "or"; rather what is used is the expression "not both" or "not possibly both." Unlike the Stoïcs, who call the relation expressed by "not both" a negation, al-Fārābī considers it as a kind of disjunction, because it expresses a conflict (*'inād*) between the two elements. "Not both" is used to express the accidental incomplete disjunction; while "not possibly both" corresponds to the essential incomplete disjunction.

In the complete disjunctive syllogism, when premise 2 is one disjunct, the conclusion is the opposite of the second disjunct and if premise 2 is the opposite of one disjunct, the conclusion is the other disjunct itself.

As illustrations, al-Fārābī presents the following concrete syllogisms:

I/ "- This number is either even [p] or odd [q]
- But it is even [p]
- Therefore it is not odd [not q]" ([15], 139, symbols added).

This corresponds to the indemonstrable IV in the Stoïc logic. It is valid only when the disjunction is exclusive, since if the disjunction were inclusive, one could not deduce "not q" from both "p or q" and "p", given that "p" and "q" could be true together.

Al-Fārābī provides the following variant too:

"- This number is either even [p] or odd [q]
- But it is odd [p]
- Therefore it is not even [not p]" ([15], 139, symbols added)

The other disjunctive syllogism is the following:

II/ "- This number is either even [p] or odd [q]
- But it is not even [not p]
- Therefore it is odd [q]" ([15], 139, symbols added)

Another variant is expressed as follows:

"- This number is either even [p] or odd [q]
- But it is not odd [not q]
- Therefore it is even [p]" ([15], 139, symbols added).

This corresponds to the indemonstrable V, which is valid both with an exclusive and with an inclusive disjunction.

It could also contain three disjuncts as in the example below:

"- The water is either cold [p] or hot [q] or middling [r]
- But it is cold [p]
- Therefore it is neither hot [not q] nor middling [not r]" ([15], 139, symbols added).

Here too, the syllogism would not be valid with an inclusive disjunction.

One can also have complex syllogisms of this kind:

"- This number is either greater [p] or smaller [q] or equal [r] (to another)
- But it is not smaller [not q] (to the other)
- Therefore it is either greater [p] or equal [r] (to the other)
- But it is not greater (to the other) [not p]
- Therefore it is equal (to the other) [r]" ([15], 139).

Here, two deductions are made successively by applying II. The syllogism is valid whatever disjunction is used.

When the disjunction is incomplete, we have the following:

- Zayd is not both at home [p] and in the market [q]
- But he is at home [p]
- Therefore he is not in the market [not q].

Or we could have the following:

- Zayd is not both at home [p] and at school [q]
- But he is at school [q]
- Therefore he is not at home [not p].

However, this kind of syllogism is not conclusive if the second premise is negative, for if we say:

"- This colour is not both white [p] and black [q]

- But it is not white [not p]

we can not deduce:

- Therefore it is black [q]" ([15], 139, symbols added).

because the color could be neither white nor black, in which case the conclusion would be false, hence not deductible.

Now what is the difference between the two kinds of incompatibility in both disjunctions? In the complete disjunction, the incompatibility is essential in all cases, but the elements are either analytically incompatible (as "even" and "odd") or physically incompatible (as "cold", "middling", and "hot"), while in the incomplete disjunction, the incompatibility is either essential or accidental. In the first case, the disjuncts are physically incompatible, but both of them could be not satisfied, as in the example involving the colors, which are contrary to each other; in the second case, the incompatibility is not essential but merely factual and chancy as is shown by the example "Not (Zayd is present and 'Amr is talking)," given that the two propositions involved in this example could be true together, for there is no essential reason to claim that whenever Zayd is present, 'Amr is not talking or vice versa. So this last kind is not really a semantic incompatibility but rather a factual one. It may be interpreted as saying: "it is not the case that both p and q," in which case, the incompatibility would be related to the truth values of the elements (not really to their meanings).

Does this mean that this second kind of disjunction is truth-functional? al-Fārābī does not evoke this notion of truth-functionality, which is a modern notion, but we might consider that in this particular kind of disjunction, where there is no real semantic, physical, or essential incompatibility, and where the disjunction is expressed in terms of a negated conjunction, it would not be absurd to talk about truth-functionality, given that the truth conditions of the conjunction are intuitively clear to every one.

5.1.2.2 The Hypothetical Syllogisms in Avicenna's Frame

Avicenna is much more prolific than al-Fārābī (and later Averroes) in his analysis of the hypothetical logic, for he constructs a whole system involving the conditional and disjunctive syllogisms that he presents especially in *al-Qiyās* and summarizes in other treatises such as *al-Najāt* and *al-Ishārāt wa al-Tanbīhāt*. This system is, at least in its first part which involves the conditional propositions, parallel to the classical categorical logic, for it contains and proves the same moods, the only difference being that the terms are replaced by whole propositions, and that the quantification ranges over situations (or times, depending on the interpretation chosen) rather than on objects. However, Avicenna talks also about the classical hypothetical syllogisms like those that we can find in the Stoïc logic and which we just analyzed in al-Fārābī's frame. These syllogisms are evoked in the very last part of *al-Qiyās*, but since they are close to what al-Fārābī presents, we will talk about them at the beginning of this section.

Let us first define what Avicenna means by a hypothetical syllogism. The hypothetical syllogisms may be either *sharṭī* (literally "conditional") or *istithnā'ī*, translated as "exceptive" by Tony Street in ([137], 546), the latter being the usual Stoïc syllogisms. As to the first kind, it reproduces the categorical syllogism, except that the premises and the conclusion contain two propositions related either by an implication or a disjunction.

Both categorical and *sharṭī* syllogisms are called *iqtirānī* (literally "conjunctive"), presumably because the two premises are conjoined but none is detached. In this first kind, the conclusion does not occur in the premises; but the second kind uses the "*istithnā*", which we could call the detachment, for in this kind of syllogisms, "[the conclusion or its contradictory] occurs explicitly [in the premises]" (Avicenna, *al-Ishārāt*, cited in [137], 546). The word "*istithnā*" itself is used by al-Fārābī, Avicenna and Averroes in the same meaning: the proposition which is said to be "*mustathnāt*" (detached) is the one introduced by the word "but" ("*lākin*"). It could be the antecedent of the first proposition, or its consequent, but it could also be the contradictory of the antecedent or of the consequent. This proposition is assumed to be true on its own; this is why it is detached.

The system that resembles the one presented by al-Fārābī is what we could call the exceptive hypothetical logic. It is called PL1 by Wilfrid Hodges (see [79]) and does not contain any temporal condition, unlike the system with quantified hypothetical propositions which will be developed afterward, and where the quantification ranges over times or situations. So it seems relatively independent from all other Avicennan systems, whether hypothetical or categorical, because of this absence of quantification. This might be the reason why Avicenna presents it at the very end of his *al-Qiyās*. This absence of quantification is sometimes criticized by Avicenna, for instance, in the following passage:

> "In our native country we came to know a long annotated book on this subject which we have not seen since we left our country and travelled around to look for a means of living. However, it might still be there. After we obtained this part of knowledge nearly 18 years ago we came across a book on conditional (propositions and syllogisms) attributed to the most excellent among later (scholars). It seems to be wrongly imputed to him. It is neither clear nor reliable. It neither gives an extensive survey of the subject nor does it achieve its purpose. It gives a mistaken exposition of *hypothetical* propositions, of a large number of syllogisms which accompany them, of the reasons for productivity and sterility, and of the number of moods in the figures. The student should not pay any attention to it - it is distracting and misleading. For the author did not know what makes conditional propositions affirmative, negative, *universal*, *particular*, and *indefinite*; nor did he know how conditional propositions *oppose* or *contradict* each other. He also did not know how one conditional proposition can *be the subaltern* of the other." (*Qiyās* vi.6, p. 356, trans. [132], 159, my emphasis)

In this passage, Avicenna is talking about a book that seems to have been attributed (wrongly, according to him) to al-Fārābī, which talks about *hypothetical* propositions, but does not consider their oppositional relations (he evokes contradiction, subalternation, and opposition in general), and does not either distinguish, among these hypothetical propositions, between *universal*, *indefinite*, and *particular* ones. So he seems to blame the author of this book for not using quantification in this system of hypothetical logic. This criticism seems astonishing since usually the hypothetical logic is not quantified, for instance, in the Stoic system and in al-Fārābī's one and Avicenna's own *exceptive* (= *istithnā-ī*) hypothetical logic is not quantified either. So why is he blaming this author, since he himself presented a non-quantified hypothetical logic? And how should one consider his own exceptive hypothetical logic? Is it less important than the quantified hypothetical logic?

A possible answer might be that the exceptive hypothetical logic should be considered as a first preliminary system that has to be completed by a quantified hypothetical one, which enters into more details in analyzing the propositions and provides a greater number of arguments and moods. Maybe also he thinks that Aristotelian logic should be the main basis in logic, even in hypothetical logic. In that case, his opinion would be contrary to the modern one, according to which propositional logic is more fundamental than predicate logic, since predicate logic is based on it.

The exceptive hypothetical syllogisms are the usual Stoïc ones, that is, the *Modus Ponendo Ponens*, the *Modus Tollendo Tollens* ([34], 391 + 395), as well as the *Modus Ponendo Tollens*, which is stated as follows:

"- This number is either even or odd
- But it is even
- Therefore it is not odd" ([34], 401)

and the *Modus Tollendo Ponens*:

"- This number is either even or odd
- But it is not even (or it is not odd)
- Therefore it is odd (or it is even)" ([34], 401).

He states also the following:

"- This thing is not both an animal and a tree
- But it is a tree (or it is an animal)
- Therefore it is not an animal (or it is not a tree)" ([36], 452).

Avicenna also states the following:

"- Either this is not an animal or it is not a plant
- But it is an animal (or –but it is a plant)
- Therefore it is not a plant (or –it is not an animal)" ([36], 451).

He also presents a syllogism whose disjunctive premise contains three elements, which is reminiscent of al-Fārābī's examples. This syllogism is the following:

"- This number is either greater or smaller or equal [to another one]
- But it is equal [to another one]
- Therefore it is neither greater nor smaller [to that other one]" ([34], 401).

But he says that in this syllogism, there are two conclusions, not only one, for the conjunction that we find in the conclusion of this syllogism relates two negative propositions, each of which is a discourse on its own. Unlike al-Fārābī, however, he does not go further in the deduction so that to arrive at a single conclusion by applying the same indemonstrable.

Now in the first and the second syllogistic moods (*Modus Ponens* and *Modus Tollens*), the first premise must be a *real* implication (*luzūm*), where the consequent really follows from the consequent "*muttasilāt luzūmīya*" ([34], 390). This real implication is distinguished from the "*ittifāq*". The word "*ittifāq*" is thus opposed to the word "*luzūm*" by the fact that it does not express a real entailment, which the *luzūm* expresses. But what does this word mean exactly? And how could one translate it? In Arabic, the word "*ittifāq*" has several meanings, which are related to at least two main concepts: The first meaning is related to agreement, which can be between people, or between opinions, it then expresses the fact that there is a harmony between these people or these opinions. In addition, *ittifāq* can also mean an agreement by convention and "*ittifāqī*" would then have the meaning of "conventional". The second main meaning is related to the concept of chance, accident, or coincidence, in which case it is opposed to necessary; thus understood, *ittifāqī* means accidental, coincidental, not necessary. It is opposed to the *luzum* because the latter requires a necessary link between the antecedent and the consequent, and because this link is intensional, i.e., related to the meanings of the antecedent and the consequent. As we will see below, unlike the *luzūm*, *ittifāq* might even be considered in some way as extensional and truth-functional.

The second meaning has been privileged by Nabil Shehaby who has translated "*ittifāq*" as "chance connection" in his book *The propositional logic of Avicenna, a translation from al-Shifā al-Qiyās*, while the first meaning has been privileged by Wilfrid Hodges, who criticizes N. Shehaby by saying that Shehaby's account of *ittifāq* does not agree with Avicenna's examples. Let us consider both interpretations.

N. Shehaby's interpretation of "*ittifāq*" as being a kind of chance connection seems to be based on Avicenna's distinction between a strong relation of entailment exemplified by the *luzūm*, where the consequent really *follows from* the antecedent and where the sentence is true only because of this entailment, whether the antecedent and the consequent are true or not, and a conditional where the link between the antecedent and the consequent is not strong and does not express a relation of entailment. For this reason, it would be chancy because the truth of the consequent does *not depend on* that of the antecedent. For instance, if one says "If every man is speaking, then every donkey is braying" ([34], 267), there is no link between the antecedent and the consequent, which are true independently of each other. In that case, the relation expressed by "if…then" is not a real implication, it is rather an *ittifāqī* or an accidental relation, because both propositions *happen to be true* together, *by accident*; the truth of the sentence does not depend on the fact that its consequent *follows from* its antecedent.

But this interpretation is criticized by Wilfrid Hodges, who says that Avicenna's "examples do not support the interpretation of *ittifāq* as a kind of 'chance connection'," for "'chance' is irrelevant to them" ([92], 257).

According to him, as he says in the passage below, the word "*ittifāq*" should be understood in the light of its first meaning as a kind of agreement, namely, an agreement, not between opinions or people, but rather between a discourse and the fact related to it. This is what he says:

> "**History 14.2.7** In this passage it seems that Ibn Sīnā understands an *ittifāqī* sentence (a; mt)(p; q) to be one which is taken to be true on the basis that 'Always q' is true. If so, then describing an (a, mt) sentence as *ittifāqī* has nothing to do with the meaning of the sentence; it refers to the way the sentence has been introduced in an argument. When he moves on to (e, mt) sentences, he speaks of these as 'denying the *ittifāq*' (*sālibata l-muwāfaqa, Qiyās* 299.4); though the facts are not completely clear, he can be read as saying that a sentence (e, mt)(p, q) 'denies the *ittifāq*' if it is taken as true on the grounds that q is always false." ([92], 257)

So for him, the conditional proposition "Whenever p then q" [an (a, mt) sentence] expresses an *ittifāq* when it is true on the basis of the truth of its consequent (q), while it would express a *luzūm* if it is true on the basis of the fact that q *follows from* p or *is entailed by* p. This is why for him, "*ittifāq*" expresses a kind of agreement between a sentence and a fact, since the whole conditional is considered true only because its consequent is true (i.e., is in agreement with the reality), not because of the link between its consequent and its antecedent.

To verify this opinion, let us consider the example that Avicenna is talking about at pages 296ff. This example raises a problem which is the following:

Avicenna considers the two premises: "Whenever 2 is odd, then 2 is a number, and whenever 2 is a number, then white is a colour" ([34], 297.16–17). From these premises, could one deduce the conclusion "whenever 2 is odd, then white is a colour"? Avicenna's answer is that our knowledge of the sentence "white is a colour" does not follow from our knowledge that "2 is a number," nor from the premises of the syllogism. For there is no link between the color white and the fact that 2 is a number or that it is even or any premise of the above syllogism. Besides that, if one changes the antecedent, one does not change anything in the truth of the consequent ("white is a colour") ([34], 297.18), which does not depend on its antecedent at all. For when one says, "whenever 2 is even, then white is a colour," the sentence does not express an entailment either; does it express an *ittifāq*? And what would be the meaning of *ittifāq* in this context? Avicenna says that the two sentences "2 is a number" and "white is a colour" are "known by themselves," their knowledge is due to an agreement with the facts, not to any kind of entailment or of dependence. This is why "*ittifāq*" can in some way be seen as an agreement of the sentences with the corresponding facts. In this case, there is an agreement between the antecedent and its corresponding mathematical fact and between the consequent and its corresponding perceptual fact. But what about the sentence "whenever 2 is odd, then white is a color"? Does the antecedent, which is impossible, express an agreement? It is clear that this antecedent does not agree with any mathematical fact. So if *ittifāq* means agreement, what entities would this agreement involve? If we interpret "*ittifāq*" as an agreement with the facts involved, then it would concern only the consequent, in that case. This kind of *ittifāqī* propositions with impossible or false antecedents is examined also at page 270, where Avicenna says that the sentence "If every donkey is talking then every human is talking" is true just because its consequent is true, the antecedent being a hypothesis (*iftirāḍ*) that need not be true, since it is only supposed ([34], 270.14–15). But if the consequent is also false, the *ittifāq* is not true for Avicenna says that when both the consequent and the antecedent are false, then the proposition is true only when it expresses a real implication ([34], 282.6–7).

If we interpret *ittifāq* as being a kind of chance connection, the answer is not clear, for there is no connection at all between "2 is odd" and "white is a color" apart from the fact that they are linguistically related by the expression "if…then". They seem thus to be linked artificially to each other, without any explicit reason. Could we then consider that *ittifāq* is comparable to a material conditional, where there is no semantic link between the antecedent and the consequent too? It does not seem to be so, for the truth conditions of *ittifāq* are different from those of both the *luzūm* and the material conditional. The material conditional is true when its antecedent and its consequent are true, when they are both false and when the antecedent is false while the consequent is true. It is false when the antecedent is true and the consequent is false, while *ittifāq* is true only when its consequent is true (hence in two cases, not in three cases), and it is false when both its elements are false and when its antecedent is true while its consequent is false. But it involves in some way the notion of a hypothesis expressed by the antecedent, which is supposed true. As to the *luzūm* or implication, it is true when both its elements are true, when the consequent is true, and in *some cases* (depending on their meanings) when its two elements are false. It is false when the antecedent is true while the consequent is false.

This being so, it seems that the *luzūm* is *not* truth-functional, since its truth value depends on the meanings of its elements, not only on their truth values, while the material conditional is truth-functional, since its truth value depends only on the truth or falsity of its elements. As to *ittifāq*, since it cannot be true when the antecedent is false, it is different from the material conditional, despite the presence of "if…then", but it seems truth-functional too, since its truth conditions are the same as those of its consequent. These truth conditions can then be determined easily and they are different from those of the *luzūm*, of the material conditional, *and* of the conjunction, which is true only when its two elements are true.

As a result, it seems that if *ittifāq* means a kind of agreement with the facts, this agreement involves only the consequent, since when the consequent is false, the *ittifāq* is false. As to the antecedent, it can be false or even impossible and it does not express an agreement with the facts in the *ittifāq*; rather it expresses a hypothesis (*faraḍ*) ([34], 272). In the *ittifāq*, the antecedent is supposed by hypothesis, but this hypothesis does not have any incidence on the consequent and cannot consequently warrant the truth of the whole conditional sentence when the consequent is false. For this reason, the *ittifāqī* sentences are not used in the *istithnā'ī* syllogisms, since they don't provide any new knowledge. This last feature is stressed by Avicenna in the following quotation where he distinguishes between the *luzūm* and the *ittifāq* (or "*muwāfaqa*"):

> "If you say: If A is B, then J is D, and you detach: but A is B, and you already know that J is D, then you don't learn anything new when stating J is D. But if you ignore that J is D, and all you know is that it follows from [the known] A is B, then if A is B is true, so will be J is D" ([34], 390).

This means that Avicenna uses the *luzūm*, which is an intensional implication involving either conceptual entailments or causal relations, in his hypothetical *istithnā'ī* syllogisms rather than *ittifāqī* sentences, which as he says do not provide any additional knowledge, given that the consequent of these *ittifāqī* sentences is already known to be true. So although the *ittifāqī* sentences can be seen as truth-functional since their truth conditions are those of their consequents and are determined precisely, they are not very useful since they do not add any new knowledge, and for this reason they are not needed.

On the other hand, Avicenna does not allow for complete conditionals which would be convertible and would admit more inferences than the incomplete ones. For he criticizes al-Fārābī in this respect by saying what follows:

> "When we say: 'If A is B, then J is D', and if this is a premise of our syllogism, we must consider what this means by considering its form, and decide what follows from this form. Saying that its consequent may be converted with its antecedent depends on something that is not the form of the premise, rather it has to do with the matter of the premise. This is like

asking if the predicate of the universal affirmative is identical with its subject or not" ([34], 391.16–17–392.1–3).

This shows that Avicenna privileges the form of the syllogism and its premises. With regard to the conditional, this form is related to the order of its elements. In this respect, in every conditional, the order considered must be the same so that the antecedent must always *precede* the consequent and there should be no difference between the complete ones and the incomplete ones, as in al-Fārābī's frame. Avicenna's logic seems thus to be more formal than al-Fārābī's one, even if it is not entirely formalized like modern logic. He justifies his rejection of the complete implication by the comparison he makes between this conditional proposition and the universal affirmative in categorical logic. For his argument is the following: if we have to interpret the universal affirmative in two ways, then the conversion of **A** would lead to a particular proposition in some cases and to a universal one in other ones, that is, when the copula expresses an identity or an equivalence. In this latter case, i.e., when the conversion leads to a universal affirmative, then the syllogistic moods would be different for we would have an **AAA** mood in the third figure instead of *Darapti*. But this "mood" has never been admitted by any logician before and al-Fārābī himself does not admit such a mood. So why does he talk about a complete implication? Since the two cases are comparable and since the only valid conversion of the universal affirmative in categorical logic is the one that leads to a particular affirmative, the conditional in the hypothetical logic should also be viewed in a unique way, that is, the way that makes the antecedent always precede the consequent and not the other way round.

Generally speaking, the conditional is true when both its antecedent and its consequent are true; it is false if the antecedent is true while the consequent is false; but the cases where the antecedent is false are not determined strictly in Avicenna's frame. For the conditional *may* be true when its two elements are false as when we say: "If men are stones, men are inert" ([34], 260); it *may* also be true when only its antecedent is false as when we say: "If men are stones, they are bodies" ([34], 260). The conditional can even be true when the truth values of both elements are not known as in the following example: "If Abdullah is writing, then he is moving his hand" ([34], 361). But the truth of the conditional in all these cases is *not warranted*, for we cannot say that whenever the antecedent is false, then the conditional is true. This being so, one *cannot* say that the conditional is truth-functional in Avicenna's frame, for as put by Edgington "Non-truth-functional accounts agree that 'If *A*, *B*' is false when *A* is true and *B* is false; and they agree that the conditional is sometimes true for the other three combinations of truth values for the components; but they deny that the conditional is always true in each of these three cases" (see [68], Sect. 2.1). Avicenna's account of the conditional is thus non-truth-functional.

In *al-Qiyās*, Avicenna mentions incidentally the counterfactual conditional which can be used to deduce the impossible from the impossible. This kind of conditional is illustrated by the sentence "If men were not animals, they would not be sensitive" ([34], 238). But this kind is not really used in the hypothetical moods that will be presented afterward in the same treatise.

As to the disjunction, it is expressed by the word "*immā*" (= or) and is said to have three meanings in *al-Qiyās*. These meanings are the following: 1. The complete disjunction, i.e., the exclusive disjunction which is true only when one element is true while the other one is false as in "either this number is even or it is odd" ([34], 242); 2. A meaning which is close to the first one but admits more than two disjuncts as in "this thing is either a plant or is inert (*jamād*)." This does not exhaust all the cases for the thing can be an animal, for instance; 3. The third is expressed as "not p or not q" and is true when both disjuncts are true together ([34], 247).

In *al-Qiyās*, the first meaning, called the real (*ḥaqīqiyya*) disjunction, is defined as follows:

> "This is what you indicate when you say, for example, 'It is exclusively (*lā yakhlū*) one of the cases'. This is the same as saying 'Either this number is even or it is odd'. In this case your aim is to indicate that these…are conflicting things and the thing [i.e. the number] is exclusively one of them." (*al-Qiyās*, translated in [132], 44)

while the second meaning, which equals "not (p and q)," is characterized as follows:

> "The second sense is a modification of what the former indicates… If someone says: 'this thing is inanimate and animal' we answer him by saying: 'Either it is inanimate or animal' and by this, we mean that these two are in conflict with each other and therefore the thing cannot be both." ([132], 45)

As to the third meaning, it seems to be equivalent to "$\sim p \lor \sim q$", and is illustrated as follows:

> "As if someone said: 'This thing is inanimate and animal' and we answered him saying: 'Either it is not inanimate or not animal'" ([132], 45)

However, Avicenna does not seem to realize that there is no real difference between the second and the third meanings in *al-Qiyās*. He will correct this "error" in *al-Ishārāt*, where he clarifies the difference between the meanings 2 and 3, as Ṭūsī puts it in his commentary.

As explained by Ṭusī, the exclusive disjunction prevents the conjunction (*jam'*) *and* the vacuity (*khulūw*) [i.e., p and q are neither true together, nor false together], while in (2), it prevents *only* the conjunction [i.e., p and q are not true together] and in (3), it prevents *only* the vacuity [i.e., p and q are not false together]. Consequently, we have now three *distinct* senses, which are the following:

Sense 1: "$p \veebar q$": Either p or q, but not both
Sense 2: "$\sim(p \land q)$": Not both p and q
Sense 3: "$p \lor q$": Either p or q or both (i.e., without exclusion).

Avicenna seems thus to admit both the exclusive *and* the inclusive meanings of the disjunction, together with a medium meaning, which corresponds to the negation of a conjunction. So the third meaning and the second one are really different. In this respect, Avicenna's opinion about the two last kinds of disjunctions in *al-Ishārāt* is less confused than it was in *al-Qiyās*.

This admission of the inclusive disjunction is relatively new in the Arabic tradition, for as we saw above, al-Fārābī tended to associate all kinds of disjunctions with the notion of conflict, while Avicenna, unlike al-Fārābī, talks explicitly about a non-conflictual kind of disjunction—the third one—which we now call an inclusive disjunction, and illustrates it by some examples even in *al-Qiyās*. In this third meaning, the disjunction may be true even when both disjuncts are true as Avicenna says in the following quotation:

> "The word '*immā*' (or) can mean that the thing does not admit more than two things but *possibly both of them*... as when one says: 'the savant either adores God or is generous with people'" ([34], 245–3, italics added).

These characterizations of the three kinds of disjunctions have been considered by some people as meaning that the disjunction is truth-functional in Avicenna's frame ([132]), given that the restrictions expressed by the words "*khulūw*" and "*jam'*" seemed to determine the truth conditions of these different disjunctions.

But the truth-functionality of the disjunction is not obvious, because all the examples contain elements which are semantically related. Consequently, we cannot say that the disjunction is defined *solely* by its truth conditions. Rather it depends on the meanings of the disjuncts and has a modal connotation as we will see below.

The disjunction means the incompatibility only in the first two senses, but not in the third one, for in the third sense, the two propositions are compatible, i.e., possibly true together.

The incompatibility of the first two senses is either complete (sense 1) or incomplete (sense 2). Generally speaking, the kinds of disjunctions admitted are the following:

1. "$p \veebar q$", which is equivalent to "$\sim p \equiv q$" and also to "$p \equiv \sim q$" (and to "$\sim(p \equiv q)$') for if "a number is either even or odd, but not both," then we can deduce that "when it is odd, it is not even and vice versa."
2. "$\sim p \vee \sim q$", which is equivalent to "$p \supset \sim q$" and also to "$q \supset \sim p$".
3. "$\sim p \vee q$", which is equivalent to "$p \supset q$" for "When they say: 'Not A is B or C is D' ... it is undoubtedly (*lā maḥāla*) a conditional (*shartiyya*), [...], it thus resembles the following hypothetical: 'If A is B, then C is D'..." ([34], p. 251.16–17).
4. "$p \vee q$", which is equivalent to "$\sim p \supset q$" for the word '*immā*' (or) does not only mean an explicit conflict (*'inād*), but also that the second [is true] (*kā'inun*) when the first is not" ([34], 244.16).

The sense (1) is called a complete disjunction and admits the following rules:

- $p \veebar q$ but p therefore not q
- $p \veebar q$ but q therefore not p
- $p \veebar q$ but not p therefore q
- $p \veebar q$ but not q therefore p

The indemonstrable III of the Stoïc logic is replaced by the following:

- $\sim p \vee \sim q$ but p therefore $\sim q$
- $\sim p \vee \sim q$ but q therefore $\sim p$.

This suggests that he holds the second De Morgan's law, that is, $\sim(p \wedge q)$ $(\sim p \vee \sim q)$. Since he has also applied the first De Morgan's law in both his categorical logic, when stating the contradictories of the special assertorics, and in his modal logic, when stating the contradictories of the bilateral possible propositions, we can say that he holds both De Morgan's laws quite clearly, even if they are not explicitly stated as they are in modern logic. This can be considered as a significant advance in logic.

These meanings include both the exclusive meaning and the inclusive one and they show that the disjunctions can always be translated in terms of a conditional (or a biconditional) and vice versa. This translatability is very helpful for the analysis of the syllogistic moods involving the disjunctive propositions.

The meaning of the disjunction is intensional, for it is related to the intensional implication. It seems modal, because Avicenna uses quantifications to characterize it and distinguishes between universal and particular disjunctions. We will evoke these quantifications together with the quantified conditionals in the next section, where we will analyze Avicenna's system of hypothetical propositions and moods.

What about Averroes' opinion about these hypothetical moods? This will be examined in what follows.

5.1.2.3 The Hypothetical Syllogisms in Averroes' Frame

Averroes holds the indemonstrables I, II, IV, and V with the variants of the latter two inferences. He classifies them in two categories, which are the "conjunctive" or "conditional" (*muttaṣil*) syllogisms and the disjunctive (*munfaṣil*) syllogisms. The first kind (*al-Qiyās al-muttaṣil*) includes two moods (*Modus Ponens* and *Modus Tollens*) which are stated as follows:

(1) - If the sun is up, then it is daytime
- But the sun is up
- Therefore it is daytime ([26], 235.2–5)

(2) - If the sun is up, then it is daytime
- But it is not daytime
- Therefore the sun is not up ([26], 235.2–6).

The second kind (*al-Qiyās al-munfaṣil*) contains the disjunctive propositions and includes four other moods which are stated as follows:

(3) - This time is either daytime or nighttime
- But it is daytime
- Therefore it is not nighttime

(4) - This time is either daytime or nighttime
- But it is nighttime
- Therefore it is not daytime

(5) - This time is either daytime or nighttime
- But it is not daytime
- Therefore it is nighttime

(6) - This time is either daytime or nighttime
- But it is not nighttime
- Therefore it is daytime ([26], 235.8–13).

Unlike al-Fārābī, he does not use the so-called complete implication in the two first moods, for here, the implication is not convertible, despite the fact that the examples used are exactly the same examples that al-Fārābī used to illustrate the complete (i.e., the convertible) implication. He does not therefore distinguish between two kinds of implications: the complete one and the incomplete one as al-Fārābī did. In this respect, he seems closer to Avicenna. However, his theory is far less developed than Avicenna's theory, since Avicenna develops a whole system of hypothetical logic that we don't find in any theory before him or after him, as we will see below.

Averroes's analysis of these syllogisms is minimal, for it presents the initial Stoic indemonstrables and some variants without adding any further suggestion.

As to the disjunctive syllogisms, he seems to admit only the moods that use the exclusive kind of disjunctions, for he does not mention the indemonstrable III, whose first premise is $\sim(p \wedge q)$, maybe because he does not consider it as a disjunction, given that it is the negation of a conjunction, as it is stated. So his list of moods is even shorter than the classical Stoic list of indemonstrables. This also means that he does not distinguish between several kinds of disjunctions as was the case with al-Fārābī and Avicenna. According to him, the disjunction expresses a strong conflict in all cases, and this conflict is obvious when one considers the semantic or physical incompatibility between the two disjuncts in the examples given. It is thus an intensional, non-truth-functional exclusive disjunction.

He does not add new inferences at least in his *Kitāb al-Qiyās* and devotes to them only a few pages in that treatise. But he does use them in the *reductio ad absurdum*. In these proofs, they play a significant role.

His opinion is that the hypothetical syllogism does not have the same nature as the categorical syllogism; this is why he criticizes al-Fārābī who holds a different opinion.

For he says: "As to the two premises in the hypothetical syllogism, they do not need to combine (*ta'līf*) to conclude what follows from them, because the implication (*luzūm*) is one of the premises; for this reason it does not enter into the class of the syllogisms as Abū Naṣr [that is, al-Fārābī] thought because what is concluded in the categorical syllogism is produced by the two premises, while it is part of the [premises] admitted in the hypothetical syllogism" ([26], 236).

According to him, there is a difference between the hypothetical syllogism and the categorical one, for the latter needs to combine the two premises in order to deduce *from both of them* the conclusion, which is new and different from them, while in the former, the conclusion is already inside one of the premises, so that it does not add anything new to what is already stated.

However, by saying that, he seems to defend the opinion that the second premise of the hypothetical syllogism is not indispensable to deduce the conclusion given the passage where he says that the two premises "do not *need* to *combine* to conclude what follows from them." This sentence does not take into account the fact that the deduction of the conclusion could not be made *without* the adjunction of the second premise to the first one, for the first premise *alone does not lead to* the conclusion. In this respect, *there is and there must be* a *combination* between the two premises, unlike what Averroes says, despite the fact that the conclusion is indeed part of one of the premises.

His distinction between the hypothetical syllogisms and the categorical ones is not as precise as the distinction introduced by Avicenna between the *istithnā'ī* syllogisms and the *iqtirānī* ones, for Averroes does not analyze the two kinds of syllogisms with enough precision, and he does not seem to give much importance to the hypothetical kind, unlike Avicenna who develops his distinction by entering into much more details.

5.1.3 The Compound Syllogisms

These hypothetical syllogisms are used in some complex arguments and applied in other parts of the systems considered. For instance, al-Fārābī applies his hypothetical logic in his metaphysica and uses them to deduce some metaphysical theses as we will see in the sequel, while Avicenne constructs a whole hypothetical system containing several kinds of hypothetical propositions and combines between these propositions and the categorical ones to prove many more moods than the usual categorical and hypothetical moods.

Let us first consider al-Fārābī's treatment of the more complex syllogisms and how he combines between several syllogisms in one and the same argument. Then we will turn to Avicenna's system. As to Averroes, we will talk about his theory of reduction ad absurdum, for he does not really develop a hypothetical logic like his predecessors and seems to privilege the categorical logic over all other kinds of logics.

5.1.3.1 The Compound Syllogisms in al-Fārābī's Frame

Al-Fārābī mentions some of the combined syllogisms but does not study them at great length. These syllogisms can be made of many hypothetical ones successively, so that the conclusion of the first one becomes the first premise of the second one, which, along with a second added premise, gives rise to another conclusion and so on.

They can also combine many kinds of syllogisms as in the following example:

"The world is either eternal (*qadīm*) [P] or created (*moḥdath*) [Q]. If it is eternal [P], it is not attached (*muqārin*) to the phenomena [∼R]. But it is attached to the phenomena [R], given that it is a body [S], and if the body is not attached to the phenomena [∼R], then it does not contain them (*khālin minhā*) [∼T], and that which does not contain the phenomena [∼T] is not produced (*mu'allaf*) [∼U] and cannot move [∼V], but this is impossible; therefore the world is created [Q]" ([16], 172, symbols added).

We could express the whole argument with variables in the following way:

1. $P \veebar Q$ (first premise),
2. $P \supset \sim R$ (assumption),
3. But S (assumption),
4. And $S \supset R$ (true premise),
5. Therefore R (MP, 3–4),
6. Aut $\sim R \supset \sim T$ (assumption),
7. and $\sim T \supset (\sim U \wedge \sim V)$ (true premise),
8. $P \supset \sim T$ (2, 6 + transitivity),
9. $P \supset (\sim U \wedge \sim V)$ (7, 8 + transitivity),
10. $P \supset \sim V$ (9 + def of conjunction),
11. But V (for $\sim V$ is false and impossible),
12. $\sim P$ (10, 11 + MT), and
13. Therefore Q (1, 12 + disjunctive syllogism).

The whole argument starts with an exclusive disjunction and proves by using several assumptions and what follows them that the first disjunct cannot be true; therefore, it is the second one that must be admitted as true. The steps include the various moods admitted by al-Fārābī, which are the *Modus Ponens*, the *Modus Tollens,* and the Disjunctive Syllogism. We have added the law of transitivity, which is a kind of *Barbara* mood applied to hypothetical logic, since it is strongly presupposed in steps 2–6 and 7–9. Inside the reasoning, the *reductio ad absurdum* is used for $\sim V$ in step 11. Since P leads to a false proposition, then P itself is false by *Modus Tollens*. Therefore, Q is true, since it is strongly incompatible with P.

However, al-Fārābī does not develop a whole system of hypothetical logic similar to the Stoic system or to Avicenna's system.

By contrast, Avicenna devotes more than a hundred pages of his *al-Qiyās* to analyze the hypothetical moods. He also presents some moods containing hypothetical as well as categorical propositions. In the sequel, we will present this system and will analyze the moods considered by Avicenna as valid and those that he rejects because of their alleged invalidity. For this purpose, we need first to examine the quantified hypothetical propositions which Avicenna presents in this part of *al-Qiyās* and to interpret them in a way that makes it possible to validate the moods considered by Avicenna as valid and to invalidate what he considers as invalid moods.

This will be the topic of the next section.

5.1.3.2 Avicenna's Hypothetical Figures and Moods

To begin with, let us consider the hypothetical propositions as they are presented by Avicenna in his hypothetical logic. According to Avicenna, the hypothetical propositions, like the categorical ones, may be singular, indefinite, or quantified, whether they are conditional or disjunctive. The syllogistic moods contain the quantified hypothetical propositions as is the case with the categorical syllogistic.

The words expressing the quantity of these different kinds of propositions are the following: "when", "if…then" when the propositions are conditional but not quantified and the words "either…or" (*immā*) whey they are disjunctive but not quantified. When they are quantified, they may be universal, in which case they contain the word "whenever" when they are conditional or the words "always" or "permanently" when they are disjunctive. The particular conditionals contain the expression "It happens that" (*qad yakūn*)[2] when they are affirmative. The negation expressed by "laysa" may be added to the affirmative propositions to obtain the negative ones. Thus, the four quantified conditional propositions are expressed in the following way by Avicenna:

$\mathbf{A_C}$: Whenever (*kullamā*) A is B then H is Z ([34], 265).
$\mathbf{E_C}$: Never (= *laysa al-battata*) (if A is B then H is Z) ([34], p. 280).
$\mathbf{I_C}$: It happens that (*qad yakūn*) if every A is B then every H is Z ([34], 278).
$\mathbf{O_C}$: It happens that $\sim$ (if… then…) (*qad lā yaqūn*) ([36], 235, note).

Note that $\mathbf{O_C}$ may also be expressed by "Not whenever (if….then)," in which case it is simply the contradictory negation of $\mathbf{A_C}$.

One could ask: where do these propositions come from, since in the Stoïc system as well as the hypothetical logic of al-Fārābī, no proposition is quantified? Could one find in Greek logic some propositions that resemble those stated by Avicenna? If we look at the works of Greek commentators, we find in Theophrastus's system some propositions called "prosleptic propositions," which are stated in a comparable way, as the quotation below from Susanne Bobzien's article "Ancient Logic" shows:

> "Theophrastus introduced the so-called prosleptic premises and syllogisms (Theophrastus fr. 110 Fortenbaugh). A prosleptic premise is of the form:
>
> For all X, if $\Phi(X)$, then $\Psi(X)$'
>
> where $\Phi(X)$ and $\Psi(X)$ stand for categorical sentences in which the variable X occurs in place of one of the terms. For example:
>
> (1) *A* [holds] of all of that of all of which *B* [holds].
> (2) *A* [holds] of none of that which [holds] of all *B*.

[2]The expression "*qad yakūn*" has been translated in several ways. W. Hodges translates it as "sometimes", K. El-Rouayheb translates it as "Once".

Theophrastus considered such premises to contain three terms, two of which are definite (*A*, *B*), one indefinite ('that', or the bound variable *X*). We can represent (1) and (2) as

$\forall X(BaX \rightarrow AaX)$
$\forall X(XaB \rightarrow AeX)$" ([45], Sect. 3.2)

These propositions contain also an explicit quantification as Avicenna's ones (see below). But the quantifier in these prosleptic propositions does not range over situations or times as is the case with Avicenna's hypothetical propositions, and the variable X which is used in Bobzien's quotation is not a time variable or a variable that stands for a situation. Rather, as Bobzien says in the quotation above, "it occurs in place of one of the terms." In these formulas, for instance, this one: "$\forall$X (BaX $\rightarrow$ AaX)," the universal affirmative categorical proposition "*a*" occurs. S. Bobzien says that the formula is read as follows: "A [holds] of all of that of all of which B [holds]." So it seems to say something like: "For all X, if B [holds] of all X, then A [holds] of all these Xs." The use of the verb "to hold" is comparable to what Avicenna is saying in his hypothetical propositions, since one can read Avicenna's $\mathbf{A}_\mathrm{C}$ propositions as saying something like "For all situations s, if 'A is B' is true in s [i.e. holds in s], then 'H is Z' is true in s [i.e. holds in s]."

So there is some closeness with Theophrastus' propositions, but Avicenna's propositions are not exactly the same as those of Theophrastus, because of the quantification on tiles or situations, which is not present in Theophrastus' prosleptic propositions. This relation between Theophrastus and Avicenna is suggested by Maróth in ([114]), who considers as one referee says that Avicenna's hypothetical logic is "in relation to the aims of Theophrastus in this area."

These aims of Theophrastus' probably have to do with the syllogistic hypothetical moods, for instance, the following mood presented by S. Bobzien in her article:

"A [holds] of all of that of all of which B [holds]
B holds of all C
Therefore, A holds of all C" ([45], Sect. 3.2)

This syllogism mixes prosleptic propositions with categorical ones, and we will see below that Avicenna too mixes between hypothetical propositions and categorical ones in some of his hypothetical moods. So maybe there is some closeness here too, but as we will see, Avicenna's system is more complex and more elaborate than Theophrastus' theory about prosleptic propositions.

As to the disjunctive quantified propositions, they are expressed as follows by Avicenna:

$\mathbf{A}_\mathrm{D}$: Always either A is B or C is D.
$\mathbf{E}_\mathrm{D}$: Never either every A is B or every C is D ([34], 283).
$\mathbf{I}_\mathrm{D}$: It happens that (*qad yakūn*) either this or that or … ([34], 288).
$\mathbf{O}_\mathrm{D}$: It happens that not either A is B or C is D (or "Not always either A is B or C is D").

Now the problem is to determine what Avicenna means exactly by these quantified propositions. If we consider the quantified conditional propositions, there is some consensus in their interpretation, although people disagree on some specific points and in particular about the entities over which the quantifiers range, for some authors like Nicholas Rescher say that the quantifiers range over times while others like Zia Muvahed say that they range over situations. This disagreement, however, is not a real problem for it can be handled by saying that Avicenna considers the situations in a very large meaning so that they include times too. This opinion is suggested by Wilfrid Hodges, who considers that Avicenna's times include situations as well. We can also say, to be fair to N. Rescher, that he has based his opinion on only two treatises, which are *al-Ishārāt wa al-Tanbīhāt* and *al-Najāt*, as noted by one reviewer. The large amount of information that we can find in *al-Qiyās* was not available to him when he published his article.

The usual formalizations of the quantified conditionals is the following:

"**Ac**: $(\forall s)$ $(Ps \rightarrow Qs)$
Ec: $(\forall s)$ $(Ps \rightarrow \sim Q)$
Ic: $(\exists s)$ $(Ps \ \& \ Qs)$
Oc: $(\exists s)$ $(Ps \ \& \ \sim Qs)$" ([119], 12–13).

Thus formalized, the universal conditional propositions contain an implication, while the particular ones contain conjunctions. This is so because of the following equivalences:

Ac: $(\forall s)(Ps \supset Qs)$; therefore $\sim$**Ac** $= \sim(\forall s)(Ps \supset Qs) = (\exists s)(Ps \wedge \sim Qs) =$ **Oc**
Ec $= (\forall s)(Ps \supset \sim Qs) = \sim(\exists s)(Ps \wedge Qs) = \sim$**Ic**; therefore **Ic** $= (\exists s)(Ps \wedge Qs)$.[3]

These equivalences are held by Avicenna for he says explicitly that "$p \supset q$" is equivalent to "$\sim(p \wedge \sim q)$", as we can deduce from the following quotation:

"When we say: 'never if A is B, then Z is H', and we mean the agreement (*almuwāfaqa*), [...], we mean that when 'A is B' is true 'H is Z' too (*ma'ahu*) is not true" ([34], 280)

This seems to assert the following equivalence: "$\sim(p \supset q) \equiv (p \wedge \sim q)$", which in turn leads to the following: "$(p \supset q) \equiv \sim(p \wedge \sim q)$".

So the two particulars are not really conditional propositions given that they express conjunctions. The two real conditional propositions are the two universals and they express real implications, whether affirmative or negative. However, we must also add that the conditional A proposition can in some cases be formalized in the same way as an A categorical proposition with import, i.e., as follows: $(\exists s)Ps \wedge (\forall s)(Ps \supset Qs)$. This formalization is needed to validate **A**-conversion, and also some moods held in hypothetical logic too, which are the hypothetical *Darapti* and *Felapton* which are admitted by Avicenna in this system too. Besides that, the logical relations of the square of oppositions, which are stated by Avicenna in his

[3]See [70, 71] for a different interpretation of these conditional propositions and [91, 132, 142] for a discussion of these interpretations.

hypothetical logic too require the **A** propositions to "have an import," i.e., to presuppose the truth of their antecedent in order to be themselves true, as appears in the following quotation:

"When we say: 'If A is B, then H is Z', we assume from this (*nūjibu min hāḍha*) that at any time where 'A is B' ***is the case*** and when A is B then H is Z, as if the fact that H is Z follows the fact that A is B, ***in so far as in effect*** A is B (*min hayṭhu huwa kā'inun A [huwa] B*)" ([34], 263.8–9, my emphasis).

We will return to this particular point below, when analyzing the moods held by Avicenna.

As to the disjunctive quantified propositions, we could say that they can be interpreted in different ways, for there is no real consensus on the interpretation of the **E** and **I** disjunctive propositions.

The first interpretation, which seems to be the most natural one and corresponds to the examples provided by Avicenna in his informal explanations, is the following:

$\mathbf{A_D}$: Universal affirmative: $(\forall s)\ (Ps \veebar Qs)$.
$\mathbf{I_D}$: Particular affirmative: $(\forall s)\ (Ps \vee Qs)$.
$\mathbf{E_D}$: Universal negative: $\sim(\exists s)\ (Ps \vee Qs)$.
$\mathbf{O_D}$: Particular negative: $\sim(\forall s)\ (Ps \veebar Qs)$ ([52], 190).[4]

In these formulas, the disjunction is exclusive in $\mathbf{A_D}$ and it is inclusive in $\mathbf{I_D}$. $\mathbf{E_D}$ and $\mathbf{O_D}$ are just the contradictory negations of $\mathbf{I_D}$ and $\mathbf{A_D}$, respectively.

This interpretation is in accordance with the examples provided by Avicenna which are the following:

$\mathbf{A_D}$: "Always (*dā'iman*) either this number is even or it is odd."
$\mathbf{E_D}$: "Never (*laysa al-battata*) either the sun rises or it is daytime."
$\mathbf{I_D}$: "It happens that (*qad yakūn*) either Zayd is at home or 'Amr is at home'."
$\mathbf{O_D}$: "Not always (*laysa dā'iman*) either the fever is Yellow (*safrāwīyya*) or it is blood-like (*damawīyya*)" ([125], 233).

However, these formalizations must be checked by considering the moods considered as valid and those rejected by Avicenna. In particular, the use of the inclusive disjunction in $\mathbf{I_D}$ may raise problems with regard to the validity or invalidity of the hypothetical moods, although the example illustrating $\mathbf{I_D}$ is clearly an inclusive disjunction. As to $\mathbf{E_D}$, it is not really clarified by the example provided, whose meaning is not immediately obvious. Maybe this example means what follows: "It is never the case that the two propositions 'the sun rises' and 'it is daytime' are in conflict." But then, the disjunction negated would be exclusive, which is not the case in the formalization provided given that $\mathbf{E_D}$ is just equivalent to $\sim\mathbf{I_D}$ and the disjunction in $\mathbf{I_D}$ seems inclusive.

---

[4]See also [57] for an analysis of these propositions and the conditional ones, and [134] for a modal interpretation of the disjunctive and conditional propositions.

As to the example illustrating $\mathbf{O_D}$, it is not very clear too, but we could interpret it as saying something like "It is not always the case that the fever is either Yellow or blood-like (but not both)," in which case it means that there are other kinds of fever. In this interpretation, the disjunction seems exclusive and it is negated. But the problem is that the contradictory negation of the exclusive disjunction is an equivalence, given that "$\sim(p \veebar q) \equiv (p \equiv q)$". However, "$\sim(p \veebar q)$" is also equivalent to "$\sim p \veebar q$", which is intuitively more in accordance with Avicenna's formulations, since it contains the word "or", which is more in accordance with the fact that it is supposed to express a kind of *disjunction*, not a biconditional. Anyway, the disjunction and the conditional are interchangeable and Avicenna often reminds his readers of that interchangeability by re-expressing the disjunction in terms of a conditional in his proofs of the hypothetical moods as we will see below.

Nicholas Rescher formalizes the disjunctive quantified propositions as follows:

$\mathbf{A_D}$: $(\forall t)\ (Pt \vee Qt)$,
$\mathbf{E_D}$: $(\forall t)\ \sim(Pt \vee Qt)$,
$\mathbf{I_D}$: $(\exists t)\ (Pt \vee Qt)$, and
$\mathbf{O_D}$: $(\exists t)\ \sim(Pt \vee Qt)$ ([125], 233[5]).

These formalizations, however, do not account for the exclusive character of $\mathbf{A_D}$. For since the symbol used by Rescher is the same for all kinds of propositions, either this symbol expresses the inclusive disjunction or the exclusive one in all cases. But if it expresses the inclusive disjunction, this means that the two elements of $\mathbf{A_D}$ could be true together, which is never the case in the examples provided by Avicenna. On the contrary, if it expresses the exclusive disjunction, then there is no real difference between $\mathbf{A_D}$ and $\mathbf{I_D}$ when we consider only one situation. In addition, in that case, $\mathbf{O_D}$ and $\mathbf{E_D}$ would express equivalences since in both formulas, the negation puts on the disjunction itself.

In his book entitled *Mathematical background to the logic of Ibn Sīnā*, W. Hodges provides the following interpretations for the disjunctive propositions:

"(a, mn) At all times t, at least one of p and q is true at t.
14.3.5 (e, mn) At all times t, if p is true at t then q is true at t.
(i, mn) There is a time at which p is true and q is not true.
(o, mn) There is a time at which neither p nor q is true" ([92], 263).

These interpretations correspond to what he calls Option A (Option α in a recent article). But he does consider also another option which he calls option B (Option β in that same article), where the propositions $\mathbf{E_D}$ and $\mathbf{I_D}$ are interpreted in another way. The interpretations of the propositions in both options α and β are stated clearly in the article entitled "Identifying Ibn Sīnā's hypothetical sentence forms," as follows:

[5] See also ([128], 42).

"(e, mn)α (p, q) $(\forall\tau)\,(p(\tau) \to q(\tau))$
(i, mn)α (p, q) $(\exists\tau)\,(p(\tau) \wedge \neg q(\tau))$
(e, mn)β (p, q) $(\forall\tau)\,(q(\tau) \to p(\tau))$
(i, mn)β (p, q) $(\exists\tau)\,(q(\tau) \wedge \neg p(\tau))$" (W. Hodges, [91], 28).

Here, it is clear that in the second option (= β), $\mathbf{E_D}$ and $\mathbf{I_D}$'s formulations permute the propositions "p" and "q", so that "q" becomes the antecedent of the formula in $\mathbf{E_D}$, while "p" is its consequent. In $\mathbf{I_D}$'s first formulation (α) "q" is negated and "p" is not, while in the second option (β), it is the proposition "p" that is negated, while "q" is affirmative. As to $\mathbf{A_D}$ and $\mathbf{O_D}$, they don't change from an option to another, except that sometimes $\mathbf{A_D}$ is interpreted as an exclusive disjunction, which Avicenna calls a "strict disjunction," in which case $\mathbf{O_D}$, which is in all cases its contradictory, must be formalized accordingly.

These formalizations disagree with Rescher's formalizations with regard to $\mathbf{I_D}$ and $\mathbf{E_D}$. $\mathbf{E_D}$ is expressed directly in terms of a conditional, while $\mathbf{I_D}$ is a conjunction containing an affirmative proposition and a negative one. Although they do not use the word "or" and do not seem, for this reason, to express a kind of *disjunction*, we will see that Avicenna himself seems to corroborate them in his proofs.

Now it should be clear that Avicenna himself does not distinguish explicitly between these two formulations α and β of both $\mathbf{E_D}$ and $\mathbf{I_D}$, since he does not use a specific symbolism or the modern one. It is presumably by formalizing the moods and checking the best formal interpretations that could validate them that Professor Hodges came to test these formalizations. I myself thought about interpretation β when considering one of the moods in subgroup VI-a, because Option α did not work for it.

Now one could try to understand Avicenna's own reasoning by considering the following features: the proposition $\mathbf{I_D}$ is a *particular* disjunctive proposition. As a particular proposition, it could be true in some cases. So it could be true when one of its elements is true but not the other one. But the true element is not necessarily the same one. Consequently, we could say that $\mathbf{I_D}$ could be formalized in two ways: Either as 1. "$P \wedge {\sim}Q$" (α interpretation), or alternatively as 2. "$Q \wedge {\sim}P$" (β interpretation). In both cases, only one proposition is true, which makes the whole disjunction true, for we could read the two formulas as follows:

1. Either P or Q = P is true but not Q [or else "when not Q, then P"].
2. Either P or Q = Q is true but not P [or else "when not P, then Q"].

These readings may be considered as *intuitively* admissible, even if they do not say the same thing and do not correspond to a *unified formal* interpretation.

In this interpretation, $\mathbf{I_D}$ would say only *one part* of what $\mathbf{A_D}$ says, given that $\mathbf{A_D}$ could be interpreted as follows:

$\mathbf{A_D}$: Either P or Q = (P is true but not Q) or (Q is true but not P).

[In case $\mathbf{A_D}$ is inclusive, one could add a third element which is "or both are true"].

Now, how can one choose between (1) and (2) above, whenever $\mathbf{I}_D$ is concerned? In which case should we interpret it as (1) and in which case the right interpretation should be (2)? Presumably, this depends on the mood in which we find $\mathbf{I}_D$, if we take into account Avicenna's developments and the proofs he provides. This is not very satisfying from a strictly formal point of view, for we have two different interpretations of the same proposition and consequently two different contradictories, given that the interpretation of $\mathbf{I}_D$ influences the interpretation of $\mathbf{E}_D$, which is its contradictory. So $\mathbf{E}_D$ would be interpreted either as "$\sim P \vee Q$" or as "$\sim Q \vee P$", depending on $\mathbf{I}_D$'s formulation. The problem, here, is not that much the fact that we have two interpretations instead of a unique one; it is rather that there are no clear criteria that help distinguish between these two interpretations. Of course one could also test systematically the two interpretations. According to one referee, this would be a reliable solution.

As to $\mathbf{A}_D$, it can be interpreted in terms of an inclusive disjunction or as an exclusive one when it is considered as "strict". Avicenna himself tells his readers when the exclusive meaning should be used and when it is the inclusive meaning that is intended.

In all cases, the couples of contradictories are the following: $\mathbf{A}_C/\mathbf{O}_C$ and $\mathbf{E}_C/\mathbf{I}_C$ for the conditional propositions and $\mathbf{A}_D/\mathbf{O}_D$ and $\mathbf{E}_D/\mathbf{I}_D$ for the disjunctive ones. So the formalization of the disjunctive propositions, whatever it is, should validate these contradictory relations.

Let us now examine the moods presented by Avicenna in his hypothetical logic and test the best formalizations that validate them and invalidate those that are said to be invalid by Avicenna. We will first consider the system presented in *al-Qiyās* vi.1 (pp. 295–304) [A] which uses only conditional hypothetical propositions, then *al-Qiyās* vi.2 (pp. 305–318) [B] which mixes between conditional and disjunctive hypothetical propositions, and *al-Qiyās* vi.3 (pp. 319–324) [C] which mixes between different disjunctive propositions. Then in Sect. 5.1.3.3, we will consider those where Avicenna mixes between all kinds of hypothetical propositions and categorical ones, that is, *al-Qiyās* vi.4 (pp. 325–336) [D] which deals with moods in Fig. 1, *al-Qiyās* vi.5 (pp. 337–348) [E] which deals with such combinations in all figures, and finally *al-Qiyās* vi.6 (pp. 349–357) [F] which deals with what Avicenna calls "the divided syllogism" (*al-Qiyās al-muqassam*). Let us start by the first hypothetical system.

A/ *al-Qiyās*, section vi.1

Avicenna starts by the syllogistic involving the conditional propositions. This syllogistic is exactly parallel to the categorical one and the moods admitted are exactly the same except that the premises are not simple propositions containing a subject and a predicate, but two propositions related by a logical connective.

The figures are defined as follows:

- Figure 1: The middle is the consequent in one premise and the antecedent in the other one.
- Figure 2: The middle is the consequent in both premises.
- Figure 3: The middle is the antecedent in both premises.

The rules are also the same, for no syllogism is valid with two particulars, nor with two negatives, nor with a negative minor premiss and a major particular ([34], 295). The proofs use the same methods.

The moods provided are just those of categorical logic, for instance, *Barbara* is expressed as follows:

"Whenever A is B, then C is D
Whenever C is D, then H is Z
Therefore Whenever A is B then H is Z" ([34], 296.3–4).

This mood is valid without any doubt, for if P: A is B, Q: C is D, and R: H is Z, and if one considers only one situation (for convenience), then the mood is formalized as follows: "$[(P \supset Q) \wedge (Q \supset R)] \rightarrow (P \supset R)$". This formula is valid by the transitivity of the implication. If we express the propositions by using the quantifiers, as Avicenna presumably would have done, we would have the following formula:

$$[(\forall s)(Ps \supset Qs) \wedge (\forall s)(Qs \supset Rs)] \rightarrow (\forall s)\,(Ps \supset Rs).$$

This formula would be analyzed as follows if we consider two situations (S_1 and S_2) and not only one situation as above:

$$\{[(Ps_1 \supset Qs_1) \wedge (Ps_2 \supset Qs_2)] \wedge [(Qs_1 \supset Rs_1) \wedge (Qs_2 \supset Rs_2)]\} \rightarrow [(Ps_1 \supset Rs_1) \wedge (Ps_2 \supset Rs_2)].$$

In all cases, the mood is valid without any doubt, as one can check easily. Given this validity, we won't always add the formulations with two situations to express the moods below, since adding them does not change anything in the validity of the considered mood. But since Avicenna uses quantifications and since these quantifications are important for him, we will express some of the moods with two situations, especially those that contain both universal and particular propositions, just to show the effect of the quantification on the mood considered.

As to *Celarent*, it is expressed as follows:

"Whenever A is B, then C is D (= **A**c)
Never (if C is D then H is Z) (= **E**c)
Therefore Never (if A is B, then H is Z) (= **E**c) ([34], 296.5–7).

If P: A is B, Q: C is D, and R: H is Z, then the formula corresponding to it is the following: "$[(P \supset Q) \wedge (Q \supset \sim R)] \rightarrow (P \supset \sim R)$".[6] This formula is obviously valid. With the quantifiers, it is expressed thus: "$[(\forall s)(Ps \supset Qs) \wedge (\forall s)(Qs \supset \sim Rs)] \rightarrow (\forall s)(Ps \supset \sim Rs)$".

Darii is expressed as follows:

"It happens that if A is B, then C is D (Ic)
Whenever C is D then H is Z (Ac)
Therefore It happens that if A is B, then H is Z (Ic)" ([34], 296. 8–10).

If A is B: P, C is D: Q, and H is Z: R, then the formula representing it is the following: "$[(P \wedge Q) \wedge (Q \supset R)] \rightarrow (P \wedge R)$". This formula is valid.

If we formalize the mood by using the two quantifiers used, we will get the following formula:

$$[(\exists s)(Ps \wedge Qs) \wedge (\forall s)(Qs \supset Rs)] \rightarrow (\exists s)(Ps \wedge Rs)$$

If we consider two situations (S_1 and S_2), the formula is analyzed as follows: $\{[(Ps_1 \wedge Qs_1) \vee (Ps_2 \wedge Qs_2)] \wedge [(Qs_1 \supset Rs_1) \wedge (Qs_2 \supset Rs_2)]\} \rightarrow [(Ps_1 \wedge Rs_1) \vee (Ps_2 \wedge Rs_2)]$

As to *Ferio* it is expressed as follows:

"It happens that if A is B then C is D (Ic)
Never if C is D, then H is Z (Ec):
Therefore, Not (Whenever A is B, then H is Z) (Oc)" ([34], 296.11–13).

Likewise, if A is B: P, C is D: Q, and H is Z: R, then the formula representing this mood is the following: "$[(P \wedge Q) \wedge (Q \supset \sim R)] \rightarrow \sim(P \supset R)$". This formula is also valid. With the quantifiers, it is expressed thus: "$[(\exists s)(Ps \wedge Qs) \wedge (\forall s)(Qs \supset \sim Rs)] \rightarrow \sim(\forall s)(Ps \supset Rs)$".

In the same way, the moods of the second figure are valid for they reproduce all the categorical moods of that figure. Avicenna states these moods in the following way:

Cesare

"Whenever A is B then C is D (Ac)
Never if Z is H then C is D (Ec)
Therefore Never if A is B then H is Z" (Ec) ([34], 300.15–16).

This is expressed by the following valid formula: "$[(P \supset Q) \wedge (R \supset \sim Q)] \rightarrow (P \supset \sim R)$". With the quantifiers, it is expressed thus: "$[(\forall s)(Ps \supset Qs) \wedge (\forall s)(Rs \supset \sim Qs)] \rightarrow (\forall s)(Ps \supset \sim Rs)$".

[6]This formula and all formulas below are expressed by considering only *one* situation. But they are also valid with two or more situations.

Camestres

"Never if A is B, then C is D (**E**$_C$)

Whenever H is Z, then C is D (**A**$_C$)

Therefore Never if A is B, then H is Z (**E**$_C$)" ([34], 301.3–4).

When formalized, it becomes: "$[(P \supset \sim Q) \wedge (R \supset Q)] \rightarrow (P \supset \sim R)$" and is valid. With the quantifiers, it is expressed thus: "$[(\forall s)(Ps \supset \sim Qs) \wedge (\forall s)(Rs \supset Qs)] \rightarrow (\forall s)(Ps \supset \sim Rs)$".

Festino

"It happens that when A is B, C is D (**I**$_C$)

Never if H is Z then C is D (**E**$_C$)

Therefore Not whenever A is B then H is Z (**O**$_C$)" ([34], 301.8–10).

This is rendered by the following valid formula: "$[(P \wedge Q) \wedge (R \supset \sim Q)] \rightarrow \sim (P \supset R)$". With the quantifiers it is expressed thus: "$[(\exists s)(Ps \wedge Qs) \wedge (\forall s)(Rs \supset \sim Qs)] \rightarrow \sim (\forall s)(Ps \supset Rs)$".

Baroco

"Not whenever A is B then C is D (**O**$_C$)

Whenever H is Z then C is D (**A**$_C$)

Therefore Not whenever A is B, then H is Z (**O**$_C$)" ([34], 301.11–12).

This may be expressed formally as follows: "$[\sim (P \supset Q) \wedge (R \supset Q)] \rightarrow \sim (P \supset R)$". With the quantifiers it is expressed thus: "$[\sim (\forall s)(Ps \supset Qs) \wedge (\forall s)(Rs \supset Qs)] \rightarrow \sim (\forall s)(Ps \supset Rs)$".

All these moods are proved by the usual methods already used in categorical logic, that is, conversion and *ekthesis* as well as *reductio ad absurdum*. The hypothetical *Baroco*, in particular, is proved by *ekthesis* "by postulating (*tu'ayyana*) a particular situation (*ḥāl*) where A is B is the case while C is D is never the case. Let this happen when K is T. Then "Never if K is T then C is D" is true and "Whenever H is Z then C is D." It follows that "Never if K is T then H is Z." If we add that "It happens that when A is B, K is T, it follows that Not whenever A is B then H is Z" ([34], 301.12–302.1).

We can rewrite this proof as follows, with "A is B": P, "C is D": Q, "K is T": S, "H is Z": R:

1. $S \supset \sim Q$ (**E**$_C$) (assumption),
2. $R \supset Q$ (**A**$_C$) (major premise),
3. $S \supset \sim R$ (**E**$_C$) (from 1, 2, by *Camestres*), and
4. But $P \wedge S$ (**I**$_C$) (assumption).

Therefore $\sim (P \supset R)$ (**O**$_C$) (from 3, 4, by *Ferio*).

In the third figure, however, the two moods *Darapti* and *Felapton* raise problems as was the case in categorical logic. *Darapti*, for instance, is formalized as follows when we consider one situation: "$[(P \supset R) \wedge (P \supset Q)] \rightarrow (Q \wedge R)$". Unfortunately, this formula is not valid, unless one adds a second element to (at least) the first premise. To validate it, we must write it as follows: "$\{[(P \wedge (P \supset R)] \wedge (P \supset Q)\} \to (Q \wedge R)$". In this rewriting, we have added "P" to "$P \supset R$" to make the first part true only in case the antecedent "P" is true.[7] Only in that case, the mood could be valid; otherwise, i.e., when "P" is false, it is invalid because the two premises could be true while the conclusion is false, given that the two conditionals could be true when both their elements are false, while the conjunction is false in that case. The same problem affects *Felapton*, which is valid only when $\mathbf{A}_\mathbf{C}$ is interpreted in that new way.

Now what does Avicenna say about this interpretation of $\mathbf{A}_\mathbf{C}$? If we go back to the text, we find the following passage, where Avicenna seems to defend the above interpretation, given that he stresses the idea that the antecedent of $\mathbf{A}_\mathbf{C}$ *must* be true:

"When we say: 'If A is B, then H is Z', we assume from this (*nūjibu min hāḏha*) that at any time where 'A is B' is the case and when A is B then H is Z, as if the fact that H is Z follows the fact that A is B, in so far as in effect A is B (*min hayṯhu huwa kā'inun A [huwa] B*)" ([34], 263.8–9).

This adjunction in $\mathbf{A}_\mathbf{C}$ is called an "augment" by Wilfrid Hodges in his book *Mathematical background to the logic of Ibn Sīnā* and several other writings. This augment is needed to validate *Darapti* and *Felapton*, among other things such as $\mathbf{A}_\mathbf{C}$-conversion, for instance, for $\mathbf{A}_\mathbf{C}$-conversion would be invalid without it, since a conditional does not entail a conjunction. However, no augment of that kind is needed for $\mathbf{E}_\mathbf{C}$, which remains a single implication.

The other third figure moods are as valid as the categorical third figure ones. These moods are expressed as follows:

Datisi

"It happens that when C is D, H is Z (Ic)
Whenever C is D then A is B (Ac)
Therefore It happens that when H is Z, A is B" (Ic) ([34], 303.3–5).

With C is D: P, H is Z: Q, and A is B: R, we obtain "$[(P \wedge Q) \wedge (P \supset R)] \to (Q \wedge R)$". This formula is valid. If we use quantifications, we would have the following:

$[(\exists s)(Ps \wedge Qs) \wedge (\forall s)(Ps \supset Rs)] \to (\exists s)(Qs \wedge Rs).$

Disamis

"Whenever C is D then H is Z (Ac)
It happens that when C is D, A is B (Ic)
Therefore It happens that when H is Z, A is B" (Ic) ([34], 303.6–8).

The formula corresponding to it is the following: "$[(P \supset Q) \wedge (P \wedge R)] \to (Q \wedge R)$". With the quantifiers, it is expressed thus: "$[(\forall s)(Ps \supset Qs) \wedge (\exists s)(Ps \wedge Rs)] \to (\exists s)(Qs \wedge Rs)$".

[7] We could also add "p" to "$p \supset q$", since it is also a universal proposition and for the symmetry of the whole formula, but this addition is not required to validate the mood, for the "augment" (as Professor Wilfrid Hodges calls it) added to the first proposition is sufficient to validate it.

Bocardo

"Whenever C is D, then H is Z (**A**c)

Not whenever C is D, then A is B (**O**c)

Therefore Not whenever H is Z, then A is B (**O**c)" ([34], 303.9–10).

When C is D: P, H is Z: Q, and A is B: R, this is rendered by the following valid formula: "$[(P \supset Q) \wedge (P \wedge \sim R)] \rightarrow (Q \wedge \sim R)$". With the quantifications, the formula is the following: "$[(\forall s)(Ps \supset Qs) \wedge (\exists s)(Ps \wedge \sim Rs)] \rightarrow (\exists s)(Qs \wedge \sim Rs)$". Note that "$(\exists s)(Ps \wedge \sim Rs)$" is equivalent to "$\sim (\forall s)(Ps \supset Rs)$" and that "$(\exists s)(Qs \wedge \sim Rs)$" is equivalent to "$\sim (\forall s)(Qs \supset Rs)$".

This mood is also proved by *ekthesis* in Avicenna's text ([34], 303.11–15).

The last valid mood is *Ferison* and is expressed as follows:

Ferison

"It happens that when C is D, H is Z (**I**c)

Never if C is D, then A is B (**E**c)

Therefore Not whenever H is Z, then A is B (**O**c)" ([34], 304.1–3).

Its corresponding formula is the following: "$[(P \wedge Q) \wedge (P \supset \sim R)] \rightarrow \sim (Q \supset R)$". This formula is valid. With the quantification, it is formalized as follows:

$$[(\exists s)(Ps \wedge Qs) \wedge (\forall s)(Ps \supset \sim Rs)] \rightarrow \sim (\forall s)(Qs \supset Rs)$$

Now what happens when one adds the disjunctive propositions? What moods are valid in that case? And how are they expressed?

B/ *al-Qiyās*, section vi.2

In this system, Avicenna mixes the conditional propositions with the disjunctive ones in section vi.2 ([34], 305–318). In this section, as we will see below, several modifications are introduced by Avicenna, due to the use of disjunctive propositions, which significantly change the nature of the moods, in particular, when the negative elements are the antecedents of the propositions and when they are taken into account as such, i.e., as negative. According to Wilfrid Hodges, this introduction of negative elements "makes a huge difference to the logic (see [79] in [67])" since it leads to the introduction of new rules different from those already held in the categorical syllogistic as well as in the hypothetical system of section **vi.1**. The rules make it possible to account for the negation even when it is present in the antecedents (= what corresponds to the subjects in categorical logic) of the premises involved, which was not possible in the previous system and in the syllogistic in general. They are made necessary because of the equivalences between some conditional propositions and some disjunctive ones, for as we saw above, Avicenna holds the equivalence between "$p \supset q$" and "$\sim p \vee q$", and also between "$q \supset p$" and "$p \vee \sim q$" (or "$\sim q \vee p$") and all the equivalences that follow from them, e.g., "$p \supset \sim q$" and "$\sim p \vee \sim q$", "$\sim p \supset q$" and "$p \vee q$", "$p \supset q$" and "$\sim (p \wedge \sim q)$", and so on. He also holds the principle of contraposition, namely,

"$(p \supset q) \equiv (\sim q \supset \sim p)$". All these rules and principles make it possible to re-express the moods in several ways depending on the connectives they contain. We will also see in the sequel that Avicenna sometimes introduces the negation in front of either element of the conclusion when it is conditional and one of the premises is disjunctive and that the conclusion of a mood is said to be an $\mathbf{A_C}$ proposition, despite the presence of the negation in its consequent, as is the case with the very first mixed *Barbara* mood that we will consider below. The same happens with the $\mathbf{I_C}$ propositions in other moods, which relativizes the general distinction between affirmative and negative propositions.

Let us now analyze these moods presented by Avicenna in section vi.2. Since these moods are analyzed systematically by Avicenna with a precise schedule organizing the whole frame, let us first present this schedule, and then we will consider the moods by considering the different conditions stated, which determine their groups and the sub-conditions which determine the subgroups inside each group.

To determine Avicenna's schedule, we will follow the text by using the following method: whenever Avicenna says "the combinations containing such and such *(al-ta'līfāt al-kā'ina...)*," this means that he is talking about one particular group, and whenever he says "the moods of this.... *(ḍurūbu dhālika....)*," this means that he is considering a subgroup pertaining to that particular group.

So his schedule, as it is presented in *al-Qiyās* (pp. 305–318) is the following:

Group I: starts p. 305.6

Sub-groups: I-a: p. 305.8; I-b: p. 306.3; I-c: p. 306.14

Group II: starts p. 306.18

Sub-groups: II-a: p. 307.2–5; II-b: p. 307.14; II-c: p. 308.7; II-d; p. 308.13; II-e: p. 309.1

Group III: starts p. 309.5–8

Sub-groups: III-a: p. 309.8-10; III-b: 310.8; III-c: p. 311.1

Group IV: starts p. 311.7

Sub-groups: IV-a: 311.9; IV-b: 311.15; IV-c: 312.3

Group V: starts p. 312.6

Sub-groups: V-a: p. 312.7; V-b: p. 312.13; V-c: p. 313.1, V-d: p. 313.3

Group VI: starts p. 313.5

Sub-groups: VI-a: p. 313.7; VI-b: p. 313.15; VI-c: p. 314.3

Group VII: starts p. 314.8

Sub-groups: VII-a: p. 314.9; VII-b: p. 314.16; VII-c: p. 315.1

Group VIII: starts p. 315.7

Sub-groups: VIII-a: p. 315.8; VIII-b: p. 315.15; VIII-c: p. 316.1

Group IX: starts p. 316.9

Sub-groups: IX-a: p. 316.10; IX-b: p. 317.1; IX-c: p. 317.4

Group X: starts p. 317.12

Sub-groups: X-a: p. 317.14; X-b: p. 318.1; X-c: p. 318.4

The first group is determined by the following conditions:

Group I/ *The conditional proposition is the minor premise, and the disjunctive proposition is a "real disjunction"; the shared premise is the consequent of the conditional* (*al-Qiyās, p. 305.6–7*).

I-a: In this subgroup, the following condition is added: "*the two premises are affirmative*" (305.5).

This leads to the first mood which is the following *mixed Barbara*:

Mood 1 (mixed *Barbara*)

"Whenever H is Z, then C is D (A_C)

Always either C is D or A is B (A_D)

Therefore Whenever H is Z, then not A is B (A_C)" ([34], 305.8–10).

With H is Z: P, C is D: Q, and A is B: R, and if we consider only one situation, then this mood is formalized thus: "$[(P \supset Q) \wedge (Q \veebar R)] \rightarrow (P \supset \sim R)$". This formula is valid *only* when the disjunctive premise is an *exclusive* one as Avicenna himself said from the start. As a matter of fact, if it is not exclusive, then we have a case of falsity under the main implication precisely when P, Q, and R are all true. In that case, the antecedent of the whole mood is true if the disjunction is inclusive while the conclusion is false.

However, in his proof, Avicenna says that the disjunctive proposition "becomes (*taṣīru*) 'Whenever C is D, then not A is B'" ([34], 305.10). As a matter of fact, when the major premise is expressed by the conditional in that way (i.e., as "$Q \supset \sim R$"), the mood is valid by the transitivity of implication. But "$Q \veebar R$" and "$Q \supset \sim R$" are *not* equivalent. Rather, the former implies the latter, but not conversely. Furthermore, the mood is also valid if one reduces "$Q \veebar R$" to "$R \supset \sim Q$" rather than to "$Q \supset \sim R$". It follows from these remarks that Avicenna relies intuitively *both* on the incompatibility between the two disjuncts *and* on the usual structure of *Barbara* in stating the conditional proposition "$Q \supset \sim R$", for it is this structure of *Barbara* that makes him privilege "$Q \supset \sim R$" over "$R \supset \sim Q$", despite the fact that they *both* validate the mood.

He then considers a mood fulfilling the same conditions (*minor conditional and real disjunctive as the major*), but with a particular minor and says that this too is valid without stating it explicitly ([34], 305.11–12). This would correspond to a mixed *Darii*, and *could be* stated as follows:

Mood 2 (mixed *Darii*)

It happens that when H is Z then C is D (I_C)

Always either C is D or A is B (A_D)

Therefore It happens that when H is Z then not A is B (I_C)

This is formalized as follows: "$[(P \wedge Q) \wedge (Q \veebar R)] \rightarrow (P \wedge \sim R)$" [= "$I_C$, $A_D \rightarrow I_C$"]. When it is stated in this way, i.e., with a negative element in the conclusion, this formula is valid.

However, some questions should be raised in this context. For one thing, in both cases, whether the mood is a mixed *Barbara* or a mixed *Darii*, the conclusion contains a *negative* consequent, which makes it equivalent to a negative proposition, not really to an $\mathbf{A}_C$ or an $\mathbf{I}_C$. Given these negative consequents in both the conclusions of *Barbara* and of *Darii*, the moods should rather be, respectively, "$\mathbf{A}_D$, $\mathbf{A}_C$ therefore $\mathbf{E}_C$" and "$\mathbf{A}_D$, $\mathbf{I}_C$ therefore $\mathbf{O}_C$." So why does Avicenna consider these conclusions as affirmative? The most plausible answer is that he relies basically on the usual categorical syllogistic even in this part of his logic. This explains why in this kind of logic, the distinction between affirmative and negative propositions becomes less strict, since the affirmative disjunctive propositions ($\mathbf{A}_D$, for instance) become negative when they are re-expressed as conditional ones. His replacement of disjunctive propositions by conditional ones allows him to state the conditional conclusion that follows from this replacement and to state a mood parallel to those of categorical syllogistic. But this method prevents him from stating a valid mood which contains $\mathbf{A}_C$ and $\mathbf{A}_D$ as premises but does not have the usual structure of *Barbara*. For this reason, it seems that the categorical syllogistic narrows his investigations because it makes him miss or at least not mention some valid moods which do not have the structures of the usual categorical moods.

We can add that in both these cases, he rightly says that when the disjunctive proposition is not a "real disjunction," that is, if it is an inclusive one, then the mood is not productive ["*lam tuntij*" ([34], 305.12]. This can be shown by the invalidity of the following two formulas corresponding, respectively, to the *Barbara* and *Darii* moods above but with inclusive disjunctions as major premises:

Combination 1 (mixed *pseudo – Barbara*)

$$[(P \supset Q) \wedge (Q \vee R)] \rightarrow (P \supset \sim R)$$

and

Combination 2 (mixed *pseudo – Darii*)

$$[(P \supset Q) \wedge (Q \wedge \sim R)] \rightarrow (P \wedge \sim R)$$

However, this second combination uses the first option of $\mathbf{I}_D$ propositions, that is, $\mathbf{I}_D$ = "$(\exists s)$ $(Ps \wedge \sim Qs)$" stated above, which is presumably considered by Avicenna.

But Wilfrid Hodges shows in his notes about Avicenna's hypothetical logic that this combination is valid if one uses Option β instead, for then we would have the following (formalized) mood:

Mood 3: (*Darii* mixed and new)

$$[(P \supset Q) \wedge (\sim Q \wedge R)] \rightarrow (\sim P \wedge R)$$

This would be a new mood in the first figure.

I-b/ The second subgroup with this same set of conditions (*minor: conditional, major: real disjunction*) is the one where the "*conditional proposition is negative (and convertible)*" ([34], 306.3). Inside this subgroup, Avicenna states the following mood:

Mood 4 ($A_D E_C E_D$, new)

"Never if H is Z then C is D (E_C)

Always either C is D or A is B (A_D)

Therefore Never either H is Z or A is B (E_D)

Rather (*bal*) 'whenever H is Z then A is B', and also 'never (*laysa al-battata*) if H is Z, then not A is B'" ([34], 306.3–5).

Here, he states explicitly the equivalences between the disjunctive and the conditional propositions, for the conclusion of this mood is said to be equivalent to "whenever H is Z then A is B" (= "$P \supset R$") and also to "never if H is Z, then not A is B" (= "$\sim(P \wedge \sim R)$"). This being so, the conclusion "never either H is Z or A is B" should be rendered as "$\sim P \vee R$", which is equivalent to the two formulas above. This reveals the real structure of $\mathbf{E_D}$ in his frame, which should be rendered in general as "not (p and not q)" or as "not p or q", and confirms Wilfrid Hodges' α-interpretation of that kind of sentences. If we use quantifications, its formalization would be the following (where H is Z: P, C is D: Q, and A is B: R):

'$[(\forall s)(Ps \supset \sim Qs) \wedge (\forall s)(Qs \veebar Rs)] \rightarrow (\forall s)(\sim Ps \vee Rs)$.'

If we consider only one situation, this formalization becomes the following:

"$[(P \supset \sim Q) \wedge (Q \veebar R)] \rightarrow (\sim P \vee R)$."

This formula is valid. Given its validity and the correspondences Avicenna establishes explicitly between the possible formulations of the conclusion, we can deduce that in his frame, $\mathbf{E_D}$ is expressible as "not p or q", where "or" is inclusive (for if "or" is exclusive in the above conclusion, then the formula is no more valid, given that the exclusive disjunction can be false when its two elements are both true). We will test this interpretation in the other moods offered by Avicenna. Note also that it would not be valid if the disjunctive premise $\mathbf{A_D}$ were inclusive.

This mood, however, does not correspond to the usual *Celarent* of the first figure, for its negative premise is the minor one, so that it is an $\mathbf{A_D E_C E_D}$ mood. This mood is thus a new one, since we do not find it in any of the systems above or in Aristotle's syllogistic.

In addition, Avicenna considers two parallel moods where one premise is particular, first a conditional particular and second a disjunctive particular, for he says: "the mood is conclusive if the particular is conditional, and it is conclusive too with a particular disjunctive" ([34], 306.9–10). He thus considers that the same structure would be valid too if one of the premises is particular. So let us test the two options mentioned in the quotation above. The mood with a disjunctive particular premise should be formally expressed as follows:

Mood 5 ($A_D I_D I_C$)

$[(P \wedge \sim Q) \wedge (Q \veebar R)] \rightarrow (P \wedge R)$

This formula is valid and contains the following propositions: $\mathbf{A_D}$, $\mathbf{I_D}$ therefore $\mathbf{I_C}$. This shows that, generally speaking, $\mathbf{I_D}$ is equivalent to $\mathbf{O_C}$, for they are both expressible by "p and not q." The mood is a *Darii* $\mathbf{A_D I_D I_C}$.

As to the second combination, it may contain $\mathbf{A_C}$ (with a negative consequent, since the initial proposition is $\mathbf{A_D}$) as the minor premise, $\mathbf{O_C}$ as the particular conditional and what follows from them. If the particular is $\mathbf{O_C}$, then the two premises are the following: "$[(P \supset \sim Q) \wedge (Q \wedge \sim R)]$". The question is: what is the conclusion that follows from these premises? According to Avicenna, the conclusion should be an $\mathbf{O_D}$ proposition, for he says: "For the conditional is convertible, so that its converse implies 'whenever C is D, then not H is Z'. If one adds 'it happens that when C is D, then not A is B; it follows: 'it happens that when not H is Z, then not A is B; and *this implies* (*yalzimu*) 'not always (*laysa dā'iman*) either H is Z or A is B''" ([34], 306.12–13, my emphasis). So the whole mood would be expressed thus:

Mood 6

'Whenever C is D, then not H is Z (A_C)

It happens that when C is D, then not A is B (O_C)

Therefore It happens that when not H is Z, then not A is B (O_C)'

Which in turn implies the following:

'Not always either H is Z or A is B (O_D)'.

If we follow what is said in the quotation, and if we have the following correspondences: C is D: Q; H is Z: P; and A is B: R, then the formula should be the following, when we consider the conclusion as an $\mathbf{O_C}$ with a negative antecedent: "$[(Q \supset \sim P) \wedge (Q \wedge \sim R)] \rightarrow (\sim P \wedge \sim R)$" (= "$\mathbf{A_C}$, $\mathbf{O_C} \rightarrow \mathbf{O_C}$", where $\mathbf{O_C}$ (the conclusion) would have a negative antecedent and would be expressed by "not P and not R"). This formula is valid.

According to Avicenna, the conclusion in this formula implies a disjunctive **O**, that is, $\mathbf{O_D}$, expressed by "not always either P or R." The question is thus the following: How should we formalize $\mathbf{O_D}$? The conclusion in this mood has the form "$\sim p \wedge \sim q$", and this is said to be equivalent to an $\mathbf{O_D}$ proposition, that is, to the contradictory negation of an $\mathbf{A_D}$ proposition. It thus corresponds to $\sim(p \vee q)$, i.e., $\sim \mathbf{A_D}$ where the disjunction in $\mathbf{A_D}$ is inclusive. So he is saying that $\sim(p \vee q)$ (= $\sim \mathbf{A_D}$) is equivalent to "$\sim p \wedge \sim q$" (= $\mathbf{O_D}$). In other words, he is claiming an equivalence which *is in fact* one of the De Morgan's laws [= $\sim(p \vee q) \equiv (\sim p \wedge \sim q)$]. If so, we can assume that Avicenna is in some way aware of this De Morgan's law. As a matter of fact, even the other De Morgan's law [i.e., "$\sim(p \wedge q) \equiv (\sim p \vee \sim q)$"] was assumed by Avicenna in his modal logic as we saw above, even if it is not explicitly expressed. It seems thus that Avicenna applies the second De Morgan's law almost intuitively in his hypothetical system, as he has applied the first De Morgan's law intuitively in his modal logic, even if he does not, in both cases, state them explicitly. This also shows a real novelty in Avicenna's hypothetical system (as well as in his modal logic), since the De Morgan's laws were not known by his predecessors, or even by some of his successors.

I-c: The third subgroup with these same conditions: *minor conditional, major: real disjunction, shared element: the consequent of the conditional,* is the one where "*the disjunctive alone is negative*" ([34], 306.14). The first mood considered is said to be not productive by Avicenna because, according to him, it leads sometimes to a negative conclusion and other times to an affirmative one. But as we will see below, unlike what Avicenna says, the mood can be valid in some interpretations. Thus, if the disjunction is negative, then the two available possibilities are $\mathbf{E_D}$ and $\mathbf{O_D}$. If the minor is conditional, it should be an $\mathbf{A_C}$ proposition or an $\mathbf{I_C}$ one, depending on the quantity of the disjunctive premise. Now let us consider the case where the conditional proposition is universal, which corresponds to the concrete example provided by Avicenna, i.e., the following "Whenever this is even, then it is a number; never either it is a number or it is a plurality divisible by two equals (*kathratun munqasima bi mutasāwiyayni*); then sometimes an affirmative universal is true, while other times a negative universal is true" ([34], 306.15–17). This example contains an $\mathbf{A_C}$ proposition and an $\mathbf{E_D}$ one. If we formalize it, we get the following combinations (where "this is even": P; "this is a number": Q, "this is divisible by two equals": R):

First combination(= Mood 7)

$$[(P \supset Q) \wedge (\sim Q \vee R)] \rightarrow (P \supset R) \; [= \mathbf{A_C}, \mathbf{E_D} \rightarrow \mathbf{A_C})$$

Or the following:

Second combination (= Mood 8)

$$[(P \supset Q) \wedge (\sim Q \vee R)] \rightarrow (\sim P \vee R) \; [= \mathbf{A_C}, \mathbf{E_D} \rightarrow \mathbf{E_D})$$

These combinations are both valid, unlike what Avicenna thinks, given that "$\sim Q \vee R$" is equivalent to "$Q \supset R$", which means that the two moods have the structure of *Barbara*. Besides that, both conclusions are equivalent since the disjunction in the conclusion of the second mood is inclusive and since "$p \supset q$" is equivalent to "$\sim p \vee q$".

Now could we consider the disjunction in the premise $\mathbf{E_D}$ as exclusive, since the general conditions stated by Avicenna evoke exclusive disjunctions without any other precision? If we consider this possibility, the mood would be valid too, for then we would have "$\sim Q \veebar R$" in both cases, and the two moods would be stated as follows:

$$(7 \text{ bis}) \; [(P \supset Q) \wedge (\sim Q \veebar R)] \rightarrow (P \supset R)$$

and

$$(8 \text{ bis}) \; [(P \supset Q) \wedge (\sim Q \veebar R)] \rightarrow (\sim P \vee R)$$

These two formulas are valid.

So Avicenna's claim according to which these moods are not valid is wrong. This claim was justified by the fact that two different conclusions follow from the same premises, which is the way by which Avicenna, following Aristotle, shows the invalidity of a given mood. However, in this case, the two conclusions amount to the same, even if one of them is affirmative while the other one seems negative. So how could we explain Avicenna's error? According to Wilfrid Hodges, Avicenna considers these moods as invalid because he has chosen option β for the proposition $\mathbf{E_D}$ rather than option α. As a matter of fact, when one formalizes the two moods with option β, one gets the following formulas:

(7′) $[(P \supset Q) \wedge (\sim R \vee Q)] \supset (P \supset R)$

and

(8′) $[(P \supset Q) \wedge (\sim R \vee Q)] \supset (\sim P \vee R)$.

In this case, the combinations are *invalid indeed* when the disjunctions of the premises are *not real*. If so, we could assume that Avicenna takes the $\mathbf{E_D}$ proposition as inclusive and interprets it in option β.

As a matter of fact, if only $\mathbf{A_D}$ can be strict just as only $\mathbf{A_C}$ can contain an augment, $\mathbf{O_D}$ too should contain a strict disjunction just as $\mathbf{O_C}$ contains the augment. For when $\mathbf{A_D}$ has the form "$P \underline{\vee} Q$", $\mathbf{O_D}$ would have the following forms: "$\sim (P \underline{\vee} Q)$" = "$P \underline{\vee} \sim Q$" = "$\sim P \underline{\vee} Q$".[8] If so, the various disjunctive propositions would be stated as follows:

$\mathbf{A_D}$ (inclusive): $(\forall s)(Ps \vee Qs)$; its contradictory is $\mathbf{O_D}$ (inclusive): $\sim(\forall s)(Ps \vee Qs) = (\exists s) \sim(Ps \underline{\vee} Qs) = (\exists s)\ (\sim Ps \wedge \sim Qs)$.
$\mathbf{A_D}$ (exclusive): $(\forall s)(Ps \underline{\vee} Qs)$; its contradictory is $\mathbf{O_D}$ (exclusive): $\sim(\forall s)\ (Ps \underline{\vee} Qs) = (\exists s) \sim(Ps \underline{\vee} Qs) = (\exists s)(\sim Ps \underline{\vee} Qs) = (\exists s)(Ps \underline{\vee} \sim Qs)$.
$\mathbf{E_D}$ (option α): $(\forall s)(\sim Ps \vee Qs)$; its contradictory is $\mathbf{I_D}$ (option α): $\sim(\forall s)(\sim Ps \vee Qs) = (\exists s) \sim(\sim Ps \vee Qs) = (\exists s)(Ps \wedge \sim Qs)$.
$\mathbf{E_D}$ (option β): $(\forall s)(\sim Qs \vee Ps)$; its contradictory is $\mathbf{I_D}$ (option β): $\sim(\forall s)(\sim Qs \vee Ps) = (\exists s) \sim(\sim Qs \vee Ps) = (\exists s)(Qs \wedge \sim Ps)$.

So we have 2 $\mathbf{A_D}$s and their corresponding $\mathbf{O_D}$s and 2 $\mathbf{E_D}$s and their corresponding $\mathbf{I_D}$s.

At the very end of page 306, Avicenna says that the "sterility" [of the above combination] is even "more obvious" if we have a "particular" ([34], 306.16–17) proposition, instead of the universal one. The first combination could contain $\mathbf{A_C}$ and $\mathbf{I_D}$ and could be stated as follows:

$[(P \supset Q) \wedge (Q \wedge \sim R)] \supset (?)\ (P \wedge \sim R)$.

This combination is not valid.

[8] All these are equivalent to "$P \equiv Q$" too, but Avicenna does not mention that in this part of the text.

However, if we consider the following combination, where we have $\mathbf{A_C}$ and $\mathbf{O_D}$, we get what follows:

$$[(P \supset Q) \wedge (\sim Q \wedge \sim R)] \rightarrow (\sim P \wedge \sim R).$$

This combination is valid and leads from $\mathbf{A_C}$ and $\mathbf{O_D}$ to $\mathbf{O_D}$. It would have been invalid if the conclusion were an $\mathbf{I_D}$ proposition, though.

Group II/ The second set of conditions is the following: *minor: conditional, major: not real disjunction, shared element: the consequent of the conditional* ([34], 306.18–307.1).

II-a: The first subgroup is where the disjunctive proposition can contain "*an affirmative element and a negative one, and the shared element is the affirmative one*" ([34], 307.2–5).

These conditions are illustrated by a "mood" which contains "two affirmative universal propositions" ([34], 307.5) and is said to be "not productive" (*lā yuntiju*) ([34], 307.7). Its two premises are the following:

Whenever H is Z then C is D ($\mathbf{A_C}$).
Always either C is D or not (A is B) ($\mathbf{A_D}$).

If H is Z: P, C is D: Q, and A is B: R, then the premises are rendered as follows: "$(\forall s)(Ps \supset Qs)$" and "$(\forall s)(Qs \vee \sim Rs)$", that is, "$(P \supset Q) \wedge (Q \vee \sim R)$" when we consider one situation. Since "$Q \vee \sim R$" = "$R \supset Q$", which is *not* equivalent to "$Q \supset R$", we *cannot* deduce "$P \supset R$", for this is not the structure of *Barbara*. It follows that Avicenna is right when he says that these premises are not productive. His own argument is a counterexample involving concrete terms. This example is the following: "whenever this is a man, then it is an animal; and always either it is an animal or it is not a bird" ([34], 307.7–8). In this example, the disjunctive proposition is clearly an inclusive one (or a "not real" one, as Avicenna says), for obviously the two elements of the sentence "it is an animal or it is not a bird" can be true together, given that many animals are not birds.

If the disjunctive is particular, then the mood is not productive, according to Avicenna, for it can have two different conclusions, depending on the premises chosen. He justifies that by a concrete example, where the first option is the following: "Whenever he is walking, then he is willing (*murīdan*), and it happens that either he is willing or he is not moving" ([34], 307.10–11), while in the second option, we would have the following disjunctive premise "it happens that either he is willing or he is not at rest, that is, willing to be at rest." In these two cases, the conclusions would be "contrary (*ḍidd*)" to each other ([34], 307.11–13). As a matter of fact, if we formalize these premises by taking: P for "He is walking," Q for "He is willing," and R for "He is moving," we will first have the following set of premises: $(P \supset Q) \wedge (Q \wedge \sim \sim R)$ with option α, since "it happens that either P or Q" = "$P \wedge \sim Q$", which means that "it happens that either P or $\sim Q$" = "$P \wedge \sim \ \sim Q$ = "$P \wedge Q$". The premises do not lead to any conclusion, so Avicenna is right in considering this combination as sterile. In the other case considered by Avicenna, we would have the following premises: "$(P \supset Q) \wedge (Q \wedge \sim R)$", since "he is at rest" = "he is not moving," so "he is not at rest = he is moving". Thus, we would have "it happens that either he is willing or he is at rest" = "$Q \wedge \sim R$" in option α. In this case too, the combination is sterile.

But what happens if we choose option β for the disjunctive premise? In that case, the two combinations would be the following:

(1) $(P \supset Q) \wedge (\sim Q \wedge \sim R)$.
(2) $(P \supset Q) \wedge (\sim Q \wedge \sim \sim R)$ [$= (\sim Q \wedge R)$].

Among these sets of premises, only the first one can lead to a conclusion, for we would have the following valid combination:

(1) $(P \supset Q) \wedge (\sim Q \wedge \sim R) \rightarrow (\sim Q \wedge \sim P)$,

where the conclusion would say something like "it happens that either he is willing or he is not walking," and would also be interpreted in option β. As to the second set of premises, it is not productive in option β too.

II-b: In this second subgroup, where the "*conditional is negative*" ([34], 307.14) under these same conditions (*minor: conditional, major: not real disjunction, shared element: the consequent of the conditional*), we can have the following mood, where both premises are universal:

Never if H is Z then C is D (E_C)
Always either C is D or not (A is B) (A_D)
Therefore Never if H is Z then A is B (E_C) ([34], 307.15–17).

If H is Z: P, C is D: Q, and A is B: R, then the formula corresponding to this mood is the following: "$[(\forall s)(Ps \supset \sim Qs) \wedge (\forall s)(Qs \vee \sim Rs)] \rightarrow (\forall s)(Ps \supset \sim Rs)$". Here too, Avicenna proves the mood by reducing the disjunctive proposition to its corresponding conditional one. With one situation, we can express the mood as follows: "$(P \supset \sim Q) \wedge (Q \vee \sim R) \rightarrow (P \supset \sim R)$". This formula is valid, for the disjunctive major premise is equivalent to "$R \supset Q$", itself equivalent to "$\sim Q \supset \sim R$", which means that this mood is valid by the transitivity of implication. This mood has the structure $\mathbf{A_D E_C E_C}$, which is new too as it is expressed by Avicenna, i.e., with two universal negative conditionals.

Avicenna adds that if the conditional is particular, the mood is productive too. We would then have the following formula:

'$[(\exists s)(Ps \wedge \sim Qs) \wedge (\forall s)(Qs \vee \sim Rs)] \rightarrow (\exists s)(Ps \wedge \sim Rs)$'

With one situation, this formula becomes

$[(P \wedge \sim Q) \wedge (Q \vee \sim R)] \rightarrow (P \wedge \sim R)$.

This formula is valid, which means that the mood is productive. It has the structure $\mathbf{A_D O_C O_C}$.

Next Avicenna considers the configuration where the disjunctive is particular. He says that in that case, the mood is productive too, "as was the case with its analogous (*naḍīratihā*) with a real disjunction" ([34], 308.5–6). Let us verify this claim. If the disjunctive is particular, it can be interpreted either in option α or in option β, while the conditional premise would remain the same in both cases. So we would have the following combination:

Never if C is D then H is Z ($\mathbf{E_C}$)
It happens that either C is D or (A is B) ($\mathbf{I_D}$)
Therefore Not always either H is Z or A is B ($\mathbf{O_D}$)

With C is D: P, H is Z: Q and A is B: R, we would have the following two formulas:

1. $[(\forall s)(Ps \supset \sim Qs) \wedge (\exists s)(Ps \wedge \sim Rs)] \rightarrow (\exists s)(\sim Qs \wedge \sim Rs)$ [option α].
2. $[(\forall s)(Ps \supset \sim Qs) \wedge (\exists s)(\sim Ps \wedge Rs)] \rightarrow (\exists s)(\sim Qs \wedge \sim Rs)$ [option β].

The first one is valid while the second one is not. This means that Avicenna is choosing option α here too. This mood pertains to the third figure and has the structure $\mathbf{I_D E_C O_D}$. The conclusion, which is an $\mathbf{O_D}$ must be non-real (i.e., it must have the structure "not (p or q)" where "or" is an inclusive disjunction) in order for the mood to be valid. This mood is a new one and pertains to the third figure.

II-c: With these same conditions (*minor: conditional, major: not real disjunction, shared element: the consequent of the conditional*), the third subgroup is the one where the "*disjunctive is negative*" ([34], 308.7). In this subgroup, no mood is conclusive, according to Avicenna ([34], 308.7), but as we will show below, this opinion can be challenged. The concrete examples he gives can have two distinct conclusions, when the premises are universal. This concrete example is the following:

"Whenever this is an accident, then it has a carrier (*kāna lahu ḥāmilun*), and Never either it has a carrier or it is not a substance" ([34], 308.8–9)

If we change the second premise, we have the following:

"Never either it has a carrier or not every quantity is finite (*lā yakūnu kull miqdārun mutanāhiyan*)" ([34], 308.10).

Let us verify the first two premises. If "this is an accident": P, "it has a carrier": Q, and "it is a substance": R, then the two first premises can be formalized thus (with one situation):

$[(P \supset Q) \wedge (\sim Q \vee \sim R)] \rightarrow (\sim P \vee \sim R)$ (with option α for $\mathbf{E_D}$) (i.e., if "never either p or q" = $\sim p \vee q$, in which case "never if p or $\sim q$" = $\sim p \vee \sim q$).

This formula is valid, unlike what Avicenna says. Since the proposition formalized by "$(\sim Q \vee \sim R)$" is equivalent to "$Q \supset \sim R$", and the one expressed by $(\sim P \vee \sim R)$ is equivalent to "$P \supset \sim R$"; if we apply it to the concrete example given by Avicenna, the premise would say: "whenever something has a carrier, then it is not a substance" and the conclusion would say: "whenever this is an accident, it is not a substance" and both would be true.

However, if we apply option β to the example above, we would have the following formalization for the disjunctive proposition: "never either it has a carrier or it is not a substance" = "never either Q or $\sim$R" = "$\sim\sim$R $\vee$ Q" = "R $\vee$ Q", which would lead to the following formula for the whole combination:

$$[(P \supset Q) \wedge (R \vee Q)] \rightarrow (R \vee P)$$

This formula is not valid just like what Avicenna says. If we apply this formalization to the concrete example given by Avicenna, and since the proposition "R $\vee$ Q" is equivalent to "$\sim$R $\supset$ Q", this means that the disjunctive premise in Avicenna's concrete example should be interpreted as "whenever this is not a substance, then it has a carrier," and would be true, and the conclusion "whenever this is not a substance then it is an accident" can be true too. But when we change "it is a substance" by "every quantity is finite," then we would have the following: the premise would be "whenever every quantity is not finite then it (?) has a carrier" and the conclusion would be "whenever every quantity is not finite then it (?) is an accident," which would be false, according to him. So it seems that Avicenna's choice of the alternative premise seems to show that he has chosen option β, even if his particular example is not very convincing, because the proposition "every quantity is not finite" is not related to the rest of the propositions of the mood. The calculus shows anyway that in that option, the combination is not valid whatever concrete example we may give.

In case the premise is particular, no combination is valid.

II-d The next subgroup (p. 308.13–17) contains the same general conditions (*minor: conditional, major: not real disjunction, shared element: the consequent of the conditional*), but "*the combination contains two affirmatives* (*al-ta'līf min mūjibatayni*)" ([34], 308.13).

The premises of the first combination given by Avicenna are the following:

"Whenever H is Z then not (C is D) (A$_{C}$)
Either not (C is D) or A is B (A$_{D}$)" ([34], 308. 13–14).

Formalized, the premises become "$[(P \supset \sim Q) \wedge (\sim Q \vee R)]$". Do these premises lead to a universal disjunctive (**A**$_{D}$)? If the conclusion is disjunctive, we would have the following formula:

"$[(P \supset \sim Q) \wedge (\sim Q \vee R)] \nrightarrow (\sim P \vee R)$" (conclusion: **A**$_{D}$).

This formula is not valid, for it is possible for the two premises to be true while the conclusion is false. If the conclusion is conditional, the conclusion is false too. This means that Avicenna is right when he says that these premises are not conclusive.

In this same group IId, he adds that when in such combinations one of the propositions is particular, the combinations are not productive either. Let us verify this by taking first a particular conditional (with a negative element as in the universal given by Avicenna) and a universal disjunctive. We would have the following formula, where the conclusion is a particular conditional **I**$_{C}$:

$$[(P \wedge \sim Q) \wedge (\sim Q \vee R)] \supset ?(P \wedge R).$$

This formula is not valid (and it would also be not valid with "$P \wedge \sim R$" (= $\mathbf{I_D}$) in the conclusion), which confirms Avicenna's claim.

If the disjunctive premise is particular while the universal is conditional, we would have the following formula:

$$[(P \supset \sim Q) \wedge (\sim Q \wedge \sim R)] \supset ?\,(P \wedge \sim R).$$

This is not valid too and confirms Avicenna's claim.

II-e: This subgroup is the one where under the same general conditions (*minor: conditional, major: not real disjunction, shared element: the consequent of the conditional*), we have the case where "*the disjunctive is negative*" ([34], 309.1).

In this case, Avicenna says that the combinations are not productive and gives a concrete example illustrating this invalidity, which is the following: "whenever this is an accident then it is not a substance, and never either this is not a substance or it is in a subject. And never either it is [not][9] a substance or the quantity is not actually finite" ([34], 309.1–4). Here too, he gives two distinct conclusions, which are supposed to prove that the combination is not valid, since sometimes the conclusion is true and sometimes false. Given that one of the premises is negative, the conclusion should be negative too. Let us consider the case where the conclusion is negative and conditional and formalize accordingly the example:

If "this is an accident": P, "this is a substance": Q, and "this is in a subject": R, then we have the following formalized premises, where $\mathbf{E_D}$ is interpreted in option α:

$$[(P \supset \sim Q) \wedge (\sim \sim Q \vee R)].$$

In this formula, if we reduce the premise $\mathbf{E_D}$ to a conditional proposition, we would have the following conditional "$\sim Q \supset R$", which joined to the $\mathbf{E_C}$ proposition, would lead to "$P \supset R$", by transitivity, since the formula would be the following:

$$[(P \supset \sim Q) \wedge (\sim Q \supset R)] \rightarrow (P \supset R).$$

This formula is valid since it has the structure of *Barbara* with a negative element "$\sim Q$" as predicate in the first premise and subject in the second one.

However, if the $\mathbf{E_D}$ premise is interpreted in option β, it would have the following form: $(\sim Q \vee \sim R)$, which when reduced to a conditional proposition would be the following: "$Q \supset \sim R$". With these two premises, the whole combination would be the following: $[(P \supset \sim Q) \wedge (Q \supset \sim R)]$. This combination has the structure $\mathbf{E_C E_C}$, and as such, it does not lead to any conclusion, since both premises are negative universal conditionals. This justifies Avicenna's claim that this combination is sterile. It seems then that he is taking $\mathbf{E_D}$ as being of option β.

"Group III/ In this third group, the set of conditions is stated as follows: (*minor conditional, major disjunctive*; *shared element*: *the antecedent of the conditional premise*" ([34], 309.8–9).

---

[9] Here the word 'not' is added by N. Shehaby.

III-a. This is the subgroup where the disjunctive is a "*real disjunction*" (*munfaṣila haqīqiya*) and "*the premises are affirmative*" ([34], 309.8–10). The first mood presented is the following:

"Whenever H is Z then C is D (**A**c)
Always either H is Z or A is B (**A**D)
Therefore It happens that if C is D then not (A is B) (**I**c)" ([34], 309.10–12).

This seems to be a mixed *Darapti* (the third figure being evoked explicitly by Avicenna, at 309.14) despite the negative element in the conclusion, which suggests that it should rather be an $\mathbf{O}_C$ proposition. But if one states it without adding the "augment" in the conditional, it is not valid. For the formula corresponding to it is the following, with "H is Z": P, "C is D": Q, and "A is B": R:

$$[(P \supset Q) \wedge (P \veebar R)] \rightarrow (Q \wedge \sim R).$$

This formula is not valid, for it is possible for the premises to be true while the conclusion is false. Only with the augment it becomes valid. So the formula corresponding to it is in fact the following: $\{[P \wedge (P \supset Q)] \wedge (P \veebar R)\} \rightarrow (Q \wedge \sim R)$.

However, in his proof, Avicenna says what follows: "It can be proved by considering that the disjunctive becomes 'whenever H is Z then not A is B'" ([34], 309.13–14). But "$P \vee R$" is *not* equivalent to "$P \supset \sim R$", as the passage quoted suggests, although the former implies the latter. The mood is valid only when the disjunction is exclusive.

According to Avicenna, another (presumably second figure) mood may have a universal conclusion if the first premise is "the converse of the contradictory of the conditional" ([34], 309.15). This mood contains $\mathbf{E}_C$, $\mathbf{A}_C$ as premises and leads to $\mathbf{E}_C$. It is expressed as follows:

"Never if not C is D then H is Z (**E**c)
Whenever not (A is B) then H is Z (**A**c)
Therefore Never if not C is D then not (A is B) (**E**c)" ([34], 310.1–4).

The formula corresponding to this mood is the following (C is D: Q, H is Z: P; A is B: R): "$[(\sim Q \supset \sim P) \wedge (\sim R \supset P)] \rightarrow \sim(\sim Q \wedge \sim R)$". This formula is valid.

Inside this same subgroup, he considers next the two cases with a particular conditional and a particular disjunctive. If the conditional is particular, we would have the following combination, where "H is Z": P, "C is D": Q, and "A is B": R:

$$(\exists s)(Ps \wedge Qs) \wedge (\forall s)(Ps \veebar Rs)] \rightarrow (\exists s)(Qs \wedge \sim Rs).$$

This combination is valid and corresponds to a mixed *Datisi* $\mathbf{A}_D\mathbf{I}_C\mathbf{I}_D$. With one situation, it becomes $[(P \wedge Q) \wedge (P \veebar R)] \rightarrow (Q \wedge \sim R)$.

When the disjunctive is particular, then we would have the following combinations, where the minor is conditional and universal, and the disjunctive can be either of option α or of option β:

Option α: $[(P \supset Q) \wedge (P \wedge {\sim}R)] \rightarrow (Q \wedge {\sim}R)$ (*Disamis* $\mathbf{I_D A_C I_D}$).
Option β: $[(P \supset Q) \wedge ({\sim}P \wedge R)] \supset\ ?\ ({\sim}Q \wedge R)$ ($\mathbf{I_D A_C I_D}$).

The first combination is valid, even without the augment. But the second one is not valid. Since Avicenna thinks that the mood is valid, he seems to have chosen option α. This is confirmed by what he says according to which the disjunctive premise would be reduced to "It happens that if C is D then not (A is B)" ([34], 310.5) which corresponds to "$Q \wedge {\sim}R$" in the first formula.

III-b. The second subgroup in this group is where "*the conditional is negative and convertible*" ([34], 310.8).

In this subgroup, the first mood considered is stated as follows:

"Never if H is Z, then C is D ($\mathbf{E_C}$)
Always either H is Z or A is B ($\mathbf{A_D}$)
Therefore Never if C is D then not (A is B) ($\mathbf{E_C}$)" ([34], 310.8–10).

If we formalize it with "H is Z": P, "C is D": Q, and "A is B": R, we get the following formula: ${\sim}(\exists s)(Ps \wedge Qs) \wedge (\forall s)(Ps \veebar R) \rightarrow {\sim}(\exists s)(Q \wedge {\sim}R)$. With one situation, this becomes $[{\sim}(P \wedge Q) \wedge (P \veebar R)] \rightarrow {\sim}(Q \wedge {\sim}R)$.

This formula is valid and corresponds to $\mathbf{A_D E_C E_C}$. This also is a new mood **AEE** in the third figure. It does not correspond to any known third figure mood of categorical logic, since we do not find a valid **AEE** mood in the third figure. So it is one of the moods that the use of disjunctive propositions introduces. What is also interesting is that Avicenna expresses the conclusion in three ways: First as "never when C is D then not (A is B)" [$= {\sim}(\exists s)(Qs \wedge {\sim}Rs)$], second as "whenever C is D then A is B" [$= (\forall s)(Qs \supset Rs)$], and third as "never either C is D or A is B" [$= (\forall s)({\sim}Qs \vee Rs)$]. This confirms the fact that he is considering option α for the conclusion, if it were to be interpreted as a disjunctive proposition. However, the disjunction in the premise is a real disjunction.

Avicenna proves this mood by converting the $\mathbf{E_C}$ proposition, which becomes "$(\forall s)(Qs \supset {\sim}Ps)$" leading to a mood where the shared element is "the consequent" ([34], 310.12), i.e., to a mood of the first figure.

A variant of the above $\mathbf{A_D E_C E_C}$ mood contains a particular conditional and is expressed as follows:

"It happens that if not (C is D) then H is Z ($\mathbf{I_C}$)
Whenever H is Z, then not (A is B) ($\mathbf{A_C}$)
Therefore it happens that when not (C is D) then not (A is B) ($\mathbf{I_C}$)" ([34], 310.13–15).

The formula expressing this mood is the following:

$$(\exists s)({\sim}Ps \wedge Qs) \wedge (\forall s)(Qs \supset {\sim}Rs) \rightarrow (\exists s)(({\sim}Ps \wedge {\sim}Rs).$$

With one situation, it becomes "$[({\sim}P \wedge Q) \wedge (Q \supset {\sim}R)] \rightarrow ({\sim}P \wedge {\sim}R)$".

This formula is valid and the mood has the structure of *Darii* of the first figure, but it contains some negative elements either as antecedents or as consequents.

If the mood contains a disjunctive particular proposition, then the disjunctive "could be reduced to a conditional one" ([34], 310.16), while the universal negative conditional could be replaced by "an affirmative conditional" ([34], 310.15–16), so that the mood is expressed as follows:

"Whenever H is Z, then not (C is D) (**A**$_C$)

It happens that when H is Z then not (A is B) (**I**$_C$)

Therefore It happens that when not (C is D) then not (A is B) (**I**$_C$)" ([34], 310.17–18).

This seems to have the structure of *Disamis* $\mathbf{I_C A_C I_C}$ from the third figure, the only difference between the usual *Disamis* being the use of negative elements inside the conditional propositions. The formula expressing it is the following:

"$[(\forall s)(Ps \supset \sim Qs) \wedge (\exists s)(Ps \wedge \sim Rs)] \rightarrow (\exists s)(\sim Qs \wedge \sim Rs)$."

With one situation, it becomes "$(P \supset \sim Q) \wedge (P \wedge \sim R) \rightarrow (\sim Q \wedge \sim R)$". This formula is valid.

III-c. Under these same conditions, i.e., "*minor conditional, major disjunctive*; *shared element*: *the antecedent of the conditional premise,*" we have the third subgroup where "*the disjunctive proposition is negative*" ([34], 311.1). The first combination contains the following premises:

"Whenever H is Z then C is D" (**A**$_C$)

and "Never either H is Z or A is B" (**E**$_D$).

These premises are said to be "not productive" ([34], 311.2).

As a matter of fact, if $\mathbf{E_D}$ is formalized as "$\sim P \vee R$", then the whole formula would be symbolized thus, assuming that the mood's conclusion is disjunctive, negative, and universal: $[(\forall s)(Ps \supset Qs) \wedge (\forall s)(\sim Ps \vee Rs)] \rightarrow (\sim Qs \vee Rs)$. With one situation, it would be rendered by the following formula: $[(P \supset Q) \wedge (\sim P \vee R)] \rightarrow (\sim Q \vee R)$. This formula is not valid for it is possible for the premises to be true while the conclusion is false. But if we add the augment in $\mathbf{A_C}$ so that the first proposition becomes "$P \wedge (P \supset Q)$", then the whole formula would be expressed as follows:

$$\{[P \wedge (P \supset Q)] \wedge (\sim P \vee R)\} \rightarrow (\sim Q \vee R)$$

and it is valid indeed. So this mood has the following form: $\mathbf{E_D}, \mathbf{A_C} \rightarrow \mathbf{E_D}$, and is valid with the augment in $\mathbf{A_C}$, although Avicenna himself does not recognize its validity. However, if the conclusion is a negative conditional, as Avicenna seems to think given the concrete example he gives, which is "whenever this [number] is even, then it is divisible by two equal numbers, and never either this is even or this is a number, then it is true that: whenever this is divisible by two equal numbers, then it is a number. And if you replace 'number' by 'void', then the negative is true." ([34], 311.2–6), then it is *not valid*, with or without the augment, whether the disjunction is exclusive or inclusive. Avicenna's justification of this invalidity in the quotation above is that the two premises could produce either an affirmative or a negative conclusion depending on their matter.

Now this mood, as Professor Hodges suggests, is also productive if the disjunctive premise is interpreted in option β mentioned above. In that case, the mood would have the following structure: $[(P \supset Q) \wedge (P \vee \sim R)] \rightarrow (Q \vee \sim R)$, where the disjunction in the conclusion is not exclusive. In this case, the mood is valid too and the conclusion is equivalent to "$R \supset Q$". According to Professor Hodges, Avicenna considers the mood as invalid because he has chosen option α for the disjunctive premise, while it is option β that validates it.

Group IV: Avicenna considers *the same set of conditions (minor conditional, major disjunctive) but where the disjunction is not real and the shared element is the affirmative one* ([34], 311.7–8).

IV-a: The first subgroup is the one where "*the premises are affirmative*" (*al-Qiyās*, p. 311.9). The first mood of this subgroup, which is considered as valid by Avicenna, is the following:

"Whenever H is Z then C is D (Ac)
Always either H is Z or not (A is B) (Ad)
Therefore It happens that if C is D then A is B (Ic)" ([34], 311.9–11).

This is a mixed *Darapti* containing one disjunctive premise. But as all other *Darapti* moods it is not valid without the augment of $\mathbf{A_C}$. However, even with the augment, it is not valid if the disjunction is not exclusive, for the formulas are the following (when H is Z: P, C is D: Q and A is B: R):

1. Without the augment: $[(P \supset Q) \wedge (P \vee \sim R)] \nrightarrow (Q \wedge R)$ [this is *not* valid].
2. With the augment: $\{[P \wedge (P \supset Q)] \wedge (P \vee \sim R)\} \nrightarrow (Q \wedge R)$ [this is *not* valid too].

But his proof is interesting for it runs as follows: "And this may be proved thus: Whenever A is B then H is Z, and whenever H is Z, then C is D; it follows: whenever A is B then C is D. Then convert." ([34], 311.12–13). If we write it down we have the following *Barbara* mood:

Whenever A is B, then H is Z (Ac)
Whenever H is Z then C is D (Ac)
Therefore Whenever A is B then C is D (Ac)

Then by the conversion of the conclusion, we should obtain "It happens that if C is D then A is B," which is the real conclusion of the mood. This may seem fine, since *Barbara* is valid, but it is not quite so, given that $\mathbf{A_C}$ does not convert unless we add the augment. Now in general we don't need the augment to state *Barbara*, because it is valid by transitivity. But here, the conversion could not lead to the desired conclusion, which is a particular conditional, unless the conclusion contains the augment. We might validate the conversion by considering that all the premises of the above *Barbara* mood and its conclusion contain the augments. In that case, the mood is valid and the conversion too, and Avicenna would be right, for the formula representing the whole *Barbara* above could be the following, where A is B: R; H is Z: P; C is D: Q: $\{[R \wedge (R \supset P)] \wedge [P \wedge (P \supset Q)]\} \rightarrow [R \wedge (R \supset Q)]$. However, although this formula is valid and leads by the conversion of the conclusion to "(Q ∧

R)", it should not be considered as the right formula by which the mood is proved. For the initial mood contains a *disjunctive* proposition stated as "$P \vee \sim R$", and "$P \vee \sim R$" is *not equivalent* to "$R \wedge (R \supset P)$"; rather it is equivalent to "$R \supset P$" alone (i.e., without the augment). Consequently, the first premise above should be written thus "$R \supset P$", and the whole formula representing the *Barbara* syllogism above would be the following: "$\{(R \supset P) \wedge [P \wedge (P \supset Q)]\} \rightarrow [R \wedge (R \supset Q)]$". Unfortunately, this formula is *not* valid, for the table shows that we could find a case of falsity under the main implication as follows:

$$\{(R \supset P) \wedge [P \wedge (P \supset Q)]\} \rightarrow [R \wedge (R \supset Q)]$$

0 1 1 **1** 1 1 1 1 1 **0** 0 **0** 1 1

Although it has the general structure of Barbara, since P is the middle, R is the minor, and Q is the major, it is not valid because the premise "$R \supset P$" could be true when R is false, in which case, the conclusion would be false, since it contains the augment "$R \wedge$..." which makes the conjunction as a whole false, while it would have been true without the augment. So we must deduce that this Barbara would be valid only if the conclusion does not contain the augment, i.e., if the whole mood is expressed thus:

$$\{(R \supset P) \wedge [P \wedge (P \supset Q)]\} \rightarrow (R \supset Q).$$

But then, the conclusion would not be convertible, unless an "unexpected existential augment" is added in the conclusion, as one referee suggests, since the conclusion should be convertible.

Besides that, when we state the mood in that way, it does not represent *Darapti* either since P is not the antecedent of both premises. Rather, it is the antecedent of the second one and the consequent of the first one.

If we write again the initial mood, which seemed to be a Darapti mood, we would have the following:

"Whenever H is Z then C is D (**A**c)
Always either H is Z or not (A is B) (**A**D)
Therefore It happens that if C is D then A is B (**I**c)"

If "H is Z": P, "C is D": Q, and "A is B": R, we would have the following formula, where **A**c contains the augment and AD is not exclusive according to the conditions of this subgroup, since Avicenna says: "where the disjunctive is *not real* (*ghair ḥaqīqīya*)" ([34], 311.7):

$$\{[P \wedge (P \supset Q)] \wedge (P \vee \sim R)\} \supset (?)\, (Q \wedge R).$$

Since he equates between "Always either H is Z or not (A is B)" and "Whenever A is B then H is Z" ([34], 311.9–10 + 12) when he explains how to prove the mood. In other words, he equates between "$P \vee \sim R$" and "$R \supset P$", and the mood becomes the following, when formalized with the conditional proposition that replaces the disjunctive one:

$$\{[P \wedge (P \supset Q)] \wedge (R \supset P)\} \supset (?)\ (Q \wedge R).$$
$$1\ 1\ 1\ 1\ 1\quad \mathbf{1}\ 0\ 1\ 1\quad \mathbf{0}\quad 1\ \mathbf{0}\ 0$$

Unfortunately, this last formula is *not valid* as the truth values of the propositions show clearly. Besides that, when the disjunctive premise becomes a conditional one, the mood has not any more the structure of *Darapti*, where the antecedents of both premises are supposed to be the same proposition. Here, the antecedent of the first premise (the minor) is P, while the antecedent of the second premise (the major) is R.

So Avicenna is wrong when he considers this mood as valid.

IV-b: The second subgroup is the one where "*the conditional is negative*" ([34], 311.15). The first mood presented by Avicenna is stated as follows:

Never if H is Z then C is D (Ec)
Always either H is Z or not (A is B) (Ad)
Therefore Never if C is D then A is B (Ec) ([34], 311.15–16).

This sounds like an **AEE** mood in the third figure. If it is valid, it would add a supplementary mood to the third figure, which does not contain traditionally any **AEE** mood. If we formalize it with H is Z: P, C is D: Q, and A is B: R, we have the following formula: "$[\sim(\exists s)(Ps \wedge Qs) \wedge (\forall s)(Ps \vee \sim Rs)] \rightarrow \sim(\exists s)(Qs \wedge Rs)$". With one situation, it becomes the following: $[\sim(P \wedge Q) \wedge (P \vee \sim R)] \rightarrow \sim(Q \wedge R)$. This formula is valid even when the disjunction is inclusive, as assumed by Avicenna.

A parallel mood with a particular conditional premise is said to be valid too, for he says "and similarly when the conditional is particular. You can do what you have done with its correspondent [proposition] (*taf'alu mā fa'alta bi naḍīratihā*)" ([34], 312.1–2). This second mood would be expressed thus:

Maybe if H is Z then not C is D (Oc)
Always either H is Z or not (A is B) (Ad)
Therefore Maybe if C is D then not (A is B) (Oc)

Formalized, it becomes $[(P \wedge \sim Q) \wedge (P \vee \sim R)] \rightarrow (Q \wedge \sim R)$. Unfortunately, this formula is not valid, for the conclusion may be false when both premises are true, when the disjunction is inclusive (*ghair ḥaqīqīya*) as assumed by Avicenna, and even if it were exclusive.

IV-c: In this third subgroup, "*the disjunctive is negative*" ([34], 312.3). In this subgroup, Avicenna says that the first mood is not conclusive and illustrates this invalid combination by the following concrete example: "Whenever Zayd is drowning, then Zayd is in the water; and not either Zayd is drowning or he is not flying" ([34], 312.3–4). These two premises do not produce any conclusion, whether the conclusion is disjunctive or conditional, which shows that Avicenna is right when he says that this combination is not valid.

Group V: In this group, the same conditions are considered but "*the shared element is the negative one*" ([34], 312.6).

V-a: The first subgroup is the one where the premises "are affirmative" ([34], 312.7). In this category, the first mood presented is the following:

"Whenever not (H is Z) then C is D (A_C)
Always either not (H is Z) or A is B (A_D)
Therefore It happens that when C is D then not (A is B) (I_C)" ([34], 312.7–9).

This sounds like a mixed *Darapti* with a negative antecedent in the conditional premise and the conclusion. It may be formalized as follows, when H is Z: P, C is D: Q, and A is B: R: "$[(\sim P \supset Q) \wedge (\sim P \vee R)] \rightarrow (Q \wedge \sim R)$". This formula, however, is not valid. Even when one adds the augment, it remains invalid, as the calculus shows:

"$\{[\sim P \wedge (\sim P \supset Q)] \wedge (\sim P \vee R)\} \supset (?)\ (Q \wedge \sim R)$".
1 1 1 1 1 **1** 1 1 1 **0** 1 0 0

It would have been valid if the disjunction were exclusive, since in that case, the disjunction cannot be true when both its elements are true. But Avicenna does not say that the disjunction should be exclusive (*haqīqīya*). On the contrary, he seems to include this mood into the kind of moods where the disjunctive is "not real (*ghair haqīqīya*)" ([34], 311.7).

However, since Avicenna says at the end of the paragraph "and you can deduce the universal as you know" ([34], 312.12), one referee notes that in this case it would be valid (even without the augment) and would represent one of his new moods, for it would be expressed as follows, where "H is Z": P, "C is D": Q, and "A is B": R and where the disjunctive proposition is replaced by a conditional, as Avicenna does very frequently, since "$\sim P \vee R$" becomes "$P \supset R$":

$$[(\sim P \supset Q) \wedge (P \supset R)] \supset (\sim Q \supset R).$$

This formula is valid, but as we can see, the conclusion that follows is not the one stated by Avicenna. It would have the following structure: $\mathbf{A_D A_C A_C}$ in the third figure with negative elements in both the minor and the conclusion.

V-b. In the next subgroup, "*the conditional is negative*" ([34], 312.13). In this subgroup, the first mood is the following:

"Never if not (H is Z) then C is D (E_C)
Always either not (H is Z) or A is B (A_D)
It follows Never if not (C is D) then A is B (E_C)" ([34], 312.13–15).

Unfortunately, this mood is not valid, for the formula expressing it is the following: "$[(\sim P \supset \sim Q) \wedge (\sim P \vee R)] \rightarrow (\sim Q \supset \sim R)$". This formula is *not* valid, whether the disjunction is exclusive or inclusive. Since $\mathbf{E_C}$ is negative, there is no augment that could be added. But Avicenna says that it is valid and can be proved by the "conversion of the conditional [proposition]" ([34], 312.16). However, the text contains a logical error which is not usual in Avicenna's writings, since the conclusion is expressed thus "It follows that : Never if not (C is D) then A is B; *rather whenever C is D, then A is B*" ([34], 312.15–16, my emphasis). The passage emphasized shows that "Never if not (C is D) then A is B [that is, "$\sim Q \supset \sim R$"]" is equated with "Whenever C is D then A is B" [that is, "$Q \supset R$"]. Thus, it looks like Avicenna says that "$\sim Q \supset \sim R$" is equivalent to "$Q \supset R$". But Avicenna never makes this kind of errors and he clearly says in many occasions and writings that "$P \supset Q$" is equivalent to "$\sim Q \supset \sim P$" (not to "$\sim P \supset \sim Q$"). On the other hand, if the conclusion in this mood were "$Q \supset R$", instead of "$\sim Q \supset \sim R$", it is *indeed valid*. So it seems that either the editor has made an error in the editing process, or Avicenna himself made a confusion that he corrected afterward. Anyway, the right mood should be expressed formally as follows: "$[(\sim P \supset \sim Q) \wedge (\sim P \vee R)] \rightarrow (\sim R \supset \sim Q)$" [= "$[(\sim P \supset \sim Q) \wedge (\sim P \vee R)] \rightarrow (Q \supset R)$"].

V-c. The third subgroup is the one where "*the disjunctive is negative*" ([34], 313.1). Here, Avicenna says that the moods with this condition are not productive.

V-d. In this fourth subgroup, "*the disjunctive has two negative elements*" ([34], 313.3). This too does not contain productive moods.

Group VI: This is the group where the conditions are the following: "*Major: conditional, Middle: the antecedent, disjunctive: real*" ([34], 313.5–6).

VI-a. In the first subgroup, the premises are "*affirmative*" ([34], 313.7).

The first mood with the above conditions is a *Barbara* containing one conditional premise and a disjunctive one. It is stated as follows:

"Always either H is Z or C is D ($\mathbf{A_D}$)
Whenever C is D, then A is B ($\mathbf{A_C}$)
Therefore Whenever not (H is Z) then A is B ($\mathbf{A_C}$)" ([34], 313.7–9).

The conclusion of this mood is said to imply the following "Either H is Z or A is B" ([34], 313.9).

If we formalize it, we have the following (where H is Z: P, C is D: Q, and A is B: R):

"$[(P \veebar Q) \wedge (Q \supset R)] \rightarrow (\sim P \supset R)$". This formula is valid. But it would also be valid if the disjunction in $\mathbf{A_D}$ were inclusive, although Avicenna does not make this precision and only considers the case where the disjunction is "real" (that is, exclusive) as stipulated in the conditions above. As to the disjunctive proposition which is said to be implied by the conclusion, it *has to* be an *inclusive* disjunction; otherwise, it would not be implied by the conclusion of this mood. For "$\sim P \supset R$" implies "$P \vee R$" (and is even equivalent to "$P \vee R$"), but it *does not* imply "$P \veebar R$".

Avicenna also considers a variant of this mood where the disjunctive premise is particular ([34], 313.11). This other mood could be stated as follows:

It happens that either H is Z or C is D ($\mathbf{I_D}$)
Whenever C is D, then A is B ($\mathbf{A_C}$)
Therefore maybe when not (H is Z), then A is B ($\mathbf{I_C}$).

This mood would be equivalent to a mixed *Darii* if it were valid. How could we state $\mathbf{I}_D$ in order to validate the mood? As we saw above, $\mathbf{I}_D$ can be interpreted either as "P ∧ ∼Q" (option α) or as "∼P ∧ Q" (option β). If we follow the first interpretation, we have the following formula: "[(P ∧ ∼Q) ∧ (Q ⊃ R)] → [?] (∼P ∧ R)". Unfortunately, this formula is *not* valid, given that the conclusion could be false while the two premises are true. We have then to state $\mathbf{I}_D$ in another way. This other way could be the following: "∼P ∧ Q", which corresponds to Option β above, for in this case too, the disjunctive proposition contains a negative element and an affirmative one. Consequently, "either P or Q" is true in that case too, given that Q is true. If the conditional is particular, then the "disjunctive would be turned to a conditional" as Avicenna says in what follows: "Let the disjunctive be a conditional, then we have 'Whenever C is D, then H is Z', and add the other premise as in the third figure; it follows: 'It happens that when not (H is Z), that A is B'" ([34], 313.12–14). Thus stated, the mood is the following:

"Whenever C is D, then H is Z ($\mathbf{A}_C$)
Whenever C is D then A is B ($\mathbf{A}_C$)
Therefore It happens that when not (H is Z) then A is B ($\mathbf{I}_C$)"

This is valid by *Darapti*, provided we add the augment to the first premise $\mathbf{A}_C$, which would be stated as follows: "Q ∧ (Q ⊃ ∼P)". Thus stated the proposition is precisely equivalent to "∼P ∧ Q" [= "P ∧ ∼Q"]. So the formula corresponding to the whole mood would be the following: "[(∼P ∧ Q) ∧ (Q ⊃ R)] → (∼P ∧ R)". This formula is valid.

VI-b: The next subgroup is the one where "*the disjunctive is negative*" ([34], 313.15). The first combination considered by Avicenna precisely contains an $\mathbf{E}_D$ proposition. This combination is the following: "Never either H is Z or C is D" and [we add] "whenever C is D, then A is B," then nothing follows ([34], 313.15–16). It is said to be not productive by Avicenna.

However, if we formalize the disjunctive premise "Never either H is Z or C is D" as "∼P ∨ Q", the two premises would be rendered as follows: "(∼P ∨ Q) ∧ (Q ⊃ R)". As we can easily verify it, these two premises are indeed productive, for they just lead to "P ⊃ R" and are indeed another way of expressing a mixed *Barbara*, given that "∼P ∨ Q" is equivalent to "P ⊃ Q". The problem is then the following: Why does Avicenna say that these premises are not productive? Maybe there is no obvious answer, but we might perhaps consider the fact that the disjunctive premise is a negative one as it is stated, which could have let Avicenna search for a negative conclusion. Not having found it, he might have considered that the two premises do not lead to any conclusion. As a matter of fact, he does provide a concrete example which is supposed to function as a counterexample as appears in the following quotation: "As an example [consider] the following: 'Never either this thing is the void or it is even; and whenever it is even, then it is divisible by two equals' and replace 'the void' by 'the even of the even' (*zawju al-zawji*)" ([34], 313.16–314.2). If we make the replacement between "it is the void" and "it is the even of the even", we get the following: "Never either this thing is the even of the even or it is even, and whenever it is even, it is divisible by two equals." Thus stated, the first premise ($\mathbf{E_D}$) seems strange, for it seems to say that there is an incompatibility between being even and being "even of the even," which might be the reason why Avicenna rejects this mood.

Now there is a possible formalization that makes the premises non-conclusive as Avicenna says explicitly. This possible alternative is the formalization of $\mathbf{E_D}$ by "$\sim Q \vee P$", which is the negation of $\mathbf{I_D}$ when interpreted as "$Q \wedge \sim P$". In that case, the whole formula would be the following: "$[(\sim Q \vee P) \wedge (Q \supset R)] \nrightarrow (P \supset R)$". So maybe that was Avicenna's choice.

This possibility could be taken into account because it agrees with Avicenna's judgment. But the problem remains to determine why $\mathbf{E_D}$ should be interpreted in that way and not in another way. Avicenna's examples and analyses do not seem to clarify the matter in the sense of choosing a unique interpretation to each of these two propositions.

VI-c. The next subgroup is the one where "*the conditional is negative*" ([34], 314.3). In this subgroup, the first mood is stated as follows:

"Always either H is Z or C is D ($\mathbf{A_D}$)
Never if C is D then A is B ($\mathbf{E_C}$)
Therefore Never if not (H is Z) then A is B ($\mathbf{E_C}$)" ([34], 314.3–4).

This mood is proved by converting the disjunctive premise into a conditional one, so that it becomes "Whenever not (H is Z) then C is D." Here, the formula is the following: "$[(P \veebar Q) \wedge \sim(Q \wedge R)] \rightarrow (\sim P \supset \sim R)$." This formula is valid whether the disjunctive premise is exclusive or inclusive. But despite the fact that he talks about a "real" disjunction, Avicenna seems to equate the disjunctive premise with "whenever not P then Q," which could be formalized by "$\sim P \supset Q$" and is equivalent to "$P \vee Q$", the inclusive disjunction, not to "$P \veebar Q$" the exclusive one.

Group VII: This is the group where the conditions are the following: *"not real" ("ghair haqīqīya") disjunctives, common element between the premises: the "affirmative one"* ([34], 314.8).

VII-a. The first subgroup is the one where the premises "*are affirmative*" ([34], 314.9).

The first mood presented is the following:

"Always either not (H is Z) or C is D ($\mathbf{A_D}$)
Whenever C is D then A is B ($\mathbf{A_C}$)
Therefore if H is Z then A is B ($\mathbf{A_C}$)" ([34], 314.9–10).

This is a mixed *Barbara* with a negative element in the disjunctive premise. This disjunctive proposition becomes the following conditional one "Whenever H is Z then C is D" ([34], 314.12), which means that "$\sim P \vee Q$" is equated with "$P \supset Q$". This equivalence is quite natural and correct since the disjunction is explicitly said to be "not real", i.e., inclusive. Here, the text is clear and totally correct from the modern perspective. The mood itself is valid, since it is a mixed *Barbara* with a disjunctive minor.

However, in these particular conditions, Avicenna says that if the conditional premise is particular, then the mood is "not productive" ([34], 314.12–13). On the contrary, if the disjunctive is particular, then we would have a mixed *Darii* which is productive, according to Avicenna. But how the disjunctive particular proposition $\mathbf{I}_D$ would it be expressed? In Avicenna's wording, it would be the following: "It happens that either not (H is Z) or C is D," that is, "If not not (H is Z) then C is D," hence "When H is Z, C is D" = "$P \wedge Q$". In that case, the whole formula would be the following: "$[(P \wedge Q) \wedge (Q \supset R)] \rightarrow (P \wedge R)$" and would be valid.

VII-b. The second subgroup is the one where "*the disjunctive is negative*" ([34], 314.16). In this case, Avicenna says that the premises are "not productive" ([34], 314.16). He provides a counterexample to illustrate this invalidity. This example is the following: "Never either this is not 'not talking' or this is a human, and whenever it is a human then it is an animal. Then replace 'not talking' by 'the void'" ([34], 314.16–18).

VII-c. The third subgroup is the one where "*the conditional is negative*" ([34], 315.1). In this case, the first mood presented is valid as witnessed by the following:

"Always either not (H is Z) or C is D ($\mathbf{A}_D$)
Never if C is D then A is B ($\mathbf{E}_C$)
Therefore Never if H is Z then A is B ($\mathbf{E}_C$)" ([34], 315.1–3).

This mood may be formalized as follows, when H is Z: P, C is: Q, A is B: R, and the disjunction is not real, i.e., inclusive: "$[(\sim P \vee Q) \wedge (Q \supset \sim R)] \rightarrow (P \supset \sim R)$". Since "$\sim P \vee Q$" is equivalent to "$P \supset Q$", the mood is valid by the transitivity of implication.

Avicenna adds that the mood is valid too "when the disjunctive is particular," but "it is not valid if the conditional is particular" ([34], 315.4–5). The particular disjunctive would be an $\mathbf{I}_D$, which when it is associated to $\mathbf{E}_C$ would lead to $\mathbf{O}_C$. But how should we express $\mathbf{I}_D$, in this case? As we saw above, two alternative renderings are available for $\mathbf{I}_D$, for one could express it by "$P \wedge \sim Q$" [= sometimes when P then not Q] or alternatively by "$\sim P \wedge Q$" [= sometimes when not P then Q]. With $\mathbf{I}_D$ and $\mathbf{E}_C$, the conclusion would be $\mathbf{O}_C$, and if $\mathbf{I}_D$ is expressed as "$\sim P \wedge Q$", the mood would be expressed as follows "$[(\sim P \wedge Q) \wedge (Q \supset \sim R)] \rightarrow (P \wedge \sim R)$". This formula is valid.

Now what happens if the conditional is particular? In that case, we would have the following formula: "$[(\sim P \vee Q) \wedge (Q \wedge \sim R)] \rightarrow (P \wedge \sim R)$". This formula is not valid, which confirms what Avicenna says.

Group VIII: This group is the one where the same conditions are stated, i.e., *"not real" ("ghair haqīqīya") disjunctives* but where "*the shared element is the negative one*" ([34], 315.7).

VIII-a. In this subgroup, the premises "are affirmative" ([34], 315.8). The first mood presented is the following:

"Always either H is Z or not (C is D) ($\mathbf{A_D}$)
Whenever not (C is D) then A is B ($\mathbf{A_C}$)
Therefore Whenever not (H is Z) then A is B ($\mathbf{A_C}$)" ([34], 315.8–9).

If H is Z: P, C is D: Q, and A is B: R, we have the following formula: "$[(P \vee \sim Q) \wedge (\sim Q \supset R)] \rightarrow (\sim P \supset R)$". This formula is valid. Interestingly, Avicenna tells us that the conclusion "Whenever not (H is Z) then A is B" equals the following $\mathbf{E_D}$ "Never either not (H is Z) or A is B" ([34], 315.9–10). This could be shown by the fact that if $\mathbf{E_D}$ which usually says "Never either P or R" is formally expressed by "$\sim P \vee R$", when one replaces "P" by "$\sim P$", it becomes "Never either not P or R," that is, "$\sim \sim P \vee R$" [= "$P \vee R$" = "$\sim P \supset R$"].

Here too, Avicenna says that the corresponding mood with a particular disjunctive is valid, while it is not valid if the conditional is particular. In the first case, we will have an $\mathbf{A_C}$, $\mathbf{I_D}$, $\mathbf{I_C}$ (a mixed *Darii*) while in the second case the mood would have the form $\mathbf{I_C}$, $\mathbf{A_D}$, $\mathbf{I_C}$ (i.e., a mixed **IAI** mood, which does not exist in the first figure).

If we formalize these two combinations containing a particular premise, we have the following for the first one (mixed *Darii*):

It happens that either H is Z or not (C is D) ($\mathbf{I_D}$)
Whenever not (C is D) then A is B ($\mathbf{A_C}$)
Therefore It happens that if not (H is Z) then A is B ($\mathbf{I_C}$).

This combination is valid only if $\mathbf{I_D}$ is formalized as "$\sim P \wedge \sim Q$", when H is Z: P, C is D: Q, and A is B: R. This formalization of the particular disjunctive premise "Maybe either P or not Q" could correspond intuitively to the following: "Maybe when not P, then not Q," given that "P or Q" is generally rendered by "Maybe when not P then Q" or alternatively by "Maybe when not Q then P," where in both cases the word "when" expresses a conjunction rather than a conditional, because the conditional proposition as a whole is a particular one (= $\mathbf{I_C}$), which in Avicenna's frame is a conjunction. In this case, the whole mood is formalized thus: "$[(\sim P \wedge \sim Q) \wedge (\sim Q \supset R)] \rightarrow (\sim P \wedge R)$". This formula is valid.

What about the combination containing a particular conditional? This one is said to be invalid by Avicenna. It corresponds to the following combination:

**Always either H is Z or not (C is D) ($\mathbf{A_D}$)
It happens that when not (C is D) then A is B ($\mathbf{I_C}$)
Therefore It happens that when not (H is Z) then A is B ($\mathbf{I_C}$)**

If we formalize it, we have the following: "$[(P \vee \sim Q) \wedge (\sim Q \wedge R)] \rightarrow (\sim P \wedge R)$". This formula is *invalid* only when the disjunctive proposition $\mathbf{A_D}$ is not real, that is, inclusive. But if the disjunctive proposition is exclusive, it is on the contrary valid, for the exclusive disjunction, unlike the inclusive one, is false when its elements are both true. The inclusive character of the disjunction is in fact explicitly assumed from the start in the formulation of the specific conditions of this kind of combinations and moods ([34], 314.8). However, Avicenna does not add that in case the disjunction is "real" or exclusive, this combination would be a valid mood.

VIII-b. The second subgroup is the one where *"the disjunctive is negative"* ([34], 315.15). In this case, according to Avicenna, the combination is not valid. To illustrate this invalidity, he considers the following concrete example: "As an example, [consider] the following [premises]: 'Never either men are not bodies or they are moving, and whenever they are moving they are bodies' and replace 'not bodies' by 'the void'" ([34], 315.15–17). If "men are the void": P, "men are moving": Q. and "men are bodies: R," we have the following:

Never either humans are the *void* or they are moving (**ED**)
Whenever they are moving, they are bodies (**AC**)
Therefore Never if they are the void, then they are bodies (**EC**).

If as we saw above, $\mathbf{E_D}$ [i.e., "Never either P or Q"] is expressed as "$\sim P \vee Q$", then the formula would be the following: "$[(\sim P \vee Q) \wedge (Q \supset R)] \rightarrow (P \supset \sim R)$". This combination is not valid, as Avicenna rightly says.

VIII-c. The third subgroup is the one where *"the conditional is negative"* ([34], 316.1), then the mood is valid and is expressed as follows:

"Always either H is Z or not (C is D) ($\mathbf{A_D}$)
Never if not (C is D) then A is B ($\mathbf{E_C}$)
Therefore Never if not (H is Z) then A is B ($\mathbf{E_C}$)" ([34], 316.1–3).

If we formalize this combination, where H is Z: P, C is D: Q and A is B: R, we obtain the following formula: "$[(P \vee \sim Q) \wedge (\sim Q \supset R)] \rightarrow (\sim P \supset R)$". This formula is valid without any doubt for as Avicenna rightly says the disjunctive premise is equivalent to "Whenever not (H is Z) then not (C is D)" ([34], 316.2–3), which makes the mood valid by transitivity. This corresponds to a mixed *Celarent*.

However, if the conditional is "particular" Avicenna says that the resulting combination is not conclusive. For then we would have the following:

"Always either H is Z or not (C is D) ($\mathbf{A_D}$)
It happens that when not (C is D) then A is B ($\mathbf{I_C}$)
Therefore It happens that when not (H is Z) then A is B ($\mathbf{I_C}$)"

If we formalize this combination, we obtain the following: "$[(P \vee \sim Q) \wedge (\sim Q \wedge R)] \rightarrow (\sim P \wedge R)$". This formula is not valid, as Avicenna rightly says. However, if the disjunction is exclusive, the whole formula is valid.

Group IX: It is the group where the conditions are the following: "*Major premise: the conditional, shared element: the consequent of the conditional*" ([34], 316.9).

IX-a. The first subgroup is the one where "the disjunctive is real" ([34], 316.10). The first mood fulfilling these conditions is expressed as follows:

"Always either H is Z or C is D (A_D)
Whenever A is B then C is D (A_C)
Therefore Never if H is Z then A is B (E_C)" ([34], 316.10–12).

This seems somewhat strange, for it deduces a negative proposition from two affirmative ones, although one of the affirmatives is a disjunctive proposition. But the formula expressing it is the following: "$[(P \veebar Q) \wedge (R \supset Q)] \rightarrow \sim(P \wedge R)$" and is indeed valid, which could be explained by the intrinsic negative meaning of the disjunctive. However, if the disjunctive proposition were inclusive, i.e., "not real" in Avicenna's wording, then the whole combination would be invalid. The negative character of the conclusion is explained by Avicenna himself by the fact that the disjunctive proposition implies the following negative conditional: "Never if H is Z then C is D" ([34], 316.11–12). So if the disjunctive proposition were expressed in terms of a conditional one, the mood would have the form of *Camestres*.

IX-b. In the second subgroup, "*the disjunctive is negative*" ([34], 317.1). In this case, the combination is not productive according to Avicenna ([34], 317.1).

IX-c. In the third subgroup, "*the conditional is negative*" ([34], 317.4). In this case, the first mood is expressed as follows:

"Always either H is Z or C is D (A_D)
Never if A is B then C is D (E_C)
Therefore Never if not (H is Z) then A is B (E_C)" ([34], 317.4–6).

The formula corresponding to this mood is the following: "$[(P \veebar Q) \wedge (R \supset \sim Q)] \rightarrow (\sim P \supset \sim R)$". This formula is valid. The disjunctive is said to "convert" (*tan'aqisu*) to the following conditional: "Whenever not (H is Z) then (C is D)" ([34], 317.5–6). As a matter of fact, the disjunctive implies the conditional but not conversely because of the exclusive character of the disjunction. But if the disjunction is inclusive, it is equivalent to the conditional evoked by Avicenna and the mood would be valid too. This mood is a mixed *Cesare*.

When the disjunctive is "particular", then the conclusion is a particular negative conditional and the combination would be expressed as follows:

"It happens that either H is Z or C is D (I_D)
Never if A is B then C is D (E_C)
Therefore It happens that when not (H is Z) then not (A is B) (O_C)"

The formula expressing it is the following: "$[(\sim P \wedge Q) \wedge (R \supset \sim Q)] \rightarrow (\sim P \wedge \sim R)$". This formula is valid too, which means that the mood is a mixed *Festino* from the second figure.

In a parallel combination where the conditional is "particular", the mood would be the following:

It happens that when A is B then not (C is D) (**O**$_C$)
Whenever not (H is Z) then C is D (**A**$_C$)
Therefore Not whenever A is B, then not (H is Z) (**O**$_C$).

The proposition AC in this mood corresponds by conversion to the disjunctive proposition "Always either H is Z or C is D." The mood has the form of *Baroco* and is therefore valid. Its corresponding formula is the following: "$[(P \wedge \sim Q) \wedge (\sim R \supset Q)] \rightarrow \sim(P \supset \sim R)$". This formula is valid.

Group X: It is the one where the conditions are as follows: "*disjunctive: "not real", and "shared element: the affirmative one*" ([34], 317.12–13).

X-a. In this first subgroup, the premises "*are affirmative*" ([34], 317.14). In that case, the moods are the following:

"Always either not (H is Z) or C is D (A$_D$)
Whenever A is B then C is D (A$_C$)
Therefore Never if H is Z then A is B (E$_C$)" ([34], 317.14–16).

Here, Avicenna tells us that the disjunctive becomes the following: "Whenever H is Z, then not (C is D)" ([34], 317.15–16). So if we express the disjunctive premise in terms of the conditional provided by Avicenna, the formula is the following: "$[(P \supset \sim Q) \wedge (R \supset Q)] \rightarrow (P \supset \sim R)$". This formula is valid. If one expresses it by using a disjunction corresponding to **A**$_D$, then the formula is the following, given that "$P \supset \sim Q$" is equivalent to "$\sim P \vee \sim Q$": "$[(\sim P \vee \sim Q) \wedge (R \supset Q)] \rightarrow (P \supset \sim R)$". So if we follow Avicenna's explanations, there is something unclear in the disjunctive formula, for it seems that he understands "Always either not (H is Z) or C is D" as saying something like "not P or not Q," itself equivalent to "whenever (not not) P then not Q" (or "whenever P then not Q"). But as we will see below, this confusion may be due to an error in the editing, which is that the disjunctive premise, which is written "Although either not (H is Z) or C is D" in Avicenna's text should be rather "Always either not (H is Z) or not C is D," because it is this latter formula that corresponds to the conditional "Whenever H is Z then not (C is D)" provided by Avicenna. We will confirm the presence of this "error" in our analysis of the next (and last) mood provided by Avicenna below. However, the problem here is that in this subgroup, the middle term should be affirmative as Avicenna says and as appears in the conditions above (subgroup X-a). So there is something unclear anyway in Avicenna's text.

X-b. The next subgroup is the one where "*the disjunctive is negative*" ([34], 318.1). In this subgroup, the first combination considered is the following:

"Never either not (H is Z) or C is D
Whenever C is D then A is B" ([15], 318.1–2).

This combination is said to be not productive by Avicenna.

X-c. The last subgroup is the one where "*the conditional is negative*" ([34], 318.4). In this case, the first mood presented by Avicenna is the following:

"Always either not (H is Z) or C is D (A_D)
Never if A is B then C is D (E_C)
Therefore Never if H is Z then A is B (E_C)" ([34], 318.4–6).

In this case, the disjunctive is said by Avicenna to become the following conditional proposition "Whenever H is Z then C is D" ([34], 318.5–6) which, when formalized, becomes "$P \supset Q$". This last formalization validates the mood, which is formally expressible as follows: "$[(P \supset Q) \wedge (R \supset \sim Q)] \rightarrow (P \supset \sim R)$", or if we use the disjunctive initial premise: "$[(\sim P \vee Q) \wedge (R \supset \sim Q)] \rightarrow (P \supset \sim R)$". Since the disjunctive premise and its corresponding conditional one is formally correct, and since the disjunctive proposition concerned is exactly the same as the one used in the preceding mood, while the conditional correspondent is different, this confirms the doubts about the formulation of the disjunctive premise of the mood preceding this last one. It seems then that the right correspondence between the disjunctive premise and the conditional one is the one just provided in this last mood, that is, "$\sim P \vee Q = P \supset Q$". While the correspondence which concerns the disjunctive premise used in the mood preceding this one should rather be the following: "$\sim P \vee \sim Q = P \supset \sim Q$". Avicenna seems then to have omitted the negation in front of Q when he wrote the disjunctive premise of the next to the last mood. This omission should be corrected in order to validate the whole mood.

Let us now turn to section vi.3, where Avicenna analyzes the moods containing only disjunctive premises.

C/ *al-Qiyās*, section vi.3

At the beginning of this section, Avicenna says that no mood containing only real disjunctive premises is productive ([34], 319.1). This is so because the real disjunctive admits only two possibilities. But when one combines between a real disjunctive premise and a not real one, one may construct some conclusive moods. For instance, consider the following premises:

"Always either H is Z or C is D (A_D)
Either C is D or not (A is B)" ([34], 321.4–5).

This couple of premises is conclusive according to Avicenna for it leads to the following conclusion: "Whenever H is Z, then not (A is B)" ([34], 321.8). To prove that this conclusion ensues from the above premises, Avicenna reduces the disjunctive premises to conditional ones and deduces the conclusion as follows:

"Whenever H is Z then not (C is D) (A_C)
Whenever not (C is D), then not (A is B) (A_C)
Therefore Whenever H is Z then not (A is B) (A_C) ([34], 312.6–8)"

This looks like a *Barbara* mood with some negative elements, either as antecedents or as consequents or both. It is valid by transitivity and shows that Avicenna equates between "Always either H is Z or C is D" and "Whenever H is Z then not (C is D)," and between "Always either C is D or not (A is B)" and "Whenever not (C is D), then not (A is B)," i.e., between "$P \vee Q$" and "$P \supset \sim Q$" on the one hand and "$P \vee \sim Q$" and "$\sim P \supset \sim Q$" on the other hand. Note that if the latter formula expresses a real equivalence between an inclusive disjunction and a conditional, the former should rather express an implication between the real disjunction and the conditional. For the real equivalent of the exclusive disjunction is not only a conditional, it is rather a biconditional.

Another example is provided, where the two disjunctive premises produce a conditional one as follows:

"Either not (H is Z) or C is D
Either C is D or not (A is B)
Therefore If not (C is D) then not (H is Z)"
(And if not (H is Z) then not (A is B) ([34], 321.9))

This combination seems to be almost rejected by Avicenna, although the premises lead indeed to the first conclusion, because it is said to be non-conventional given that the two affirmative premises "lead to a negative one" ([34], 321.11), which does not conform the usual rules of syllogistic. So it seems that Avicenna, even in his hypothetical logic, restricts himself to the moods respecting the rules of the usual categorical logic and does not admit easily the arguments that do not respect them. This is a real limitation because categorical logic is not the only system containing valid inferences and its rules are not the only ones to be correct.

Now rather than the moods containing only disjunctive premises which seem to be not really interesting, Avicenna studies a more promising kind of syllogisms, which combine between the categorical propositions and the hypothetical ones. These moods will be analyzed in the next section.

5.1.3.3 The Moods Combining Hypothetical and Categorical Premises

As we noted earlier, Avicenna always uses subject–predicate propositions, even in his hypothetical logic. For the elements of the conditional or the disjunctive propositions are "H is Z" or "A is B", for instance. This device shows first how close his hypothetical logic is to the usual categorical one, which is based on this kind of subject–predicate propositions; second, it allows him to mix between the categorical propositions and the hypothetical ones in the complex syllogisms that we will now consider.

The moods studied in this part of the system are really new and very interesting from many perspectives, as we will see below. They are conclusive as are most of the other moods already studied, but they differ from the usual moods, whether categorical or hypothetical, in that the premises of a single argument are *not all explicitly stated*, although they are indeed contained in some way in the whole argument. Some moods presented in this section presuppose the presence of some more premises that could be *deduced* from the *stated ones* by means of an independent reasoning, and are needed to understand the link between the stated premises and the conclusion.

He devotes three chapters to the study of these syllogisms: the first one is chapter 4 of section 6 (pp. 325–336) and is entitled "On the syllogisms (*fī-al-qiyāsāt*) containing categorical and hypothetical [premises] in the first figure, where the categorical is the major in the three figures," and the second one is chapter 5 of the section 6 (pp. 337–348) and is entitled "On the syllogisms containing categorical and hypothetical [premises], where the categorical is common to the antecedent in the three figures." Chapter 6 of this section (pp. 349–357) is devoted to the study of the syllogisms where the first premise contains many elements related by disjunctions; it is entitled "On the divided (*muqassam*) syllogism in the fashion (*'alā namaṭi*) of the three figures."

We will consider some moods belonging to these different kinds of syllogisms but not all of them, for the sake of brevity. A further study could be devoted to the study of the whole list of moods provided by Avicenna. In what follows we will analyze section vi.4 in chapter D, section vi.5 in chapter E, and section vi.6 in chapter F.

D/ *al-Qiyās*, section vi.4.

In this section vi.4, the categorical proposition takes the "place of the major term or of the minor term" ([34], 325.5), while the shared element could be "the categorical and the consequent of the hypothetical or the antecedent of the hypothetical" ([34], 325.6–7).

If the "shared element is the consequent" of the hypothetical and "the categorical is the major" ([34], 325.6–7), then the mood would be like the following:

"Whenever H is Z, then Every C is D
and Every D is A
Whenever H is Z, then Every C is A" ([34], 326.3–4).

Here, it is clear that something is missing although it is strongly presupposed. For the consequent of the conclusion, "Every C is A" follows by *Barbara* from the consequent of the first premise "Every C is D" and the second premise "Every D is A." So the categorical *Barbara* is presupposed in the deduction of the conclusion of this mood, because it is what relates the second premise to the first one and produces the conclusion. It is also the missed premise that justifies the whole mood and warrants its validity. This premise should be stated between the second premise and the conclusion as follows:

- But if Every C is D and Every D is A, then Every C is A.

The conclusion follows by transitivity, so if "H is Z" implies "every C is D," it also implies what follows from it by *Barbara*, that is, "every C is A." The whole conclusion is justified by the transitivity of the implication according to which if "H is Z" implies a proposition (that is, "Every C is D"), it also implies what follows from it, that is, "Every C is A."

Other moods presuppose *Celarent* instead of *Barbara* in order for the conclusion to be deduced from the stated premises. For instance, the following:

"Whenever H is Z, then C is D
And No D is A
Therefore Whenever H is Z, then no C is A" ([34], 326.12–13).

It is clear here that the missing premise is the following:

- "If C is D and No D is A then No C is A"

This premise expresses *Celarent* and justifies the consequent of the conclusion, hence, the whole conclusion, since what implies a proposition implies what follows from it.

In other moods, the conditional proposition is "negative" as in the following:

"Never if H is Z then not (Every C is D and Every D is A)
It follows Never if H is Z then not Every C is A" ([34], 327.2–4).

In his explanation of this mood, Avicenna says that "the conditional implies 'Whenever H is Z, then every C is D and every D is A'. It follows that: Whenever H is Z, then every C is A." ([34], 327.4–5). This means that the missing presupposed premise is a *Barbara* syllogism, which is expressed as follows: "If every C is D and every D is A, then every C is A." The whole mood should then contain the missing premise in order to be valid. For as it is first stated, it could be formalized as follows, where "H is Z": P, Every C is D: Q, Every D is A: R, and Every C is A: S: "$\sim [P \wedge \sim (Q \wedge R)] \nrightarrow \sim (P \wedge \sim S)$."

Unfortunately, this formula is not valid. In order to validate the mood, one should then add the missing premise and formalize the whole mood as follows: "$\{\sim [P \wedge \sim (Q \wedge R)] \wedge [(Q \wedge R) \supset S]\} \rightarrow \sim (P \wedge \sim S)$."

Thus formalized, the mood is valid for there is no case where the antecedent of the whole implication is true while its consequent is false.

One might also formalize it by using the language of predicate logic and establishing the links between the subjects and predicates of the categorical propositions involved.

This mood shows that if some proposition implies a set of premises (the premises of *Barbara*) it also implies its consequence (the conclusion of *Barbara*). The premise added, just specifies that this set of premises *does indeed imply* the conclusion, which was not explicitly said in the first formulation.

Another mood is stated as follows:

"Whenever H is Z, then Every C is D

And No D is A

Therefore Whenever H is Z, then No C is A" ([34], 327.8–9).

The missing premise in this mood seems to be *Cesare* from the second figure, which Avicenna states by "converting the predicative proposition" "No D is A" and getting "No A is D." This leads to *Cesare* which could be stated as follows: "If Every C is D and No A is D, then No C is A." The whole mood is proved and restated by Avicenna as follows:

"Whenever H is Z, then C is D

No A is D

Whenever C is D and No A is D, then No C is A

It follows: Whenever H is Z, then No C is A" ([34], 327.14–17).

This shows that Avicenna has included the missing premise in his demonstration of the mood. Here the negative premise of *Cesare* is not exactly the one used in the initial mood, but it is equivalent to it by conversion. The mood itself is valid as one can easily show by formalizing it.

E/ *al-Qiyās*, section vi.5

In the following section (section 5), he studies some other kinds of arguments where one categorical proposition is the antecedent of the conditional one. In these moods, the proposition "H is Z" is the consequent of the conditional proposition, while it was the antecedent of the conditional in the moods of the previous chapter. Avicenna also says that a further condition must be added in the moods presented in this section, which is that "the antecedent [of the conditional] is true, i.e. not impossible (*lā yakūnu muḥālan*)" ([34], 337.2). We will see below that this additional condition is indispensable to validate the moods held by Avicenna.

The first mood presented is the following:

"Every C is B

And Whenever Every B is A then H is Z

Therefore It happens that when Every C is A then H is Z" ([34], 337.12–13)

In this mood, the missed premise is a *Barbara* syllogism, which is the following:

- "If Every C is B and Every B is A, then Every C is A"

If we formalize it by considering that "Every C is B": P, "Every B is A": Q, "H is Z": R, and "Every C is A": S, and after adding the missing premise, we obtain the following formula: "$\{[P \wedge (Q \supset R)] \wedge [(P \wedge Q) \supset R]\} \rightarrow (S \wedge R)$". This formula is not valid, for if Q is false, the antecedent of the whole implication would be true while its consequent would be false. This is why the condition mentioned by Avicenna is really crucial, for when taken into account, it could be rendered symbolically by adding "Q $\wedge$" to the conditional in the first element of the formula. We would then have the following formula: "$\{\{[P \wedge [Q \wedge (Q \supset R)]\} \wedge [(P \wedge Q) \supset R]\} \rightarrow (S \wedge R)$".

Thus stated, the mood is valid, which confirms Avicenna's intuitions.

The second mood is stated as follows:

"Every C is B
and Whenever No B is A then H is Z
If follows: It happens that when No C is A then H is Z" ([34], 338.3–4).

Here, the missing premise is a *Celarent*, which can be stated as follows:

"If Every C is B and No B is A then No C is A"

The condition related to the truth of the antecedent is also indispensable to validate the mood, which can be formalized as follows:

"$\{\{P \wedge [\sim Q \wedge (\sim Q \supset R)] \wedge [(P \wedge \sim Q) \supset \sim S]\} \rightarrow (\sim S \wedge R)'$"

Thus formalized, the mood is valid.

As we can see from these few examples, the missing premise is always a valid categorical mood already proved. The new moods introduced by Avicenna in these chapters are thus complex moods using both hypothetical and categorical logic. They are, though, different from the usual categorical moods and the hypothetical ones studied at the beginning of this section in that they contain implicit premises not explicitly stated in the mood but sometimes used in the proof. Without these premises, the moods as they stand are not formally valid. But with them, and sometimes with some further condition, these moods are indeed *formally valid.*

F/ *al-Qiyās*, section vi.6

Let us now turn to the last section (section 6, pp. 349–357) which presents this kind of mixed moods. In this section, the moods involved are also different in some other ways as we will see in our analysis.

These moods contain a disjunctive premise combined with one or several categorical ones, as Avicenna says at the very beginning of the section (p. 349). He even compares these moods with induction (*istiqrā*) but makes a difference between both kinds of reasoning by saying that in inductive reasonings, the "predication is not real" and "the number of premises may be incomplete," while in the present moods, "the predication is real" and "the number of premises is complete" (p. 349.6–8). He calls this kind of syllogisms "the divided syllogism" ("*al-Qiyās al-muqassam*", p. 349.8).

These moods pertain to the three classical figures too. They obey several conditions which are stated by Avicenna explicitly as follows: "The disjunctive must be affirmative, with affirmative parts, and the predicative [propositions] must be universal, with the same quality, and the conclusion is predicative" ([34], 350.1–2).

In the first figure, he presents, for instance, the following:

"Every B is either C or H or Z
and Every C and H and Z are A
Therefore Every B is A" ([34], 350.3–4).

If we formalize this mood by using the modern symbolism, we will have the following formula: $\{(x)[Bx \supset (Cx \vee Hx \vee Zx)] \wedge (x)[(Cx \vee Hx \vee Zx) \supset Ax]\} \rightarrow (x)(Bx \supset Ax)$. In other words, the mood is formalized as follows:

$(x)[Bx \supset (Cx \vee Hx \vee Zx)]$ (first premise)
$(x)[(Cx \vee Hx \vee Zx) \supset Ax]$ (second premise)
$\vdash (x)(Bx \supset Ax)$ (conclusion)

In this formula, the second premise contains inclusive disjunctions in its antecedent, which does not appear clearly in Avicenna's text. But the formula "$(x)[(Cx \vee Hx \vee Zx) \supset Ax]$" (the second premise) is indeed equivalent to that other one: "$(x)(Cx \supset Ax) \wedge (x)(Hx \supset Ax) \wedge (x)(Zx \supset Ax)$", by the laws of modern mathematical predicate logic, that is, precisely by the quantified counterpart of the following propositional law: "$[(q \vee r) \supset p] \equiv [(q \supset p) \wedge (r \supset p)]$", that we can find in B. Russell & A. N. Whitehead's *Principia Mathematica*, where it is the proposition *4.77 ([129], Section A, p. 121). So, although Avicenna did not know nor state this law, he seems to have grasped intuitively the above equivalence, for he does use "C or H or Z" as a *middle* term in his mood, thus he *equates implicitly* between "C or H or Z" with "[every] C and [every] H and [every] Z," where the conjunctions relate three different conditional propositions. This means that the consequent of the first premise, which contains disjunctions is the same as the *antecedent* of the second premise, which contains separate conjunctions in his initial formulation, which Avicenna rightly equated with inclusive disjunctions as shown above.

As a matter of fact, the mood is valid, for it has the structure of a *Barbara* syllogism with a complex middle term containing two *inclusive* disjunctions. It also shows that Avicenna uses indeed the inclusive disjunction in his hypothetical moods, given that the above syllogism would not be valid if the disjunctions were exclusive. For if the disjunction is exclusive, there is no equivalence between "$(x)[(Cx \veebar Hx \veebar Zx) \supset Ax]$" and "$(x)(Cx \supset Ax) \wedge (x)(Hx \supset Ax) \wedge (x)(Zx \supset Ax)$". Since Avicenna states explicitly his mood with conjunctions in the second premise, the middle term of his mood would not be the same term if the disjunction were exclusive in the first premise.

In the same way, the second mood he states has the structure of *Celarent* with a complex middle term. This mood is the following:

"Every B is either C or H or Z
And No C and no H and no Z is A
Therefore No B is A" ([34], 350.5–6).

Here too, the disjunction has to be inclusive in order for the whole mood to be valid, for if it were exclusive, the middle term would not be the same in both premises and the mood would not be valid.

The third and fourth moods contain a particular premise and a particular conclusion and are stated respectively as follows:

"Some B are either C or H or Z
and Every C and H and Z are A
Therefore Some B are A" ([34], 350.7-8)

"Some B are either C or H or Z
and No C and no H and no Z is A
Therefore Not every B is A" ([34], 350.9–10).

Both are valid as everyone can check by using the modern symbolism. But Avicenna says bizarrely that "if the disjunctive is particular, the predicative conclusion would not be deductible" ([34], 350.11). Since the two last moods are indeed valid, this sentence seems anachronistic and shows that something is not clear in the text.

Anyway, the moods above do not really contain hypothetical propositions, despite the presence of disjunctions in the premises, for all propositions, including those that contain disjunctions, are in fact predicative, the disjunctions being only *parts* of the propositions, namely, either their antecedent or their consequent. It also shows that there is no real separation between categorical and hypothetical logics in Avicenna's system, despite the richness of the latter.

However, Avicenna presents two other "moods" in this figure, without providing their conclusions, which tends to signify that he does not consider them valid. These two "moods", as they are stated by Avicenna, are the following:

(1) "No B is C, nor (is it) H, nor (is it) Z
and Always either A is C or H or Z" ([34], 351.1–2).
(2) "Some B is either C or H or Z
and No A is C or H or Z" ([34], 351.3–4).

The first "mood", in which conclusion is not given, seems to correspond to an analogous of *Camestres*, with a complex middle term, while the second one seems to correspond to *Festino*, with a complex middle term. In both cases, there is indeed a conclusion, which is "No B is A" for the first and "Not every B is A" for the second. If we formalize them, we obtain the following:

$$(1)\{(x)[Bx \supset (\sim Cx \wedge \sim Hx \wedge \sim Zx)] \wedge (x)[Ax \supset (Cx \vee Hx \vee Zx)]\} \rightarrow (x)(Bx \supset \sim Ax)$$

$$(2)\{(\exists x)[Bx \wedge (Cx \vee Hx \vee Zx)] \wedge (x)[Ax \supset \sim (Cx \vee Hx \vee Zx)]\} \rightarrow (\exists x)(Bx \wedge \sim Ax)$$

The first mood is indeed valid if the disjunction is *inclusive*, for it leads in that case to the conclusion, as the middle stated in the first premise is equivalent to "$\sim Cx \wedge \sim Hx \wedge \sim Zx$", which is the same as "$\sim (Cx \vee Hx \vee Zx)$", by the De Morgan's law. The latter formula is the negation of the second premise's consequent, which means that we have here a real middle term, the same for both premises, but it is affirmed in one premise and negated in the other one. So there is no reason why the mood would be invalid, since in that case it has the structure of *Camestres*, with a complex middle.

However, if the disjunction is exclusive, then "$\sim Cx \wedge \sim Hx \wedge \sim Zx$" is *not* equivalent to "$\sim (Cx \veebar Hx \veebar Zx)$", which means that the middle is no more the same, and the structure would not be that of *Camestres*, because of that difference. Since Avicenna uses the word "Always" in front of the disjunctive premise, maybe he intends the disjunction to be exclusive. This would be the reason why he does not consider the whole mood as valid. But we have to say that even in that case, the mood *is valid*, despite the fact that the middle is not shared by both premises.

As to the second mood, it is also valid and corresponds to *Festino*, with a complex middle.

So it seems that Avicenna does not always distinguish with enough clarity the validity or invalidity of the moods he presents. He may consider some moods as invalid, while they are valid and vice versa. This shows that he does not have an efficient method that makes him determine the validity of the moods without errors. His method is more intuitive than mechanical. Besides that, he tends to always rely on the results of categorical logic and apply them as much as possible in his hypothetical logic.

In the third figure, Avicenna provides an interesting mood apparently corresponding to *Darapti*, which he states as follows:

"Always either C is B or D is B
and Every C and every D is H
Therefore Some B is H" ([34], 351.10–11).

In this mood, the first premise stated (the minor premise) is really a hypothetical proposition. But the mood is not valid if the two predicative propositions in the second premise (the major premise) do not have an import, i.e., do not presuppose the existence of their subjects (namely, C and D). For the formula corresponding to this mood is the following:

$$\{[(C \text{ is } B) \vee (D \text{ is } B)] \wedge [(x)(Cx \supset Hx) \wedge (x)(Dx \supset Hx)]\} \rightarrow (\exists x)(Bx \wedge Hx).$$

This formula is valid only if the major premise "Every C and every D is H" presupposes the existence of both C and D, whether the disjunction of the minor premise is exclusive or inclusive. But the mood presupposes also some implicit relations between the two premises, for the middle is not the same in both premises. But this mood seems to be dividable into two parts as follows:

'Every C is H'	'Every D is H'
and 'C is B'	And 'D is B'
Therefore 'Some B is H' (by *Darapti*)	Therefore 'Some B is H' (by *Darapti*)

In both cases, the conclusion "Some B is H" is deductible by the categorical *Darapti*, provided the singular premises "C is B" and "D is B" are treated as universals. But this conclusion is deduced twice by the two parts of the mood; one of these parts seems therefore superfluous.

The next condition considered by Avicenna is the one where "the minor premise is predicative" ([34], 352.16), while the major premise is hypothetical. The first mood of this kind that he presents is the following:

"Every C is H and every D is Z
and Either every H is A or every Z is A
Therefore Either C is A or (when every H is A) every Z is A" ([34], 352.16–313.1).

This mood contains at least two anomalies, which could be due to the editing: first in the conclusion, the first proposition is "C is A," while it should be "every C is D," given the universality of the premises; second the second part of the conclusion is "every Z is A," which is exactly equivalent to the second disjunct of the major premise. This disjunct should rather be "every D is A," if we take into account the fact that it follows by *Barbara* from "every D is Z" and "every Z is A."

Anyway, if we formalize it as it is written, we have the following formula (where Ca and Aa are singular propositions):

$$\{[(x)(Cx \supset Hx) \wedge (x)(Dx \supset Zx)] \wedge [(x)(Hx \supset Ax) \vee (x)(Zx \supset Ax)]\} \rightarrow [(Ca \supset Aa) \vee (x)(Zx \supset Ax)].$$

This formula is valid if the two disjunctions (in the major premise and the conclusion) are inclusive; it is not valid, however, if the disjunction of the major premise is inclusive while the disjunction of the conclusion is exclusive.

When the condition is that "the predicative is the minor, and the disjunctive is the major" ([34], 353.7), the first mood is the following:

"Every C is B
and Always Every B is either H or Z
Therefore Every C is either H or Z" ([34], 353.8–9).

This mood is obviously valid for it is a *Barbara* mood, whose major term contains a disjunction. However, as many other moods already presented by Avicenna, its propositions are not really hypothetical, despite the presence of the disjunctions; they are rather predicative propositions with some complex (disjunctive) elements.

Some other moods seem to be invalid, or at least to lack some further premise(s) in order to be validated. An example is the following mood, presented at page 355:

"Whenever C is B, then H is Z
and Every Z is either D or A
Therefore Whenever C is B, then every H is either D or A" ([34], 355.12–13)

The (first) missing premise in this mood seems to be:
'If H is Z then every Z is either D or A',
From which we can deduce by *Barbara*
'Whenever C is B, then every Z is either D or A'
Then from this last premise, and
'If every Z is either D or A, then every H is either D or A'
We may deduce, by *Barbara* too:
'Whenever C is B then every H is either D or A'

which is the conclusion of the mood provided by Avicenna. So the mood is valid provided these missing premises are added. As it is presented by Avicenna, its validity does not appear clearly. But when one adds the missing premises, its validity becomes clear. So Avicenna's method of proof seems sometimes close to the mathematical methods, since it involves some implicit steps, which are not found in the same way in Aristotle's proofs or his followers' ones. This complexity has been noted by some authors, such as Wilfrid Hodges, who studies the question in a recent article entitled "Proofs as cognitive or computational: Ibn Sīnā's innovations" (see [87]). He even goes further than what we said above, by stressing the novel character of some proofs used by Avicenna, which do not contain in their premises evident support to the conclusion but are nevertheless conclusive and valid because the conclusion of such arguments cannot fail to be true. According to Wilfrid Hodges, this feature makes Avicenna's method much more modern than all the proofs used by the authors who preceded him.

Let us now turn to the section on *reductio ad absurdum*, as it is presented and developed by the three authors. We will start by al-Fārābī's analysis, then we will present Avicenna's analysis, and we will end by Averroes' analysis of this kind of proof.

5.2 The *Reductio Ad Absurdum* (*Qiyās al-Khalf*)

5.2.1 The Reductio Ad Absurdum *in al-Fārābī's Frame*

Al-Fārābī defines this kind of reasoning as follows in *Kitāb al-Qiyās*:

> "If any of the two premises of a syllogism is clearly true but the other one is doubtful, i.e. its truth or falsity are not known, and if this [combination] leads to a conclusion which is clearly false and absurd (*dāhiratu al kadhib wa al imtinā'*), this syllogism is called the *reductio ad absurdum* (*qiyās al khalf*), and it is used to prove the truth of the contradictory of the doubtful premise" ([10], 34).

To prove P, one supposes not P (the doubtful premise). If together with Q (a clearly true proposition), it leads to a clear falsehood, then not P is false, and therefore P is true.

So this kind of syllogism relies on suppositions or assumptions stated at the beginning of the argument, together with an obviously true proposition. If from these two propositions, what follows is clearly false, then, since the obvious proposition cannot in any case be false, it is the doubtful one that is false. Therefore, its contradictory will be true. This argument relies implicitly on some logical principles, such as the principle of excluded middle and the principle of non-contradiction, although al-Fārābī does not state them explicitly in *al-Qiyās*, since according to the principle of non-contradiction, the conjunction of "p" and "not p" is always false, and hence its negation is always true; while according to the principle of excluded middle, either "p" is true or "not p" is true, so that when one of them is true, the other one will be false and vice versa. Now if "not p" is the false conclusion deduced, then its contradictory, i.e., "not not p," will be true (by the definition of the negation); therefore, "p" will also be true, since it negates a falsehood. This reasoning assumes then implicitly the law of double negation as well.

Whether these principles are explicitly stated or not, they are strongly assumed in every proof by *reductio ad absurdum*, since they are at the heart of this kind of argument, which deduces from the falsity of a conclusion, the truth of its contradictory.

Al-Fārābī himself argues as follows:

> "If the conclusion is false, this means that something is false in the syllogism … either in the two premises or in one of them, but one of the premises is clearly true, and it is not possible that the falsity of the conclusion is due to the truth of that premise; it is rather due to the other doubtful one. And what implies a falsehood is itself false; therefore the doubtful premise is false; consequently, its contradictory is true. And this is what we wanted to prove from the beginning" ([10], 34)

We can summarize what al-Fārābī states in this passage as follows: To prove a proposition (say P) by *reductio ad absurdum*, we state its contradictory (= not P) by supposing that it is true. This contradictory, however, is not *known* to be true; it remains doubtful (*mashkūkun fīhi*). Then we add to this doubtful premise a clearly true premise (= Q). If this leads to a conclusion (= not R) which is clearly false (impossible, absurd), then either both premises or one of them must be false. Since the obvious premise cannot be false, it must be the other doubtful one. Given that this doubtful premise is false, its contradictory (= not not P (= P)) will be true, which is what we wanted to prove from the beginning.

This reasoning is supported by the following principle: "What implies a falsehood is itself false," together with the logical principles of excluded middle and of non-contradiction and the law of double negation already evoked above. For these principles, justify all the deductions made throughout the whole argument, even if the latter three propositional principles are not explicitly stated by al-Fārābī.

We can add that the deduction of the conclusion in that proof is syllogistic, for it is made by means of one of the valid syllogistic moods. Also, the conclusion (not R) is false but it does not necessarily express an explicit contradiction, although it is said by al-Fārābī to be "impossible (*mumtana'*) and absurd (*muḥāl*)." This impossibility is not a formal one, though. It is more related to the meaning of the proposition and may be considered as a semantic or analytic falsity, given the very meanings of the subject and the predicate, as appears in the example provided by al-Fārābī that we analyze below.

This makes al-Fārābī's account different from the modern one, for instance, from what we find in Russell and Whitehead's *Principia Mathematica*, who define the *reductio ad absurdum* by the following proposition: "* 2.01. $\vdash$: (p $\supset$ ~p) $\supset$ ~p". This proposition is called explicitly by Russell and Whitehead "the principle of the *reductio ad absurdum*" and commented as follows: "If p implies its *own* falsehood, then p is false" ([129], 100, italics added). This modern wording is more formal than al-Fārābi's because Russell and Whitehead talk about "…its *own* falsehood…" which results from the negation and appears clearly, whatever meaning the proposition can have. While al-Fārābī's conclusion in his argument by *reductio ad absurdum* is not the contradictory of the premise P itself, it is a different proposition (= R), whose falsity depends on its own meaning, not on the basis of the negation alone or on its own structure. In the contemporary account of what is now called RAA, the account is even more formal, for the conclusion deduced is an explicit and formal contradiction expressed by "P and not P," for instance, and formalized by "$\perp$".

Let us now analyze the example provided by al-Fārābī to illustrate the *reductio ad absurdum*. This example is the following:

To prove "Every man is sensitive" by *reductio ad absurdum*, we suppose that this proposition is false; hence, we start the argument by its contradictory, i.e., "Not every man is sensitive," which we suppose to be true, without being certain of its truth. Then we add an obviously true proposition, e.g., "Every man is an animal." This leads to the following syllogism:

- Not every man is sensitive,
- Every man is an animal, and
- Therefore not every animal is sensitive (by *Bocardo*).

But this conclusion is obviously false. Since its falsity cannot come from "every man is an animal" which is evidently true, it has to come from the other premise, which was supposed to be true, but appears now to be clearly false, since it leads to a falsehood. Therefore, its contradictory, namely, "Every man is sensitive," is true ([10], p. 34).

As we can see, al-Fārābī concludes to the truth of "every man is sensitive" from the falsity of "not every animal is sensitive" deduced by a valid syllogistic mood from the two premises stated. He does not add "but 'every animal is sensitive'", which together with "not every animal is sensitive" would express an explicit contradiction. It seems then that according to him, the argument does not require adding explicitly this supplementary premise, although it is strongly presupposed. The simple falsity of its conclusion is sufficient to deduce the falsity of the doubtful premise and consequently the truth of its contradictory.

According to Joep Lameer, the *reductio ad absurdum* in al-Fārābī's frame is a mixed argument which combines the hypothetical arguments and the syllogistic moods. He says: "The *qiyās al khalf* is conceived of as a complex argument, involving both a syllogistical and a hypothetical part, making it the Arabic counterpart of Aristotle's *deductio per impossibile*" ([108], 60). The propositional principles make the *reductio by absurdum* a kind of hypothetical argument rather than a pure syllogistic one, even if some syllogistic moods are used in the deduction.

As a matter of fact, al-Fārābī does use a categorical mood (*Bocardo*) in his deduction above, but his use of the hypothetical mood(s) is more informal. For starting from the falsity of the conclusion of *Bocardo* above, he says:

But "Not every animal is sensitive" is false.

Therefore, one of the premises must be false (for "what implies a falsehood must be false")

But the premise "Every human is an animal" is evidently true.

Therefore, the second premise "Not every human is sensitive" is false.

Consequently its contradictory, namely, "Every human is sensitive" is true.

This reasoning is hypothetical, for it relies on *Modus Tollens* when, from the falsity of the conclusion, it infers the falsity of one of the premises (given that the other one is true). What underlies the deduction of the last proposition is the principle of excluded middle, according to which either p is true or its contradictory negation is true (but not both). But al-Fārābī does not say explicitly that he is using *Modus Tollens*, although he had studied this hypothetical mood and the other ones in the chapter just preceding the one about *Qiyās al-khalf*. This means that his use of the hypothetical mood(s) remains implicit, which will be criticized afterward by Avicenna, who reports the whole reasoning and blames it by saying that it "hides (*iḍmār*) implicit syllogisms that were not explicitly stated (*lam yuṣarraḥ bihā*)" ([34], 410).

Now what is Avicenna's account of this same argument? This will be analyzed in the next subsection.

5.2.2 *The* Reductio Ad Absurdum *in Avicenna's Frame*

Avicenna analyzes the *qiyās al-khalf* in *al-Qiyās* (pp. 408–411) and *al-Ishārāt* (pp. 453–454) and talks about it in *al-Najāt* too (pp. 55–56). His opinion about this kind of proof is almost the same in all the treatises but the analysis that he provides in *al-Qiyās* is more developed.

In *al-Qiyās*, Avicenna says that the *reductio ad absurdum* comprises "only two hypothetical syllogisms" ([34], 408.1), even if the conclusion may be predicative. The first hypothetical syllogism is "*iqtirānī*", while the second one is "*istithnā'ī*". The "*istithnā'ī*" syllogisms correspond to the usual Stoic kind of syllogism, while the "*iqtirānī*" ones are parallel to the usual categorical syllogisms when they are simple, as we saw above.

How does Avicenna analyze this argument? In *al-Qiyās*, he presents the *reductio ad absurdum* in a way that he considers as more precise and complete than the usual one. According to him, if the aim is to prove "Not Every C is B," then

> "We say: If our sentence: "Not every C is B" is false, it follows that "Every C is B", and we add a true premise to it, that is, "Every B is A". Then from the "recombinant" that we counted as hypothetical, what follows is: If our sentence "Not Every C is B" is false, then "Every C is A". Then we say: But "Not every C is A", for it is absurd and impossible. In this way, the contradictory of the consequent is detached, which leads to the contradictory of the antecedent, i.e. "[Not] Every C is B"" ([34], 408.12–409.2).

We can write these steps as follows:

1. If "Not Every C is B" is false, then Every C is B.
2. Every B is A (true premise).
3. If [it is not the case that] "Not every C is B" then Every C is A.
4. But Not Every C is A (true because "Every C is A" is false).
5. Therefore [Not] Every C is B (from 3, 4, by *Modus tollens*).

In this proof, the *iqtirānī* syllogism is represented by premises 1, 2, and 3, while the *istithnā'ī* syllogism is expressed by premises 3, 4, and 5. The first syllogism says what follows:

If Not not every C is B, then Every C is B
and Every B is A
Therefore If Not not every C is B, then Every C is A.

The second one says what follows:

If Not not Every C is B, then Every C is A
But Not every C is A
Therefore Not every C is B.

However, the first syllogism is not formally valid. It becomes formally valid only when one adds the implicit missing premise, which is the following *Barbara* syllogism: "If Every C is B and Every B is A, then Every C is A," for it is by *Barbara* that one can deduce "Every C is A" from both "Every C is B" and "Every B is A." In his presentation of the whole proof, Avicenna did not add this implicit premise, maybe because he thought that it is so obvious that it does not need to be added. Now the missing premise is the categorical mood *Barbara*. So Avicenna's claim that the *reductio ad absurdum* contains *only two hypothetical* syllogisms is not entirely right, since the missing premise is a categorical syllogistic mood, not a hypothetical one.

The whole proof should then be stated as follows:

1. If Not not every C is B, then Every C is B (assumption).
2. Every B is A (true premise).
3. Whenever Every C is B and Every B is A, then Every C is A (*Barbara*).
4. If Not not every C is B, then Every C is A (1, 3).
5. But Not every C is A (true premise).
6. Therefore Not every C is B (from 4, 5 by *Modus Tollens*).

Even thus stated, the proof raises a problem related to step 4, which seems not to be justified by an explicit rule. This problem has been examined by Wilfrid Hodges in [85]. According to this author, the justification of step 4 is the following principle: "(12) $\eta, \psi \vdash \theta \Rightarrow \delta(\eta), \psi \vdash \delta(\theta)$" ([85], p. 9), for he says what follows: "the principle (12) holds if the inference $\eta, \psi \vdash \theta$ is any of the standard Aristotelian categorical moods and δ is either of the following two operations:

$$\delta(\psi) = \forall t\ (\phi \to \psi)$$

(13) $\delta(\psi) = \exists t\ (\phi \wedge \psi)$" ([85], 9. See pp. 9–11 for the whole developed analysis).

Anyway, the whole derivation is indeed valid if one formalizes it in the language of predicate logic and checks its validity by a classical truth table. It is even valid if one uses the language of propositional logic and checks it in the same way. For let "Every C is B" = α, "Every B is A" = β, and "Every C is A" = γ, then the formulas are the following:

1. $[(\sim\sim\alpha \supset \alpha) \wedge \beta] \to (\sim\sim\alpha \supset \gamma)$ (the *iqtirānī* syllogism).
2. - $(\sim\sim\alpha \supset \gamma)$
   - $\sim\gamma$

   $\therefore \sim\alpha$ (by *Modus Tollens*) (the *istithnā'ī* syllogism).

The second syllogism does not raise any problem because it is formally valid, but the first one cannot be considered as formally valid, *as it is stated* here. For without the missing premise, one cannot deduce "$\sim\sim\alpha \supset \gamma$" from the premises provided, given that they do not contain γ at all. This first syllogism becomes valid *only when* one adds the missing premise, which leads to the following formula:

$$\{[(\sim\sim\alpha \supset \alpha) \wedge \beta] \wedge [(\alpha \wedge \beta) \supset \gamma]\} \to (\sim\sim\alpha \supset \gamma).$$

However, we have to note that the *whole* proof *is* formally valid *with or without* the implicit premise, because of the validity of the last syllogism by *Modus Tollens* and the rule of monotony[10] which says that when an inference is valid [here the *Modus Tollens*], whatever is added to its premises does not alter its validity.

Of course, one cannot attribute to Avicenna these findings of contemporary logic, but we can see that his intuitions are confirmed by these contemporary results, at least with regard to the inferences that he is considering.[11]

This presentation of the proof is very different from al-Fārābī's one and from the usual presentation, which deduces a contradiction from the assumption and a true premise and proves the truth of the contradictory of the assumption by this (indirect) way. Unlike al-Fārābī's approach and other approaches of that argument, which use, according to Avicenna, too many informal explanations and implicit principles ([34], 410), Avicenna presents his own proof by *reductio ad absurdum* as a mere sequel of steps that are deductible by means of clear syllogisms and rules already proved. He does not arrive at any contradiction, which makes his proof comparable to direct ones. His proof is valid as we saw above, but it is not really a *mere* sequel of steps duly justified, as shown by W. Hodges. Unlike what Avicenna seems to think when he says:

> "All of these kinds of mutilation, and [these] things that are hidden and not explicit, lengthen the discussion but give us no new information. [By contrast] the account we have given is *exactly* the absurdity syllogism itself, *no more no less*." ([34], 410.9–11, translation W. Hodges, emphasis added)

there are some missing and implicit things in the proof by *reductio ad absurdum* as it is presented by Avicenna. Wilfrid Hodges, who cites this passage, says that Avicenna's claim above is "exaggerated", but contains "a grain of truth" ([85], 7). Anyway, it is certainly an original way of presenting the *reductio ad absurdum*, which when completed is perfectly acceptable.

Now as we saw in Sect. 3.3.2, Avicenna adds a proof by *reductio ad absurdum* to the other proofs in his justification of almost all the categorical syllogistic moods, unlike al-Fārābī who never uses it in his own proofs of the categorical moods. But his proofs by *reductio ad absurdum* of the categorical moods are presented in the usual way, *not* in the novel way that we just reported. For he always deduces a contradiction from the assumption and one of the premises of the mood he is proving, and deduces from this contradiction that the assumption is false, therefore its contradictory (i.e., the other premise of the mood) is true. For instance, *Cesare*, which says "Every C is B and No A is B, therefore No C is A," is proved by *reductio ad absurdum* as follows: "If the conclusion is false, then let us assume (*fa li-yakun*) Some C is A, but we had No A is B; it follows [by *Ferio*] of the first figure that Not every C is B; but we had Every C is B; which is absurd" ([34],

[10] This rule is stated as follows (for First-Order Logic) by contemporary logicians: "If a sentence φ can be inferred in FOL from a set Γ of premises, then it can also be inferred from any set Δ of premises containing Γ as a subset" (See [20]).

[11] Some followers of Avicenna have also been shown by Wilfrid Hodges to use modern "Model-Theoretic" methods. See [88] for more details.

114.8–10)[12]. In this proof, he starts from the assumption that the conclusion is false; therefore, its contradictory is true. But if it is true, and we add the negative premise, which is true, this leads to "Not every C is B." But this proposition contradicts the other premise "Every C is B." Given this contradiction, the assumption is false; therefore, its contradictory, i.e., the conclusion of *Cesare*, is true. This can be shown as follows:

If 'No C is A' (as long as it C) is false then Some C is A (as long as it is C).
So let us assume 1. Some C is A (as long as it is C)
But 2. No A is B (as long as it is A) (true premise)
Therefore 3. Not Every C is A (as long as it is C) (from 1, 2, by *Ferio*)
But 4. Every C is A (as long as it is C) (minor premise)

5. $\bot$ (3,4)
Therefore 6. No C is A (as long as it C)

So he also admits the usual proof by *reductio ad absurdum* (which he states too in the chapter devoted to the analysis of *Qiyās al-khalf*) in his logical practice. But he probably thinks that his own analysis of the proof is more adequate or more acceptable. According to Wilfrid Hodges, the reason of this preference might be that inferences involving explicit contradictions, such as "(10) $\chi, \neg\chi \vdash \bot$", are seen by Avicenna as almost unnatural, for he says "At *Qiyās*, (Ibn Sīnā 1964) 547.13f he claims that the *bāl* is incapable of accepting inputs of the form $\chi, \neg\chi$. We needn't read him as saying that (10) is an invalid inference. More likely his point is that there doesn't seem to be any way of reading (10) as a real-life inference." ([85], 8). Whether this interpretation of Avicenna's text is correct or not, it is true that Avicenna does not rely on an explicit contradiction in his own account of the *reductio ad absurdum*.

Let us now turn to Averroes' account of that same proof.

5.2.3 The Reductio Ad Absurdum *in Averroes' Frame*

As we already said, Averroes is much more Aristotelian than Avicenna, for his aim is primarily to be as faithful as possible to Aristotle and to return back to the fundamentals of the Aristotelian doctrine, without changing it by the various additions introduced by his followers. But how does he analyze the proof by *reductio ad absurdum*?

[12]Avicenna adds the conditions "as long as it is C" and "as long as it is A" to the propositions of this mood in his full analysis of the proof (pp. 114–115), but this does not alter the whole structure of the proof, which is the usual presentation of the *reductio ad absurdum*.

In *Kitāb al-Qiyās*, he says that the *reductio ad absurdum* contains both a categorical syllogism and a hypothetical one ([26], 234). The categorical syllogism shows the impossibility that ensues from a given assumption, while the hypothetical syllogism leads to the conclusion that we wanted to prove (*maṭlūb*) ([26], 234). His account is different from both Avicenna's and al-Fārābī's accounts, for he starts by a disjunctive proposition to finally prove one of the disjuncts by proving that the other one is impossible. The example starts with the following disjunction:

1. The diagonal of the square is either congruent with its side or unequal to it.
Then he supposes that the diagonal is congruent with the side.
2. If the diagonal is congruent with the side, then the ratio between the square of the diagonal and the square of the side is the same as the ratio of *one* squared number and *one* other squared number.
To prove this consequence, he uses the categorical *Barbara*, stated as follows:

3- The ratio of c^2 and a^2 is the ratio of *one* squared number and *one* squared number
4- The ratio of the square of the diagonal and the square of the side is the ratio of c^2 and a^2
5- Therefore the ratio of the square of the diagonal and the square of the side is the ratio of *one* squared number and *one* squared number
6- But the ratio of the square of the diagonal and the square of the side is the ratio of *one* squared number and the *sum* of *two* squared numbers (by the *Pythagorean Theorem*: $c^2 = a^2 + b^2$)
[c = the length of the diagonal, a = the length of one side, b = the length of the other side. In the case of the geometrical square, a = b]
7- Therefore the ratio of the square of the diagonal and the square of the side is *not* the ratio of *one* squared number and *one* other squared number (From 6).

$\perp$ (From 5 and 7)
[8- So the diagonal is not congruent with the side (From 2 and 7, by *Modus Tollens*)]*

Then returning back to premise 1, he deduces its second disjunct as follows:
1. Either the diagonal is congruent with the side or it is unequal to it (Premise)
9. But it is not congruent with the side (by 8 above)
10. Therefore it is unequal to the side (by *Disjunctive Syllogism*) ([26], 234).

We have to note, though, that the proof is not presented in Averroes' text exactly as it is presented here. For instance, the *Modus Tollens* is not explicitly stated or evoked. Averroes just says that the conclusion of the *Barbara* syllogism above shows that the diagonal "is not congruent [with the side]" ([26], 234.19). But as we can see, and unlike what he explicitly says, the proof uses not only *one* hypothetical syllogism (the *Disjunctive Syllogism* explicitly evoked) but *two* hypothetical syllogisms, given that the *Modus Tollens* is also (implicitly) used.

As to the assumption by which the proof starts, it comes here from the disjunctive premise used at the very beginning of the proof. This makes Averroes' proof different from both al-Fārābī's proof and Avicenna's one, for these two authors do not start by a disjunction; they only use an assumption.

In addition, the proof leads to a real contradiction and deduces the falsity of the assumption starting from this absurdity. It is thus very close to the usual proof by *reductio ad absurdum*, which leads to a contradiction too.

Averroes provides another example, where the *reductio ad absurdum* leads also to a clear contradiction. This example is the following:

If we want to show that "No H is A" starting from premises "Every A is B" and "No H is B," by means of the *reductio ad absurdum* we proceed as follows:

"No H is A otherwise let us assume (*fa li-takun*) Some H is A" ([26], 254)

1 Some H is A (Assumption)
2 But we had: Every A is B (true premise)
3 Therefore: Some H is B (from 1, 2, by *Darii*)
4 But No H is B (true premise)

---

5 $\perp$ (3, 4) ([26], 254)
6 Therefore, Not (Some H is A) (the assumption is false given $\perp$ in 5)
7 Consequently No H is A (by definition, from 6)

This proof is a derivation starting from an explicit assumption, expressed by "let us assume (*li-takun*)." It arrives at an explicit contradiction and infers from it the contradictory of the assumption, then its equivalent "No H is A," which was the conclusion (*maṭlūb*) to be proved. The last two steps (6 and 7) are not explicit in the text, but they are strongly presupposed.

It is a classical proof, similar to those used in categorical logic. It is different from both al-Fārābī's proof and Avicenna's one. For al-Fārābī does not state an explicit contradiction in his proof, while Avicenna proceeds in a radical different way by using rather the hypothetical moods such as the *Modus Tollens*, and avoiding the explicit contradictions in his analysis of *Qiyās al-khalf*, although he does use them in his proofs of the categorical moods.

However, we can note that Averroes uses quite explicitly the word "assumption", which was not the case with al-Fārābī, and his account is more formal than al-Fārābī's account. His proof is then closer to the classical account of RAA, although he uses mainly the categorical moods to make his derivations.

5.3 Conclusion

The three hypothetical logics appear to be very different, although we can note some common features. For as we have seen, al-Fārābī presents the classical indemonstrables and some variants, without developing a whole system, unlike

Avicenna who presents a much elaborate system where the simplest moods are parallel to the categorical ones, but the more complex ones are entirely new and different both from the Aristotelian moods and the Stoic ones.

As to Averroes, he is the one who gives the less interest to hypothetical logic, for he devotes only a few pages to its study and presents only the very basic hypothetical moods. This can be seen as a consequence of his general attitude toward Aristotelian logic, which is an explicit and radical defense of Aristotle against his followers and all those who presented altered commentaries of his texts. On the contrary, Avicenna appears to be the less faithful to Aristotle and the most innovative in this respect, while al-Fārābī remains Aristotelian but introduces some precisions and concepts that are not to be found in Aristotle's text.

As a consequence, and not surprisingly, Avicenna's system appears to be the most original and developed, although it seems at first sight to contain some errors, for all the disjunctive propositions are expressed by Avicenna in ordinary language, and some of them, in particular $\mathbf{E_D}$ and $\mathbf{I_D}$, need to be interpreted in two ways in order to validate the moods held by Avicenna. But once one formalizes them, one can choose between the two distinct interpretations the one which validates one or other of the moods held by Avicenna. Some of these moods are validated by the first interpretation while some others are validated by the second one. All in all, we can say that the system is coherent and very rich, even if in some cases the text needs some clarification. The whole system is not only very original but also very rich since it uses the tools of both categorical logic and hypothetical logic. It can be very nicely formalized by the symbolism of modern logic which is able to tell exactly which moods are valid and under which conditions.

In other complex moods, some hidden and implicit premises are presupposed that Avicenna seems to consider as so obvious that one does not need to state them, but which must be stated explicitly in order for the whole mood to be formally valid. These moods show that Avicenna, unlike al-Fārābī and Averroes, uses quite explicitly both the inclusive disjunction and the exclusive one. While the two other authors focus mainly on the exclusive disjunction, and say that the disjunction in general expresses some kind of conflict even when it is not exactly exclusive, i.e., when it is rendered by "not (P and Q)". However, in all the theories considered, even Avicenna's one, the disjunction remains non-truth-functional. It is an intensional operator in all three theories, even if its different variants are not exactly the same.

Another difference between the three authors is related to the conditional. In all three theories, the conditional is also an intensional operator, for the meanings of its antecedent and its consequent are always taken into account. However, al-Fārābī uses sometimes a biconditional in some of his hypothetical moods, which is criticized by Avicenna who says that the antecedent of the conditional should always precede its consequent, for this is the right structure of the conditional and the way by which one can recognize it. In this respect, Avicenna's theory seems more formal than al-Fārābī's, for the notion of form is given much importance, both with regard to the propositions and with regard to the inferences (on the notion of form, see [66]).

As to the proof by *reductio ad absurdum*, it is also interpreted in different ways by the three authors as we just saw. Al-Fārābī's account is the less formal one, while Avicenna's account is the most innovative since it is different from both other accounts and from the modern one, which does rely on an explicit contradiction in RAA. Averroes' account is classical, more formal than al-Fārābī's and contains some modern features, for instance, the explicit use of the word "assumption".

References

10. Al-Fārābī, Abū Naṣr. 1986. Kitāb al-Qiyās. In *Al-Manṭiq 'inda al-Fārābī*, vol. 2, ed. Rafik Al Ajam, 11–64. Beirut: Dar el Machriq.
13. Al-Fārābī, Abū Naṣr. 1988. Al-Maqūlāt. In *Al-Manṭiqiyāt li-al-Fārābi*, vol. 1, texts published by Mohamed Teki Danesh Pazuh, Edition Qom, 41–82.
15. Al-Fārābī, Abū Naṣr. 1988. Kitāb al Qiyās. In *Al-Manṭiqiyāt li-al-Fārābi*, vol. 1, texts published by Mohamed Teki Danesh Pazuh, Edition Qom, 115–151.
16. Al-Fārābī, Abū Naṣr. 1988. Al-Qiyās al-Ṣaghīr. In *Al-Manṭiqiyāt li-al-Fārābi*, vol. 1, texts published by Mohamed Teki Danesh Pazuh, Edition Qom, 152–194.
20. Antonelli, Aldo. Non-monotonic logic. In *Stanford Encyclopedia of Philosophy*, ed. E.N. Zalta. http://plato.stanford.edu/entries/logic-nonmonotonic/.
24. Aristotle. 1991. Prior Analytics. In *The complete works of Aristotle*, the Revised Oxford Edition, ed. Jonathan Barnes, vol. 1.
26. Averroes. 1982. *Talkhīṣ Manṭiq Arisṭu* (*Paraphrase de la logique d'Aristote*), volume 1: *Kitāb Al-Maqūlāt*, 3–77, *Kitāb al-'Ibāra*, 81–141, *Kitāb al-Qiyās*, 143–366, ed. Gérard Jehamy, Manshūrāt al-Jāmi a al-lubnānīya, al-Maktaba al-sharqiyya. Beirut.
34. Avicenna. 1964. *Al-Shifā', al-Manṭiq* 4: *al-Qiyās*, ed. S. Zayed, rev. and intro. I. Madkour. Cairo.
36. Avicenna. 1971. *Al-Ishārāt wa l–tanbīhāt, with the commentary of N. Ṭūsi*, intro. Dr. Seliman Donya, Part 1, 3rd ed. Cairo: Dar al Ma'arif.
45. Bobzien, Susanne. 2006. Ancient Logic. In *Stanford encyclopedia of philosophy*, ed. Edward N. Zalta. http://plato.stanford.edu/entries/logic-ancient/.
52. Chatti, Saloua. 2014. Syncategoremata in Arabic logic, al-Fārābi and Avicenna. *History and Philosophy of Logic* 35 (2): 167–197.
57. Chatti, Saloua. 2016. Avicenna (Ibn Sīnā): Logic. *Encyclopedia of logic, internet Encyclopædia of philosophy*. College Publications. www.iep.utm.edu/av-logic/.
59. Chatti, Saloua. 2017. The semantics and pragmatics of the conditional in al-Farabi's and Avicenna's theories. *Studia Humana* 6 (1): 5–17.
60. Chatti, Saloua. 2019. Logical consequence in Avicenna's theory. *Logica Universalis* 13: 101–133.
66. Dutilh Novaes, Catarina. 2011. The different ways in which logic is (said to be) formal. *History and Philosophy of Logic* 32 (4): 303–332.
67. Dutilh Novaes, Catarina, and Stephen Read. 2016. *The cambridge companion to medieval logic*. Cambridge University Press.
71. El-Rouayheb, Khaled. 2010. *Relational syllogisms and the history of arabic logic, 900–1900*. Leiden: Brill.
68. Edgington, D. 2014. Conditionals. In *Stanford encyclopedia of philosophy*, ed. E.N. Zalta. http://plato.stanford.edu/entries/conditionals/.
70. El-Rouayheb, Khaled. 2009. Impossible antecedents and their consequences: Some thirteenth-century Arabic discussions. *History and Philosophy of Logic* 30 (3): 209–225.

79. Hasnawi, Ahmed, Wilfrid Hodges. 2016. Arabic logic up to Avicenna. In *The Cambridge companion to medieval logic*, ed. Catarina Dutilh Novaes and Stephen Read, 45–66. Cambridge: Cambridge University Press.
85. Hodges, Wilfrid. 2017. Ibn Sīnā on *reductio ad absurdum*. *The Review of Symbolic Logic* 10 (3): 583–601.
87. Hodges, Wilfrid. 2018. Proofs as cognitive or computational: Ibn Sīnā's innovations. *Philosophy and Technology* 31 (1): 131–153.
88. Hodges, Wilfrid. 2018. Two early Arabic applications of model-theoretic consequence. *Logica Universalis* 12 (1–2): 37–54.
91. Hodges, Wilfrid. 2017. Identifying Ibn Sīnā's hypothetical logic I: Sentences forms. Draft, Nov 2017.
92. Hodges, Wilfrid. forthcoming. Mathematical background to the logic of Avicenna. Available at http://wilfridhodges.co.uk/arabic44.pdf.
108. Lameer, Joep. 1994. *Al-Fārābī and aristotelian syllogistics; greak theory and islamic practice*. Brill Edition.
109. Largeault, Jean. 1972. *Logique mathématique, textes*. Paris: Armand Colin.
111. Łukasiewicz. 1934. Contribution à l'histoire de la logique des propositions. French translation in J. Largeault. 1972. *Logique mathématique, textes*. Paris: Armand Colin.
114. Maróth, Miklos. 1989. *Ibn Sīnā und die peripatetische "Aussagenlogik"*. Leiden: Brill.
119. Movahed, Zia. 2009. A critical examination of Ibn Sīnā's theory of the conditional syllogism, vol. 1, no. 1, published in *Sophia Perennis*. www.ensani.ir/storage/Files/20120507101758-9055-5.pdf.
125. Rescher, Nicholas. 1963. *Studies in the history of Arabic logic*. University of Pittsburg Press; Arabic trans. Mohamed Mahrān. 1992. Cairo.
128. Rescher, Nicholas. 2006. *Studies in the history of logic*. Ontos Verlag.
129. Russell, Bertrand, and Alfred NorthWhitehead. 1973. *Principia mathematica*. Paperback edition to *56. Cambridge University Press.
132. Shehaby Nabil. 1973. *The propositional logic of Avicenna. A translation from al-Shifā al-Qiyās.* Dordrecht Holland: Kluwer, D. Reidel.
134. Street, Tony. 1995. Ṭūsī on Avicenna's logical connectives. *History and Philosophy of Logic* 16: 257–268.
137. Street, Tony. 2004. Arabic Logic. In *Handbook of the history of logic*, vol. 1, ed. Dov Gabbay and John Woods. Elsevier, BV.
142. Strobino, Riccardo. 2018. Ibn Sīnā's Logic. *The Stanford encyclopedia of philosophy*. https://plato.stanford.edu/archives/fall2018/entries/ibn-sina-logic.

Chapter 6
General Conclusion

The study of the three systems shows many differences between the three authors considered. These differences concern their categorical logics as well as their modal logics and their hypothetical logics. Let us consider these different systems one by one.

In categorical logic, al-Fārābī seems to be faithful to Aristotle in that he admits only three figures. He expresses the syllogisms in two different ways and provides proofs for both. He also starts every syllogistic mood by the minor premise introducing by doing so a kind of tradition in Arabic logic. He excludes the invalid syllogistic moods by making calculations and following the Philoponus rules rather than by providing counterexamples as was the case for Aristotle. He also introduces some novelties that are not present in Aristotle's theory or even in the theories of his Greek commentators. For instance, he proves *Baroco* and *Bocardo* by *ekthesis*, while these two moods are proved by *reductio ad absurdum* in Aristotle's theory and in his Greek followers. He is thus the first logician to introduce this kind of proofs for *Baroco* and *Bocardo*, but Avicenna will follow him and even improve one of these proofs. The reason why he uses *ekthesis* to prove these two moods might be that he never uses *reductio ad absurdum* (*Qiyās al-khalf*) in his proofs of the syllogistic moods, as we have noted. Maybe he considers that proof as an indirect one and tries to avoid it as much as possible, although he devotes a (short) chapter in his hypothetical logic to analyze it.

In addition, following Alexander of Aphrodisias, he gives definitions of the minor term and the major term, which are more formal, more precise, and more adequate than Aristotle's definitions, for he defines them by considering their places in the conclusion, which remains the same in all three figures, while Aristotle defines them by considering their extensions, which are not the same in the three figures. So al-Fārābī's definitions are applicable to all figures in the same way and appear for this reason to be more uniform than Aristotle's ones. In this respect, al-Fārābī appears to depart from Aristotle and to present a view which will be endorsed afterward by his followers such as Avicenna and by Western scholars as well, since we find it as late as the seventeenth century in Port-Royal logic, for instance.

S. Chatti, *Arabic Logic from al-Fārābī to Averroes*, Studies in Universal Logic,
https://doi.org/10.1007/978-3-030-27466-5_6

His definitions of oppositions are more precise than Aristotle's definitions and are based on the nature of each proposition, whether quantified, indefinite, or singular. He thus defines contradiction, contrariety, and subcontrariety by distinguishing among matter necessity, matter impossibility, and matter possibility. These distinctions will be admitted by Avicenna and Averroes afterward and we find them also in Western medieval treatises and even in Port-Royal logic. However, he does not explicitly evoke subalternation, which will be introduced and precisely defined afterward by Avicenna, who uses the same method in defining this last opposition.

Al-Fārābī is also the first author to endorse and defend explicitly the view according to which the affirmative *quantified* propositions have an import, while the negative ones do not, as appears in his *al-Maqūlāt,* where he provides the truth values of the quantified as well as the singular and indefinite propositions which subject exists (i.e., with import) and those where the subject does not exist (i.e., without import). Although the problem of existential import is not really raised, the existence of the subject in all kinds of propositions is discussed and leads to the view which we expressed above, when we take into account the truth values provided by al-Fārābī. This view is important since it validates the third figure moods such as *Darapti* and *Felapton* together with **A**-conversion. Both **A**-conversion and the moods evoked would be invalid if the **A** propositions were without import.

Finally, al-Fārābī seems to be influenced mostly by Alexander of Aphrodisias, whom he cites quite often, since he privileges Alexander's interpretation of the Aristotelian texts.

As to Avicenna, he presents the most original theory, for he introduces many new distinctions and precisions that are not found in Aristotle or even in his commentators' texts or in al'Farabi's ones. His analysis of what he calls the absolute propositions contains new features because he introduces many conditions to these propositions, among which we find temporal conditions. He thus distinguishes between these propositions, depending on the conditions they contain. These conditions may be temporal, such as "at some times" or "at some times but not always," but they may also be "as long as it is S," "as long as S exists," or "as long as it is P." The latter is mentioned but it is not used in the syllogistic, while the conditions "as long as it is S" and "as long as S exists" are the ones that are mostly used and that validate the syllogistic moods. The reason for introducing these last conditions is that the pure absolute propositions containing "at some times" are not convertible, according to Avicenna, which makes him avoid them in his syllogistic.

The syllogistic moods are thus stated differently and more precisely in Avicenna's theory, although they are not radically different from the Aristotelian moods. He starts by stating some general rules that must be fulfilled in order for the moods of a particular figure to be valid. Some of these rules can be found in Ancient texts, but they are much more used and more common in the medieval texts and also the 17th ones, as witnessed by their presence in Port-Royal logic.[1] His method is different from that of al-Fārābī, for he states each mood in one unique way, but provides several proofs for each mood, since he uses either conversion or *ekthesis* depending on the mood, and adds a proof by *reductio ad absurdum* in each case. Like al-Fārābī, he provides a proof by *ekthesis* of the two moods *Baroco* and *Bocardo*, but he improves the *ekthetic* proof of *Baroco*. Generally speaking, he is clearly influenced by al-Fārābī in many aspects of his theory, for he appears to clarify, to develop, and to deepen many ideas that were already but briefly stated in al-Fārābī's texts.

His theory of oppositions is very rich, for he analyzes the oppositional relations between all kinds of propositions and by taking into account the conditions they contain. He states and defines precisely all four oppositional relations, namely, contradiction, contrariety, subcontrariety as in al-Fārābī's frame, but also subalternation, which he calls "*tadākhul*", and defines precisely by means of the matter modalities of the propositions.

His analysis of the relations between the propositions containing the various conditions cited in Sect. 3.2.2 leads to the admission of what he calls the *dā'ima* propositions which contain the conditions "as long as it exists" and can be either affirmative or negative. So the contradictories of the general absolute propositions containing "at some times" are the negative propositions containing "as long as it exists." While the affirmative perpetual propositions containing "as long as it exists" contradict the negative general absolute propositions containing "sometimes not." The special absolutes containing "at some times but not permanently" are contradicted by disjunctive propositions containing "either permanently or permanently not," where the permanence is understood as being "continuity as long as the essence of the individual exists." His analysis of the general absolutes and the propositions containing the condition "as long as it exists" (either affirmative or negative or containing a disjunction between an affirmative side and a negative one) makes them parallel to the modal propositions containing "possibly", "necessarily", "impossibly", and "possibly not" together with "possibly but not necessarily." The special absolutes appear to be analogue to the two-sided possible propositions, while the general absolutes are comparable to the possible ones, and the propositions containing "as long as it exists" are comparable to the necessary or impossible ones depending on their quality. In this respect, his analysis is almost similar to the modern ones, for he finds equivalences that are stated and analyzed by modern logicians. By stating the disjunctive propositions containing "either permanently or permanently not," for instance, he seems to apply implicitly the De Morgan's laws, without stating them explicitly. However, he sometimes makes some errors as we have showed in Sect. 3.2.2. These errors are due to the fact that he does not use a symbolism comparable to the modern ones and relies mainly on his intuition. Despite the presence of these errors, however, his theory remains remarkable in

[1]John N. Martin cites many medieval authors who state explicitly similar rules, for instance, William of Sherwood, Buridan, Peter of Spain. He also talks about Port-Royal logic which contains many rules of that kind (see [115], 133–154, note 14).

many aspects and far more developed than al-Fārābī's theory and even Aristotle's and his followers' ones.

His general view about logic appears to be formal for he privileges the forms and structures of the propositions and the inferences. Despite the absence of a symbolism comparable to the modern ones, he states the propositions and the inferences in a way that privileges their form over their matter.

As to Averroes, he defends a theory based on the Aristotelian text and rejects many of the improvements made by his predecessors and even the Greek commentators. For instance, he rejects the conditions added by Avicenna and those added by Theophrastus such as the condition "as long as it is S" which is widely used by Avicenna in his syllogistic. According to him, these conditions are not stated by Aristotle and they alter the meanings of the propositions by restricting their applicability.

In his defense of the first figure moods, he relies on the *dictum de omni et de nullo*, which, according to him, justifies their validity. He provides counterexamples to rule out the invalid moods as was the case with Aristotle and admits only three figures, taking the unnaturalness of the fourth figure as an argument against it.

His aim is to return back to the Aristotelian text and to defend it against all the alterations that came through the different comments made by either the Greek commentators or the Arabic ones. Because of this faithfulness to Aristotle, his system does not contain many novelties and seems to even reject any kind of novelty.

Furthermore, he follows his two predecessors in his way of presenting the oppositions between the quantified propositions, for he also talks about matter modalities and determines the truth values of the propositions by considering them. But he remains more Fārābian than Avicennan in this respect, for he talks only about the three oppositional relations evoked by al-Fārābī, namely, contradiction, contrariety, and subcontrariety and does not mention what Avicenna calls "*tadākhul*", i.e., subalternation, although this relation is clearly determined and defined by Avicenna.

Finally, although our three main authors ignore or reject the fourth figure, because of its unnaturalness, some later authors such as Ibn al-Ṣalāh (Twelfth Century) attribute it explicitly to Galen (whom Avicenna had already evoked in his *al-Qiyās*) and accept it explicitly. This author finds five moods in the fourth figure and he proves them by the usual methods and by adding another rule, namely, the transposition of premises. He even considers that this figure should be put in the second place rather than the fourth one. So the "discovery" of the fourth figure, and its attribution to Galen can be found in the texts of some Arabic logicians.

In modal logic, the same differences between the authors can be observed, for Averroes follows Aristotle's opinions and tries to prove all the Aristotelian moods by using his rules, while Avicenna presents a different system with different conversions. As to al-Fārābī, unfortunately his modal logic is very incomplete as it stands, although we can understand from what Averroes said about him that he departed from Aristotle in some points, for instance, in his interpretation of the modal *dictum de omni et de nullo*, which made him consider some imperfect moods in Aristotle's theory as justified by the *dictum*, i.e., as perfect.

His analysis of the modal propositions and their oppositional relations is very interesting, since he states the eight modal quantified propositions and some of their oppositional relations, which we can find in the medieval logic treatises. He thus states all the contradictions and some contrarieties, subcontrarieties, and even subalternations between these quantified modal propositions, which means that all the vertices and some of the oppositional relations of the modal octagon are expressed in his theory, although he does not draw any figure. We could then consider that his theory prefigures that of Buridan[2] who draws the modal octagon and gives all the oppositional relations between its vertices. Al-Fārābī expresses also the bilateral possible singular proposition but not the quantified bilateral possible ones. However, he does not provide the exact negation of the bilateral possible singular propositions, which should be expressed in terms of a disjunction, nor does he give the negations of the quantified bilateral possible ones. In his system, all the modalities are *de re*, i.e., internal. In his logic, we can find both internal (*de re*) and external (*de dicto*) formulations of the modal propositions, and some authors (Z. Movahed, for instance) even defend the view that he anticipated the Barcan and Buridan formulas.

Avicenna improves the analysis of the modal propositions, in particular, the quantified ones and he provides all the quantified modal propositions and their negations. He thus expresses all kinds of propositions in a quite systematic and precise way, and states explicitly many of their oppositional relations, the remaining ones being deducible in his frame. Although Avicenna does not draw any figure, all the relations he is talking about lead to a dodecagon (a figure of 12 vertices), which contains several squares, hexagons, and octagons,[3] and is thus more complex than Buridan's octagon(s), as shown by Chatti ([53]). Avicenna expresses these relations mostly by using external (*de dicto*) modalities, but he also uses internal (*de re*) modalities in his modal syllogistic and distinguishes explicitly between both kinds of modalities in his *al-'Ibāra*, although he does not develop at length these distinctions, as will be the case in medieval logic. However, **E** and **O** bilateral possible propositions where the possibility is *internal* are not well expressed in his frame.

His modal syllogistic is different from Aristotle's modal syllogistic, for the conversions he admits are stated and proved differently. According to him, both **A** necessary and **I** necessary lead by conversion to **I** possible (unilateral), while in Aristotle's frame, they both lead to **I** necessary. **E** necessary leads to **E** necessary, while **O** necessary is not convertible. As to the possible propositions, **A** possible and **I** possible lead to **I** possible, but when the possibility is bilateral, the conversion leads to a unilateral possibility. As to **E** possible and **O** possible, they do not convert.

---

[2]For more details about Buridan's theory of oppositions see [124] and about medieval Western logic, see [123]. For Buridan's conception of the modalities in relation to Avicenna's one, see [86].

[3]For the theory of oppositions in general and the different figures that it leads to, see [42, 43, 116], and for the differences between Avicenna and Averroes with regard to the concept of opposition see [50].

These conversions are different from the ones held by Aristotle, which are overtly criticized by Avicenna. For according to Aristotle, **E** possible converts as itself, but Avicenna provides a counterexample saying that "If possibly no man is a writer" is true, then neither "possibly no writer is a man" nor "possibly some writers are not men" is true.

In addition, the mixed moods are stated in a very original way, since the conditions added to the absolute propositions are stated too and give rise to several kinds of moods, whose premises and conclusions are stated very precisely with all the conditions that differentiate the absolute and the modal propositions from each other. Thus, we find in his modal syllogistic all the kinds of absolute propositions that he states in his categorical syllogistic, plus several kinds of modal—necessary and possible—propositions.

As a consequence, the moods proved by Avicenna are different from those proved by Aristotle, although there are some common points. Avicenna expresses the moods containing possible propositions mostly by using unilateral possibility, but he also uses bilateral possible propositions sometimes. He admits some moods where the conclusion is necessary while one of the premises is possible or absolute. In most cases, the mood leads to only one conclusion, while in Aristotle's frame, many moods lead to two different conclusions, as in the third figure, for instance, where both **LXL** and **LXX** moods are admitted.

On the one hand, Avicenna's modal logic is simpler than Aristotle's one, because the moods lead to one conclusion in most cases, not to two ones as sometimes in Aristotle's frame, but on the other hand it is also more elaborate than Aristotle's modal logic, because of the conditions introduced inside the propositions which make the Avicennan moods much more various and complex than Aristotle's ones.

However, the *negative* bilateral possible propositions are not expressed in a correct way, which makes the validity of all the moods containing them at least problematic.

This same problem can be found in Averroes' modal syllogistic, for Averroes tries to validate all the Aristotelian modal moods and uses the bilateral possible propositions very often. Unfortunately, in his frame too, these propositions are not expressed clearly, which can invalidate the moods containing them. The moods he admits in his modal syllogistic are different from those admitted by Avicenna because of the difference between the conversions of possible and necessary propositions. Averroes admits the same conversions as Aristotle and uses the same proofs. He thus validates almost the same moods as Aristotle. However, he does not distinguish clearly between internal and external formulations of the modal propositions, which sometimes leads to some errors in his proofs. Like Aristotle, he admits some moods with different conclusions, which makes his modal syllogistic more complex and problematic than Avicenna's one. His modal syllogistic is thus much closer to Aristotle's modal syllogistic than Avicenna's one and does not contain any significant improvement. This is due to the fact, already noted, that Averroes' aim is to defend Aristotle's view as faithfully as possible, while Avicenna departs sometimes overtly from Aristotle.

Regarding hypothetical logic, Averroes presents the usual Stoic moods and al-Fārābī adds some variants to these moods, while Avicenna develops much longer the matter and presents a whole system where he introduces complex moods containing both categorical propositions and hypothetical ones. His system is very original and does not have any exact equivalent in Greek or Arabic logic or even medieval logic, although it shares with Theophrastus' developments some features. But it is more developed than Theophrastus' reflections and much richer than al-Fārābī's and Averroes' reflections on the same topic. Here too, Avicenna's system is the most innovative and creative. The moods he presents are different from the usual categorical ones, not only by their complexity but also because they contain implicitly some hidden premises that must be presupposed in order for the moods to be conclusive. These hidden premises are not explicitly expressed by Avicenna, presumably because he considers them as obvious. But they are nevertheless indispensable to validate the moods and should for this reason be explicitly stated in the mood. In these moods as well as in modal and categorical logic, Avicenna "discovers" almost implicitly and intuitively some modern laws that we can find not only in De Morgan's system but also in Russell and Whitehead's *Principia Mathematica*.

We could also say that Avicenna's method of proof and his expression of the hypothetical moods prefigure some modern developments and proofs as shown by W. Hodges, for instance (see [82, 87]). This makes his theory far more developed than all those that preceded it. Nevertheless, some of his followers such as Abū al-Barakāt use almost modern (model-theoretic) methods, which have been studied by Wilfrid Hodges in [88].

Avicenna's conception of the logical constants can be clarified by taking into account the moods held valid in his hypothetical system and the way the disjunctive and the conditional hypothetical quantified propositions are expressed in these moods. As we have shown, Avicenna seems to be almost the only author to use quite explicitly the inclusive disjunction in his system, for some moods cannot be considered valid if the disjunction were exclusive. He even defines this kind of disjunction and distinguishes it explicitly from the exclusive one in his preliminary analysis of the constants used in his system. Unlike al-Fārābī and afterward Averroes, he does not systematically relate the disjunction with the notion of conflict for according to him some disjunctions do not express any kind of conflict and can contain compatible elements.

But both the disjunction and the conditional are not truth-functional in all frames, even in Avicenna's one. These constants are always defined by taking into account not only the truth values of their elements but also and above all their meanings. In this respect, all three theories are intensional.

However, Avicenna's hypothetical logic seems more formal than the two other systems, for Avicenna gives much importance to the notion of form and says that explicitly. So despite the intensional character of his definitions of the logical constants, most of the moods held in his hypothetical theory can be shown to be valid by means of extensional methods, as we have shown when we formalized them.

Finally, the three authors' accounts of the *reductio ad absurdum* is also different as we have shown, since al-Fārābī provides a very informal analysis of this kind of proof, while Averroes provides a classical account relying on an explicit contradiction to deduce the conclusion. As to Avicenna, he analyzes the proof in a way that avoids making use of an explicit contradiction to deduce the conclusion. His reformulation of the premises enables him to use explicitly the *Modus Tollens* and deduce directly the conclusion. Here too, his account is more original and more formal than those of the two other authors. But in practice, he also relies on the classical proof by *reductio ad absurdum* in his demonstrations of the categorical moods.

The proof by *reductio ad absurdum* makes use of a categorical syllogistic mood and a hypothetical one in al-Fārābī's and Averroes' accounts, but Avicenna says that it contains two *hypothetical* moods. However, our analysis showed that there are in all cases some implicit rules or moods that are used in all three accounts of the proof without being stated explicitly. In Averroes' and al-Fārābī's accounts, this missing mood is a hypothetical one while in Avicenna's account, it is a categorical syllogism.

References

42. Blanché, Robert. 1953. Sur l'opposition des concepts. *Theoria* 19: 89–130.
43. Blanché, Robert. 1969. *Structures intellectuelles. Essai sur l'organisation systématique des concepts*, Editions Vrin, Paris.
50. Chatti, Saloua. 2012. Logical oppositions in Arabic logic, Avicenna and Averroes. In *Around and beyond the square of opposition*, ed. J. Y. Béziau and D. Jacquette. Basel: Springer.
53. Chatti, Saloua. 2014. Avicenna on possibility and necessity. *History and Philosophy of Logic* 35 (4): 332–353.
82. Hodges, Wilfrid. 2010. Ibn Sina on analysis: 1. Proof search. Or: Abstract State Machines as a tool for history of logic. In *Fields of logic and computation: Essays dedicated to Yuri Gurevich on the occasion of his 70th birthday*, ed. A. Blass, N. Dershowitz, and W. Reisig. Lecture Notes in Computer Science, vol. 6300, 354–404. Heidelberg: Springer.
86. Hodges, Wilfrid, and Spencer Johnston. 2017. Medieval modalities and modern methods: Avicenna and Buridan. *If Colog Journal of Logics and Their Applications* 4 (4): 1029–1073.
87. Hodges, Wilfrid. 2018. Proofs as cognitive or computational: Ibn Sīnā's innovations. *Philosophy and Technology* 31 (1): 131–153.
88. Hodges, Wilfrid. 2018. Two early Arabic applications of model-theoretic consequence. *Logica Universalis* 12 (1–2): 37–54.
115. Martin, John N. 2013. Distributive terms, truth and the port royal logic. *History and Philosophy of Logic* 34 (2): 133–154.
116. Moretti, Alessio. 2009. *The geometry of logical opposition*. Ph.D. University of Neuchâtel.
123. Read, Stephen, and Wilfrid Hodges. 2010. Western logic. *Journal of the Indian Council of Philosophical Research* 27: 13–45.
124. Read, Stephen. 2012. The medieval theory of consequence. *Synthese* 187: 899–912.

Chapter References

1. *Alexander of Aphrodisias on Aristotle's Prior Analytics 1.1–7*, translation J. Barnes et al., 1991.
2. Al-Fārābī, Abū Naṣr. 1960. *Şharh al-Fārābī li kitāb Aristūṭālīs fī al-'Ibāra*, 2nd edn, ed. Wilhelm Kutch and Stanley Marrow. Beirut: Dar el-Mashriq
3. Al-Fārābī, Abū Naṣr. 1968. *Iḥṣā al-'Ulūm*, edited and introduced by Uthman Amin, Maktabat al-anjelu al-misriyya, Cairo.
4. Al-Fārābī, Abū Naṣr. 1968. *al-Alfāḍ al musta'mala fi l-manṭiq*, second edition, ed. Mohsen Mahdi. Beirut: Dar el Machriq.
5. Al-Fārābī, Abū Naṣr. 1985. *Kitāb al-Tanbīh 'Ala Sabīl as-Sa'āda*, edited and introduced by Jafar Al Yasin, Dar al-Manahil, Beiruth.
6. Al-Fārābī, Abu Nasr. al-Risālah allatī ṣadara bihā al-Manṭiq (or "al-Tawṭi'a"). In *al-Manṭiq 'inda al-Fārābī*, vol. 1, ed. Rafik Al Ajam, 55–62. Beirut: Dar el Machriq.
7. Al-Fārābī, Abū Naṣr. 1986. *Al-Fuṣūl al-Khamsa*. In *al-Manṭiq 'inda al-Fārābī*, vol. 1, ed. Rafik Al Ajam, 65–93. Beirut: Dar el Machriq.
8. Al-Fārābī, Abū Naṣr. 1986. *Kitāb al-'Ibāra*. In *al-Manṭiq 'inda al-Fārābī*, vol. 1, ed. Rafik Al Ajam, 133–164. Beirut: Dar el Machriq.
9. Al-Fārābī, Abū Naṣr. 1986. *Kitāb al-Maqūlāt*. In *al-Manṭiq 'inda al-Fārābī*, vol. 1, ed. Rafik Al Ajam, 89–132. Beirut: Dar el Machriq.
10. Al-Fārābī, Abū Naṣr. 1986. *Kitāb al-Qiyās*. In *al-Manṭiq 'inda al-Fārābī*, vol. 2, ed. Rafik Al Ajam, 11–64. Beirut: Dar el Machriq.
11. Al-Fārābī, Abū Naṣr. 1986. Kitāb al-Qiyās al-Ṣaghīr 'alā ṭarīqati al-mutakallimīn. In *al Mantiq 'inda al-Fārābī*, vol. 2, ed. Rafik Al Ajam, 65–93. Beirut: Dar el Machriq.
12. Al-Fārābī, Abū Naṣr. 1988. *Sharh al-'Ibāra*". In *al-Manṭiqiyāt li-al-Furābī*, vol. 2, texts published by Mohamed Teki Danesh Pazuh, Edition Qom, 1409 of Hegira.
13. Al-Fārābī, Abū Naṣr. 1988. Al-Maqūlāt. In *Al-Manṭiqiyāt li-al-Fārābi*, vol. 1, texts published by Mohamed Teki Danesh Pazuh, Edition Qom, 41–82.
14. Al-Fārābī, Abū Naṣr. 1988. *al-Qawl fī al-'Ibāra*, in *al-Manṭiqiyāt li-al-Fārābi*, volume 1, texts published by Mohamed Teki Danesh Pazuh, Edition Qom, 83–114.
15. Al-Fārābī, Abū Naṣr. 1988. *Kitāb al Qiyās*. In *al-Manṭiqiyāt li-al-Fārābi*, vol. 1, texts published by Mohamed Teki Danesh Pazuh, Edition Qom, 115–151.
16. Al-Fārābī, Abū Naṣr. 1988. *al-Qiyās al-Ṣaghīr*. In *al-Manṭiqiyāt li-al-Fārābi*, vol. 1, texts published by Mohamed Teki Danesh Pazuh, Edition Qom, 152–194.
17. Al-Fārābī, Abū Naṣr. 1988. *Kitāb al-Burhān*. In *al-Manṭiqiyāt li-al-Fārābī*, vol. 1, texts published by Mohamed Teki Danesh Pazuh, Edition Qom, 267–349.

S. Chatti, *Arabic Logic from al-Fārābī to Averroes*, Studies in Universal Logic,
https://doi.org/10.1007/978-3-030-27466-5

18. Al-Fārābī, Abū Naṣr. 1988. *Mā yanbaghī an yuqaddama qabla ta'allum al-falsafa*. In *al-Mantiqiyyāt li-al-Fārābī*, texts published by Mohamed Teki Danesh Pazuh, Edition Qom, 1–10.
19. Al Fārābī, Abū Naṣr. 1990. *Kitāb al Ḥurūf*, Dar al Machriq, Beirut.
20. Antonelli, Aldo. 2008. Non-monotonic logic. In *Stanford encyclopedia of philosophy*, ed. E. N. Zalta. http://plato.stanford.edu/entries/logic-nonmonotonic/.
21. Aristote. 1971. *Premiers Analytiques*, Translated by J. Tricot, Librairie philosophie J. Vrin, Paris.
22. Aristotle. 1991. *Categories*. In *The complete works of aristotle*, vol. 1, ed. Jonathan Barnes. The Revised Oxford Edition.
23. Aristotle. 1991. *De Interpretatione*. In *The complete works of aristotle*, vol. 1, ed. Jonathan Barnes. The Revised Oxford Edition.
24. Aristotle. 1991. *Prior analytics*. In *The complete works of aristotle*, vol. 1, ed. Jonathan Barnes. The Revised Oxford Edition.
25. Arnault, Antoine and Pierre Nicole. 1970. *La logique ou l'art de penser*. Editions Flammarion.
26. Averroes. 1982. *Talkhīṣ Manṭiq Arisṭu* (*Paraphrase de la logique d'Aristote*), vol. 1: *Kitāb Al-Maqūlāt* (pp. 3–77), *Kitāb al-'Ibāra* (pp. 81–141), *Kitāb al-Qiyās* (pp. 143–366), edited by Gérard Jehamy, Manshūrāt al-Jāmi a al-lubnānīya, al-Maktaba al-sharqiyya, Beirut.
27. Averroes. 1982. *Kitāb al-Jadal*. In *Talkhiṣ Manṭiq Arisṭu*, vol. 2, ed. Gérard Jehamy, Manshūrāt al-Jāmi a al-lubnānīya, al-Maktaba al-sharqiyya Beirut, 499–661.
28. Averroes. 1983. *Middle commentary on Aristotle's prior analytics*, Critical edition by M. Kassem, completed, revised and annotated by C. E. Butterworth, and A. Abd al-Magid Haridi, Cairo.
29. Averroes. 1983. *Maqālāt fī al-manṭiq wa-al-'ilm al-ṭabī'ī* [Essays on logic and natural science], ed. Jamāl al-Dīn al-'Alawī. Casablanca.
30. Averroes. 2001. *Kitāb al-Muqaddamāt fi al-Falsafa, al-Masā'il fi-al Manṭiq wa al 'ilm al-ṭabī'ī wa al-Ṭibb*, ed. Assad Jemaa. Tunis: Markez al-Nashr al-Jāmi'ī.
31. Avicenna. 1910. *Manṭiq al-Mashriqiyīn*, Muḥyī al-Dīn al Khatīb and 'Abdelfattāh al Qatlane, Cairo.
32. Avicenna. 1938. *al-Najāt*, Muḥyi al-Dīn Sabrī al-Kurdī, second edition, Library Mustapha al Bab al Hilbi, Cairo. Avicenna.
33. Avicenna. 1959. *al- Shifā', al-Manṭiq* 2: *al-Maqūlāt*, ed. G. Anawati, M. El Khodeiri, A.F. El-Ehwani, S. Zayed, rev. and intr. by I. Madkour, Cairo.
34. Avicenna. 1964. *al-Shifā', al-Manṭiq* 4: *al-Qiyās*, ed. S. Zayed, rev. and intr. by I. Madkour. Cairo.
35. Avicenna. 1970. *al-Shifā', al-Manṭiq 3*: *al-'Ibāra*, ed M. El Khodeiri, rev and intr. by I. Madkour, Cairo.
36. Avicenna. 1971. *Al-Ishārāt wa l–tanbīhāt, with the commentary of N. Ṭūsi*, intr by Dr. Seliman Donya, Part 1, third edition, Cairo: Dar al Ma'arif.
37. Avicenna. 1982. *Manṭiq al-Mashriqiyīn*, ed. Shokri Najjar, Dār al Ḥadātha, Beirut.
38. Avicenna. 2017. *Al-mukhtaṣar al-'awsaṭ fī al-manṭiq*, ed. Seyyed Mahmoud Yousofsani, Muassasah-i Pizh ūhishī-i Ḥikmat va Falsafan-i Īrān, Tehran (hij. 1396).
39. Bäck, Allan. 1992. Avicenna's conception of the modalities. *Vivarium* XXX (2): 217–255.
40. Badawi, Abderrahman. 1980. *Manṭiq Arisṭu*, vols. 1 and 2, Dar al Kalam, Beirut.
41. Black, Dedorah. 1998. *Logic in Islamic philosophy*. Routledge. http://www.muslimphilosophy.com/ip/rep/H017.htm#H017SECT2.
42. Blanché, Robert. 1953. Sur l'opposition des concepts. *Theoria* 19: 89–130.
43. Blanché, Robert. 1969. *Structures intellectuelles. Essai sur l'organisation systématique des concepts*, Editions Vrin, Paris.
44. Blanché, Robert. 1970. *La logique et son histoire, d'Aristote à Russell*. Paris: Armand Colin.

45. Bobzien, Susanne. 2006. Ancient Logic. In *Stanford encyclopedia of philosophy*, ed. Edward N. Zalta. http://plato.stanford.edu/entries/logic-ancient/.
46. Brumberg Chaumont and Julie. 2016. The legacy of ancient logic in the middle ages. In *The Cambridge Companion to Medieval Logic*, ed. Catarina Dutilh Novaes and Stephen Read, 19–44. Cambridge University Press.
47. Buridan, Jean. 1985. *Jean Buridan's logic, the treatise on supposition, the treatise on consequences*, trans. P. King. Dordrecht and Holland: D. Reidel.
48. Burnett, Charles. 2004. The translations of Arabic works on logic into Latin in the middle ages and the renaissance. In *Handbook of the history of logic*, vol. 1, ed. Dov Gabbay and John Woods, 597–606. Elsevier BV.
49. Carnap, Rudolf. 1988. *Meaning and necessity, a study in semantics and modal logic*, 2nd edn. Chicago and London: Midway Reprint Edition, The University of Chicago Press.
50. Chatti, Saloua. 2012. Logical oppositions in Arabic logic, Avicenna and Averroes. In *Around and beyond the square of opposition*, ed. J. Y. Béziau, and D. Jacquette. Basel: Springer.
51. Chatti, Saloua, and Fabien Schang. 2013. The cube, the square and the problem of existential import. *History and Philosophy of Logic* 34 (2): 101–132.
52. Chatti, Saloua. 2014. Syncategoremata in Arabic logic, al-Fārābi and Avicenna. *History and Philosophy of Logic* 35 (2): 167–197.
53. Chatti, Saloua. 2014. Avicenna on possibility and necessity. *History and Philosophy of Logic* 35 (4): 332–353.
54. Chatti, Saloua. 2015. Les carrés d'Avicenne. In *Le carré et ses extensions. Approches théoriques, pratiques et historiques*, ed. Hmaid Ben Aziza and Saloua Chatti. Publications de la Faculté des Sciences Humaines et Sociales de Tunis.
55. Chatti, Saloua. 2016. Existential import in Avicenna's modal logic. *Arabic Sciences and Philosophy* 26 (1): 45–71 (Cambridge University Press).
56. Chatti, Saloua. 2016. Les oppositions modales dans la logique d'al-Fārābī. In *Soyons logiques / Let's be Logical*, ed. Amirouche Moktefi, Alessio Moretti, and Fabien Schang. Collection: *Cahiers de logique et d'épistémologie* (ed. Shahid Rahman and Dov Gabbay). London: College Publications.
57. Chatti, Saloua. 2016. Avicenna (Ibn Sīnā): Logic. In *Encyclopedia of logic, internet encyclopedia of philosophy*. College Publications, www.iep.utm.edu/av-logic/.
58. Chatti, Saloua. 2017. On the asymmetry between the four corners of the square. Available online in the site 'Cercle Ferdinad de Saussure', proceedings of the workshop 'The Arbitrariness of the Sign', organized by Prof. J-Y. Beziau in the congress *Le Cours de Linguistique Générale*, 1916–2016. L'émergence, Geneva, 9–13 janvier 2017. https://www.clg2016.org/contribution/281.html.
59. Chatti, Saloua. 2017. The semantics and pragmatics of the conditional in al-Farabi's and Avicenna's theories. *Studia Humana* 6 (1): 5–17.
60. Chatti, Saloua. 2019. Logical consequence in Avicenna's theory. *Logica Universalis* 13: 101–133.
61. Chatti, Saloua. 2019. The logic of Avicenna, between *al-Qiyās* and *Manṭiq al-Mashriqiyyīn*. *Arabic Sciences and Philosohy* 29 (1): 109–131.
62. Couturat, Louis. 1901. *La logique de Leibniz*, Georg Olms Verlagsbuchhandlung Hildesheim, New Edition (1969).
63. Czeżowski, Tadeusz. 1955. On certain peculiarities of singular propositions. *Mind* 64: 287–308.
64. D'Ancona, Cristina. 2013. Greek sources in Arabic and Islamic philosophy. In *Stanford encyclopedia of philosophy*, ed. Edward N. Zalta, http://plato.stanford.edu/entries/arabic-islamic-greek/.
65. Dadkhah, Gholamreza, and Asadollah Fallahi. 2018. *Logic in 6th/12th century Iran (Arabic texts with Persian notes and English introduction)*. Tehran: Iranian Institute of Philosophy.

66. Dutilh Novaes, Catarina. 2011. The different ways in which logic is (said to be) formal. *History and Philosophy of Logic* 32 (4): 303–332.
67. Dutilh Novaes, Catarina, and Stephen Read. 2016. *The cambridge companion to medieval logic*. Cambridge University Press.
68. Edgington, D. 2014. Conditionals. In *Stanford encyclopedia of philosophy*, ed. E.N. Zalta. http://plato.stanford.edu/entries/conditionals/.
69. Elamrani-Jamal, Abdelali. 1995. Ibn Rušd et les Premiers Analytiques d'Aristote: Aperçu sur un problème de syllogistique modale. *Arabic Sciences and Philosophy* 5: 51–74.
70. El-Rouayheb, Khaled. 2009. Impossible antecedents and their consequences: Some thirteenth-century Arabic discussions. *History and Philosophy of Logic* 30 (3): 209–225.
71. El-Rouayheb, Khaled. 2010. *Relational syllogisms and the history of arabic logic, 900–1900*. Leiden: Brill.
72. El-Rouayheb, Khaled. 2012. Post-Avicennan logicians on the subject matter of logic: some thirteenth—and fourteenth—century discussions. *Arabic Sciences and Philosophy* 22 (1).
73. El-Rouayheb, Khaled. 2016. Arabic logic after Avicenna. In *The Cambridge Companion to Medieval Logic*, eds. Dutilh Novaes, Catarina, and Read, Stephen, 67–93. Cambridge University Press.
74. Fakhry, Majid. 2002 *Al-Fārābī, founder of Islamic neoplatonism, his life, works and influence*. Oxford: Oneworld.
75. Fakhoury, Adel. 1981. *Mantiq al 'Arab min wijhati naḍar al mantiq al ḥadīth* (in Arabic) (*Arabic logic from the point of view of modern logic*), 2nd Edn., Beirut.
76. Garson, James. 2018. Modal logic. In *Stanford encyclopedia of philosophy*, ed. Edward N. Zalta. http://plato.stanford.edu/entries/logic-modal/.
77. Gutas, Dimitri. 1988. *Avicenna and the Aristotelian tradition: Introduction to reading Avicenna's philosophical works*, 1st ed. Leiden: E.J. Brill.
78. Gutas, Dimitri. 2014. *Avicenna and the aristotelian tradition: introduction to reading Avicenna's philosophical works*, 2nd ed. Leiden: Brill.
79. Hasnawi, Ahmed and Wilfrid Hodges. 2016. Arabic logic up to Avicenna. In *The cambridge companion to medieval logic*, ed. Catarina Dutilh Novaes and Stephen Read, 45–66. Cambridge: Cambridge university Press.
80. Hasse, Dag Nickolaus. 2014. Influence of Arabic and Islamic thought on the Latin West. In *Stanford encyclopedia of philosophy*, ed. Edward N. Zalta. http://plato.stanford.edu/entries/arabic-islamic-influence/#Log.
81. Hodges, Wilfrid. 2010. Ibn Sīnā on modes, '*Ibārah* ii.4'. http://wilfridhodges.co.uk/arabic07.pdf.
82. Hodges, Wilfrid. 2010. Ibn Sina on analysis: 1. Proof search. Or: Abstract State Machines as a tool for history of logic. In *Fields of logic and computation: Essays dedicated to Yuri Gurevich on the occasion of his 70th birthday*, ed. A. Blass, N. Dershowitz, and W. Reisig. Lecture Notes in Computer Science, vol. 6300, 354–404. Heidelberg: Springer.
83. Hodges, Wilfrid. 2012. 'Ibn Sīna's Modal Logic', plus 'Permanent and Necessary in Ibn Sīna', presented in the workshop *Modal Logic in the Middle Ages*, University of St-Andrews. http://wilfridhodges.co.uk/arabic20a.pdf
84. Hodges, Wilfrid. 2012. Affirmative and negative in Ibn Sīnā. In *Insolubles and consequences; essays in honour of Stephen Read*, ed. Catarina Dutilh Novaes and Ole Hjortland Thomassen. UK: College Publications, Lightning Source, Milton Keynes.
85. Hodges, Wilfrid. 2017. Ibn Sīnā on *reductio ad absurdum*. *The Review of Symbolic Logic* 10 (3): 583–601.
86. Hodges, Wilfrid, and Spencer Johnston. 2017. Medieval modalities and modern methods: Avicenna and Buridan. *IfColog Journal of Logics and Their Applications* 4 (4): 1029–1073.
87. Hodges, Wilfrid. 2018. Proofs as cognitive or computational: Ibn Sīnā's innovations. *Philosophy and Technology* 31 (1): 131–153.

88. Hodges, Wilfrid. 2018. Two early Arabic applications of model-theoretic consequence. *Logica Universalis* 12 (1–2): 37–54.
89. Hodges, Wilfrid. 2018. Nonproductivity proofs from Alexander to Abū al-Barakāt:1. Aristotelian and logical background, http://wilfridhodges.co.uk/history26.pdf.
90. Hodges, Wilfrid, and Druart Thérèse-Anne. 2018. Al-Fārābī's philosophy of logic and language. In *Stanford encyclopedia of philosophy*.
91. Hodges, Wilfrid. 2017. Identifying Ibn Sīnā's hypothetical logic I: Sentences forms. Draft, Nov 2017.
92. Hodges, Wilfrid. forthcoming. *Mathematical background to the logic of Avicenna*. http://wilfridhodges.co.uk/arabic44.pdf.
93. Hodges, Wilfrid. 2010. *Ibn Sīnā's Alethic Modal Logic*, to appear. http://wilfridhodges.co.uk/arabic47.pdf.
94. Hodges, Wilfrid. *Ibn Sina: Qiyas* ii.3, translation based on the Cairo text, ed. Ibrahim Madhkur et al. Draft. http://wilfridhodges.co.uk/arabic21.pdf.
95. Horn, Laurence. 2001. A natural history of negation. In *The David Hume series*. Stanford: CSLI, University of Chicago Press
96. Hughes, George Edward, and M.J. Cresswell. 1972. *An introduction to modal logic*. London: Methuen and Co. Ltd.
97. Ibn al-Muqaffa. 1978. *'Al-manṭiq*, ed. M. T. Dāneshpazhūh, Iranian Institute of Philosophy, Tehran.
98. Ibn al-Ṣalāh. 1966. Maqāla fī al-shakl al-rābi min ashkāl al-qiyās, In *Galen and the Syllogism*, ed. N. Rescher, pp. 76–87.
99. Ibn al-Sikkit. 1956. *Iṣlāḥ al-manṭiq*, ed. Aḥmad M. Shākir, and 'Abd-al-Salām M. Mārūn, Dār al-Ma'ārif, Cairo.
100. Ibn Khaldūn, Abdurrahmān. 2005. *Al-Muqaddima*, eds. Abdessalam Chaddadi, Beyt al Funūn wa al-'Ulūm wa al-'Ādāb, Casablanca, Morocco.
101. King, Peter. 1985. Introduction to Jean Buridan's logic. In *Jean Buridan's logic*. D. Reidel: Dordrecht.
102. Klima, Gyula. 2006. Syncategoremata. In *Encyclopedia of language & linguistics*, 2nd edn, vol. 12, ed. Keith Brown, 353–356. Oxford: Elsevier.
103. Klima, Gyula, *John Buridan: His nominalist logic, metaphysics, and epistemology*. Great Medieval Thinkers. Oxford University Press. http://www.phil-inst.hu/ ~ gyula/FILES/John-Buridan.pdf.
104. Knuutila, Simo. 2008. Medieval theories of modalities. In *Stanford Encyclopedia of Philosophy*, ed. Edward N. Zalta. http://plato.stanford.edu/entries/modality-medieval/.
105. Lagerlund, Henrik. 2009. Avicenna and Tūsi on modal logic. *History and Philosophy of Logic* 30 (3): 227–239.
106. Lagerlund, Henrik. 2010. Medieval theories of the syllogism. In *Stanford Encyclopedia of Philosophy*. http://plato.standord.edu/entries/medieval-syllogism/.
107. Lagerlund, Henrik. 2012. Arabic logic and its influence. *ul-Mukhatabat*, no. 1, pp. 175–183.
108. Lameer, Joep. 1994. *al-Fārābī and Aristotelian syllogistics; Greak theory and Islamic practice*, Brill Edition.
109. Largeault, Jean. 1972. *Logique mathématique, textes*. Paris: Armand Colin.
110. Lee, Tae-Soo. 1984. *Die Griechische Tradition der Aristotelischen Syllogistik in der Spätantike*. Göttingen: Vandenhoeck & Ruprecht.
111. Łukasiewicz. 1934. Contribution à l'histoire de la logique des propositions. French translation in J. Largeault. 1972. *Logique mathématique, textes*. Paris: Armand Colin.
112. Łukasiewicz, Jan. 1972. *La syllogistique d'Aristote dans la perspective de la logique formelle moderne*, French translation by Françoise Zaslawsky, Librairie Armand Colin, Paris (1951)
113. Malink, Marko. 2006. A reconstruction of Aristotle's modal syllogistic. *History and Philosophy of Logic* 27 (2): 95–141.

114. Maróth, Miklos. 1989. *Ibn Sīnā und die peripatetische "Aussagenlogik"*. Leiden: Brill.
115. Martin, John N. 2013. Distributive terms, truth and the port royal logic. *History and Philosophy of Logic* 34 (2): 133–154.
116. Moretti, Alessio. 2009. *The geometry of logical opposition*. Ph.D. University of Neuchâtel.
117. Moktefi, A., A. Moretti, and F. Schang (eds.). 2016. *Let's be logical*. London: College Publications.
118. Movahed, Zia. 2017. Ibn-Sīnā's anticipation of the formulas of Buridan and Barcan. In *Logic in Tehran*, ed. Ali Enayat et al., pp. 248–255. Wellesley, MA: Association for Symbolic Logic and A. K. Peters. First edition (2003).
119. Movahed, Zia. 2009. A critical examination of Ibn Sīnā's theory of the conditional syllogism, vol. 1, no. 1, published in *Sophia Perennis*. www.ensani.ir/storage/Files/20120507101758-9055-5.pdf.
120. Movahed, Zia. 2010. De re and de dicto modality in the Islamic traditional logic. *Sophia Perennis* 2 (2): 5–19.
121. Parsons, Terence. 2006. The traditional Square of Opposition. In *Stanford encyclopedia of philosophy*, ed. Edward N. Zalta. Stanford: Metaphysics Research lab, CSLI, http://plato.stanford.edu/entries/square/index.html.
122. Philoponus, John. 1905. *In Aristotelis Analytica Priora Commentaria*, ed. M. Wallies, Reimer, Berlin.
123. Read, Stephen, and Wilfrid Hodges. 2010. Western logic. *Journal of the Indian Council of Philosophical Research* 27: 13–45.
124. Read, Stephen. 2012. The medieval theory of consequence. *Synthese* 187: 899–912.
125. Rescher, Nicholas. 1963. *Studies in the history of Arabic logic*. University of Pittsburg Press; Arabic trans. Mohamed Mahrān. 1992. Cairo.
126. Rescher, Nicholas. 1966. *Galen and the Syllogism*. Pittsburgh: University of Pittsburgh Press.
127. Rescher, Nicholas. 1964. *The development of arabic logic*. University of Pittsburgh Press [Arabic translation by Mohamed Mahrān, Dar el Ma'ārif, Cairo (1985)].
128. Rescher, Nicholas. 2006. *Studies in the history of logic*. Ontos Verlag.
129. Russell, Bertrand, and Alfred NorthWhitehead. 1973. *Principia mathematica*. Paperback edition to *56. Cambridge University Press.
130. Ryle, Gilbert. 1966. *The concept of mind*. Harmondsworth: Penguin.
131. Sabra, A. I. 1980. Avicenna on the subject matter of logic. *Journal of Philosophy* 77: 746–764.
132. Shehaby Nabil. 1973. *The propositional logic of Avicenna. A translation from al-Shifā al-Qiyās*. Dordrecht Holland: Kluwer, D. Reidel.
133. Spruyt, Joke. 2007. Peter of Spain. In *The Stanford encyclopedia of philosophy*, ed. Edward. N. Zalta, http://plato.stanford.edu/entries/peter-spain/.
134. Street, Tony. 1995. Ṭūsī on Avicenna's logical connectives. *History and Philosophy of Logic* 16: 257–268.
135. Street, Tony. 2001. 'The eminent later scholar' in Avicenna's Book of the Syllogism. *Arabic Sciences and Philosophy* 11: 205–218.
136. Street, Tony. 2002. An outline of Avicenna's Syllogistic. *Archiv für Geschichte der Philosophie* 84 (2): 129–160.
137. Street, Tony. 2004. Arabic logic. In *Handbook of the history of logic*, vol. 1, eds. Gabbay, Dov, and Woods, John. Elsevier, BV.
138. Street, Tony. 2010. Appendix: Readings of the subject term. *Arabic Sciences and Philosophy* 29: 119–124.
139. Street, Tony, 2008. Arabic and Islamic philosophy of language and logic. In *Stanford encyclopedia of philosophy*, ed. Edward N. Zalta. Stanford University, http://plato.stanford.edu/entries/arabic-islamic-language/, New Edition (2013).
140. Street, Tony. 2014. Afḍal al-Dīn al-Khunājī (d. 1248) on the conversion of modal propositions. *ORIENS* 42: 454–513.

141. Street, Tony. 2016. Kātibī (d. 1277), Taḥtānī (d. 1365) and the *Shamsiyya*. *The Oxford Handbook of Islamic Philosophy* 348.
142. Strobino, Riccardo. 2018. Ibn Sīnā's logic. *The Stanford Encyclopedia of Philosophy*, https://plato.stanford.edu/archives/fall2018/entries/ibn-sina-logic.
143. Thom, Paul. 2008. al-Fārābī on indefinite and privative names. *Arabic Sciences and Philosophy* 18 (2): 193–209.
144. Thom, Paul. 2008. Logic and metaphysics in Avicenna's modal syllogistic. In *The Unity of Science in the Arabic tradition*, ed. Shahid Rahman, Tony Street, and Hassen Tahiri, pp. 361–376. Dordrecht.
145. Thom, Paul. 2010. al-Fārābī on the number of categories. http://paulthom.net/Papers-on-Arabiclogic.html.
146. Thom, Paul. 2010. Abharī on the logic of conjunctive terms. *Arabic Sciences and Philosophy* 20: 105–117.
147. Türker, Sadik. 2007. The Arabico-Islamic background of al-Fārābī's logic. *History and Philosophy of Logic* 28 (3): 183–255.
148. Versteegh, C.H.M. 1977. *Greek elements in Arabic linguistic thinking*. Leiden: Brill.
149. Versteegh, Kees. 1997. The debate between logic and grammar. In *Landmarks in linguistic thought III, The Arabic linguistic tradition*, ed. Kees Versteegh, 52–63. London: Routledge.
150. Zimmermann, F.-W. 1972. Some observations on Al-Fārābī and logical tradition. In *Islamic philosophy and the classical tradition, essays presented by his friends and pupils to Richard Walzer on his seventieth birthday*, ed. S. M. Stern, Albert Hourani and Vivian Brown, Cassirer, Oxford, 517–546.

Index of Names

Note: 'n' refers to footnote page numbers.

S. Chatti, *Arabic Logic from al-Fārābī to Averroes*, Studies in Universal Logic,
https://doi.org/10.1007/978-3-030-27466-5

Subject Index

A

Absolute (propositions), 41–45, 47, 49, 50, 102, 110, 127, 141, 170, 171, 173, 176, 177, 179, 180, 189, 190, 195, 196, 198–200, 202, 207, 210, 215, 226, 350, 351, 354
  general absolute (propositions), *see* general
  special absolute (propositions), 46, 53, 55, 55n, 57, 151, 155, 180, 351
Absurdity, 40, 117, 176, 236, 250, 341, 344
  Absurdum (*reduction ad*), *see Khalf (qiyās al-)*
Accidental, 57, 151, 196, 231, 235, 268, 270, 272, 275, 276
Affirmative, 75, 87, 108, 111, 115, 128–130, 132, 135, 136, 141, 161, 162, 166, 206, 207, 212, 221, 222, 226, 239–242, 254–256, 274, 279, 331
Alethic (modality), 5n, 56, 155, 179, 184, 185, 228
Antecedent(s), 56, 172, 267, 268, 276, 277, 297, 311, 314, 317, 326, 331, 339
Assent, 12, 19–22, 24
Assertoric (propositions), 63, 97, 125, 179, 227, 228, 259
Augment (existential), 115, 140, 141, 296, 296n, 304, 310, 311, 312, 313, 314, 316, 318

B

Barbara (mood), 84, 182–185, 206, 285, 298, 313, 314, 326
Baroco, 92, 105, 112, 129, 136, 137, 141, 176, 177, 177n, 190, 191, 193, 208, 213, 216, 219, 220, 223, 236, 249, 253, 254, 266, 324, 349, 351
Biconditional, 153, 154, 268, 282, 290, 326, 345
Bilateral (possibility), 155, 164, 205, 208, 214, 220, 221, 226, 227, 240–245, 256
Bivalent, 24, 264
Bocardo, 82, 94, 95, 105, 120, 131, 136, 137, 141, 176, 177, 201–203, 208, 222–224, 236, 243, 244, 247, 248, 255, 256, 258, 337, 338, 349

C

Categorical
  logic, 2, 16, 17, 19, 75, 173, 187, 191, 201, 202, 226, 252, 256, 258, 273, 279, 282, 284, 293, 295, 297, 311, 326, 330, 333, 344, 345, 349, 355
  propositions, 24, 28, 40, 96, 124, 225, 285, 326–328, 355
Categories, 2n, 3, 9–11, 13–15, 23, 28, 58, 71n, 124, 263, 282
Classical, 2n, 28, 42, 63, 202, 268, 273, 283, 330, 340, 344, 346, 356
Coherent, 6, 11, 83, 168, 170, 234, 345
Conception, 5n, 12, 14, 19, 20, 22, 36n, 43, 67, 92, 92n, 141, 264, 353n, 355
Conclusive *vs* non-conclusive (moods), 73, 98–100, 105, 108, 113, 120, 123, 128–130, 132, 142, 205, 215, 216, 220, 236, 238, 239, 243, 245–247, 249, 252–254, 256, 268, 272, 301, 307, 308, 315, 319, 322, 325, 327, 335, 355

S. Chatti, *Arabic Logic from al-Fārābī to Averroes*, Studies in Universal Logic,
https://doi.org/10.1007/978-3-030-27466-5

Printed in the United States
By Bookmasters